Ecological Studies, Vol. 136

Analysis and Synthesis

Edited by

M.M. Caldwell, Logan, USA
G. Heldmaier, Marburg, Germany
O.L. Lange, Würzburg, Germany
H.A. Mooney, Stanford, USA
E.-D. Schulze, Jena, Germany
U. Sommer, Kiel, Germany

Ecological Studies

Volumes published since 1992 are listed at the end of this book.

Springer
Berlin
Heidelberg
New York
Barcelona
Budapest
Hong Kong
London
Milan
Paris
Singapore
Tokyo

P.W. Rundel G. Montenegro F.M. Jaksic (Eds.)

Landscape Disturbance and Biodiversity in Mediterranean-Type Ecosystems

With 86 Figures, 3 in Color, and 57 Tables

Springer

Prof. Dr. Philip W. Rundel
Department of Biology
University of California (UCLA)
900 Veteran Avenue
Los Angeles CA 90095, USA

Prof. Dr. Gloria Montenegro
Prof. Dr. Fabian M. Jaksic
Departamento de Ecología
Pontificia Universidad Católica de Chile
Casilla 114-D
Santiago, Chile

ISSN 0070-8356
ISBN 3-540-64475-X Springer-Verlag Berlin Heidelberg New York

Libraray of Congress Cataloging-in-Publication Data.

Landscape disturbance and biodiversity in Mediterranean-type ecosystems / P.W. Rundel). G. Montenegro, F.M. Jaksic (eds.).
p. cm. – (Ecological studies / Analysis and synthesis. ISSN 0070-8356; v. 136)
Includes bibliographical references and index.
ISBN 3-540-64475-X (hardcover)
1. Mediterranean-type ecosystems. 2. Nature-Effect of human beings on. 3. Landscape ecology. I. Rundel, Philip W. (Philip Wilson) II. Montenegro Rizzardini, Gloria. III. Jaksic, F. M. (Fabian M.), 1952–. IV. Series: Ecological studies; v. 136.
QH541.5.M44L35 1998 577-dc21

Cover design: Design & Production GmbH, Heidelberg
Typesetting: Best-set Typesetter Ltd., Hong Kong

SPIN 10537279 31/3137 – 5 4 3 2 1 0 – Printed on acid free paper

Preface

Mediterranean-type ecosystems have long provided classic cases for studies of comparative ecological function because of the opportunities that they present for investigations of evolutionary convergence and divergence among the five regions of the world that share this unusual climatic regime. It is not surprising, therefore, that there has been an active history of comparative ecological studies involving researchers from throughout these mediterranean-climate regions. Past syntheses have examined comparative aspects of broad ecological function (di Castri and Mooney 1973; di Castri et al. 1981), fire ecology (Mooney and Conrad 1977; Moreno and Oechel 1994), nutrients (Kruger et al. 1983), resilience (Dell et al. 1986), plant stress response (Tenhunen et al. 1987), database syntheses (Specht 1988), plant–animal interactions (Arianoutsou and Groves 1994), biogeography (Arroyo et al. 1994), water relations (Roy et al. 1995), the function of biodiversity (Davis and Richardson 1995), and global change (Moreno and Oechel 1995). Despite the significant effects that human populations have had, and are continuing to have, throughout these regions of the world, there has not previously been an examination of the impact of human landscape disturbance on biodiversity within mediterranean-climate ecosystems.

Landscapes within the five mediterranean-climate regions of the world have experienced very different histories and intensities of human impacts. Although South Africa has likely had the longest history of human presence, it is the Mediterranean Basin itself where human urbanization and agricultural expansion first had a broad and dramatic effect. Thousands of years of such activities have had profound impacts through land degradation, altered fire frequencies, erosion, and loss of biodiversity.

Impacts by European cultures on natural landscapes in the other mediterranean-climate regions have a much shorter history of a few centuries. However, these impacts have accelerated in recent years in all five regions as populations have expanded and economic development has grown. These factors present serious environmental problems of planning and resource management at both the ecological and the political level.

Urbanization, agricultural expansion, tree plantations, grazing, exotic species invasions, and mining are all among the issues of varying relative significance within these regions.

There is much that can be learned by a comparative approach to understanding the effects of landscape degradation on broad issues of biodiversity in mediterranean-climate regions. California with its rapid rate of urbanization and South Africa with its change in political structure each face strong pressures on the sustainability of natural ecosystem function and biodiversity. The increasing urbanization and industrialization in Chile today parallels in many respects those changes that occurred in the Mediterranean Basin of Europe following World War II. Marginal farm and grazing lands are being abandoned as a formerly rural society becomes rapidly urbanized, leading to dramatic changes in vegetation cover and fire frequencies. Agricultural development in all of the mediterranean-climate regions has led to increasing fragmentation of natural habitats, and invasions of exotic species have become a growing concern.

Although the concept for this book grew out of ideas presented at the MEDECOS VI Conference in Reñaca, Viña del Mar, Chile, in October 1994, this is not a symposium volume. Authors who have contributed chapters to this volume represent both individuals present at this conference as well as others who were not there but share an interest and experience in this theme of landscape disturbance and biodiversity in mediterranean-type ecosystems. Multiple authors from each of the five mediterranean-climate regions are represented.

We have divided this volume into six groups of chapters. The first group includes three introductory chapters that broadly review the issues of landscape degradation. Part II presents six chapters that describe regional land use conflicts in each of the mediterranean-climate regions. Next, Part III provides four chapters on landscape disturbance and plant diversity, while Part IV comprises four chapters dealing with demography and ecophysiology in vegetation succession following disturbance. The theme of landscape disturbance and animal diversity is the subject of five chapters in Part V. The volume closes with a consideration of the future in Part VI with a chapter addressing issues of politics and the environment in mediterranean-climate regions.

References

Arianoutsou M, Groves RH (eds) (1994) Plant–animal interactions in mediterranean-type ecosystems. Kluwer, Dordrecht

Arroyo MTK, Zedler PH, Fox MD (eds) (1994) Ecology and biogeography of mediterranean ecosystems in Chile, California, and Australia. Springer, Berlin Heidelberg New York

Davis GW, Richardson DM (eds) (1995) Mediterranean-type ecosystems: the function of biodiversity. Springer, Berlin Heidelberg New York

Dell B, Hopkins AJM, Lamont BB (eds) (1986) Resilience in mediterranean-type ecosystems. Dr W Junk, Dordrecht

di Castri F, Goodall DW, Specht RL (eds) (1981) Mediterranean-type shrublands. Elsevier, Amsterdam

di Castri F, Mooney HA (eds) (1973) Mediterranean-type ecosystems: origin and structure. Springer, Berlin Heidelberg New York

Groves RH, di Castri F (eds) (1991) Biogeography of mediterranean invasions. Cambridge University Press, Cambridge

Kruger FJ, Mitchell DT, Jarvis JUM (eds) (1983) Mediterranean-type ecosystems: the role of nutrients. Springer, Berlin Heidelberg New York

Mooney HA, Conrad CE (tech co-ords) (1977) Symposium on the environmental consequences of fire and fuel management in mediterranean ecosystems. USDA For Serv Gen Tech Rep WO-3

Moreno JM, Oechel WC (eds) (1994) The role of fire in mediterranean-type ecosystems. Springer, Berlin Heidelberg New York

Moreno JM, Oechel WC (eds) (1995) Global change and mediterranean-type ecosystems. Springer, Berlin Heidelberg New York

Roy J, Aronson J, di Castri F (1995) Time scales of biological response to water constraints: the case of mediterranean biota. SPB Academic Publ, Amsterdam

Specht RL (ed) (1988) Mediterranean-type ecosystems: a data source book. Kluwer, Dordrecht

Tenhunen JD, Catarino FM, Lange OL, Oechel WC (eds) (1987) Plant response to stress: functional analysis in mediterranean ecosystems. Springer, Berlin Heidelberg New York

Philip W. Rundel,
Los Angeles
Gloria Montenegro
and *Fabian Jaksic*, Santiago

Summer 1998

Contents

Part I: Introduction

Part VI: The Future

Contributors

Arianoutsou, Margarita

Department of Ecology and Systematics, School of Biology, University of Athens, 15784 Athen, Greece (marianou@atlas.uoa.gr)

Aronson, James

C.N.R.S./CEPE, B.P. 5051, 1919 Route de Monde, 34033 Montpellier Cedex 01, France (aronson@cefe.cnrs-mop.fr)

Avendaño, Julia

Centro Experimental Cauquenes, INIA, Casilla 165, Cauquenes, Chile

Barton, K.P.

Science Division, Pepperdine University, Malibu, California 90263 USA

Cowling, Richard M.

Institute for Plant Conservation, University of Cape Town, Private Bag, Rondebosch 7700, South Africa (wpaisley@botzoo.uct.ac.za)

Davis, George

National Botanical Institute, Private Bag X7, Claremont 7735, South Africa (davis@nbict.nbi.ac.za)

Davis Stephen D.

Science Division, Pepperdine University, Malibu, California 90263 USA (davis@pepperdine.edu)

Del Pozo, Alejandro

Centro Regional de Investigaciones Quilamapu, INIA, Casilla 426, Chillán, Chile di Castri, Francesco

C.N.R.S./CEPE, B.P. 5051, 1919 Route de Mende, 34033 Montpellier Cedex 01, France (dicastri@cefe.cnrs-mop.fr)

Esler, Karen J.

Department of Botany, University of Stellenbosch, P.O. Box 1, Matieland, South Africa

Etienne, Michel

INRA, Unité d'Ecodéveloppement, Site Agroparc, 84914 Avignon Cedex 9, France (etienne@avignon.inra.fr)

Farina, Almo

Museo di Storia Naturale della Lunigiana, 54011 Aulla, Italy (farinams@vm.cnuce.cnr.it)

Feinsinger, P.

Department of Biological Sciences, Northern Arizona University, Flagstaff, Arizona 86011-5640, USA

Fox, Barry J.

School of Biological Science, University of New South Wales, Sydney 2052, Australia (b.fox@unsw.edu.au)

Gelderblom, Caroline

CSIR Division of Water, Environment and Forestry Technology, PO Box 320, Stellenbosch 7599, South Africa

Groves, Richard

CSIRO Division of Plant Industry, GPO Box 1600, Canberra, ACT, Australia (r.groves@plcan.pi.csiro.au)

Hobbs, Richard J.

CSIRO Division of Wildlife and Ecology, LMB 4, PO Midland, WA 6056, Australia (richard.hobbs@per.dwe.csiro.au)

Hoffmann, Alicia J.

Departamento de Ecología, Facultad de Ciencias Biológicas, Pontificia Universidad Católica de Chile, Casilla 114-D, Santiago, Chile

Hoffmann, Adriana E.

Fundación Claudio Gay, Santiago, Chile

Jaksic, Fabiau M.

Departamento de Ecología, Pontificia Universidad Católica de Chile, Casilla 114-D, Santiago, Chile (fjaksic@bio.puc.cl)

Kazaklis, Angelos

Technological Institute of Forestry, 661 00 Drama, Greece

Keeley, Jon E.

USGS Biological Resources Division, Sequoia-Kings Canyon Field Station, Three Rivers, CA 93271, USA (jkeeley@nsf.gov)

Kolb, K.J.

Science Division, Pepperdine University, Malibu, California 90263 USA

Lavin, Arturo

Centro Experimental Cauquenes, INIA, Casilla 165, Cauquenes, Chile

Le Floch, E.

C.N.R.S./CEPE, B.P. 5051, 34033 Montpellier Cedex 01, France (lefloch@cefe.cnrs.mop.fr)

Liberona, Flavia

Departamento de Ecología, Facultad de Ciencias Biológicas, Pontificia Universidad Católica de Chile, Casilla 114-D, Santiago, Chile

McDonald, David J.

National Botanical Institute, Private Bag X7, Claremont 7735, South Africa

Montenegro, Gloria

Departamento de Ecología, Pontificia Universidad Católica de Chile, Casilla 114-D, Santiago, Chile (gmonten@bio.puc.cl)

Naveh, Zev

Technion-IIT, Haifa, 32000 Haifa, Israel

Orozco-Segovia, A.

Centro de Ecología, UNAM, Apdo Postal 70-275, Mexico 04510, D.F., México (cvazquez@miranda.ecologia.unam.mx)

Ovalle, Carlos

Centro Regional de Investigaciones Quilamapu, INIA, Casilla 426, Chillán, Chile

Papanastasis, Vasilios P.

Laboratory of Range Science (236), Aristotle University 54006 Thessaloniki, Greece (vpapan@for.auth.gr.)

Quinn, Ronald D.

California State Polytechnic University, Dept. of Biological Sciences, 3801 W. Temple Avenue, Pomona, California 91768 USA (rdquinn@csupomona.edu)

Richardson, Dave M.

Institute for Plant Conservation, Botany Department, University of Cape Town, Private Bag, Rondebosch 7700, South Africa (rich@botzoo.uct.ac.za)

Rundel, Philip W.

Department of Biology, University of California, Los Angeles CA 90095 USA (rundel@biology.ucla.edu)

Samways, Michael

Invertebrate Conservation Research Centre, Department of Zoology and Entomology, University of Natal, Private Bag X01, Scottsville 3209, South Africa (samways@zoology.unp.ac.za)

Trinder-Smith, Terry H.

Bolus Herbarium, Botany Department, University of Cape Town, Rondebosch 7700 South Africa van Wilgen, Brian W.

CSIR Division of Water, Environment and Forestry Technology, PO Box 320, Stellenbosch 7599, South Africa

Vázquez-Yanes, Carlos

Centro de Ecología, UNAM, Apdo Postal 70-275, Mexico 04510, D.F., Mexico (cvazquez@miranda.ecologia.unam.mx)

Walter, Hartmut S.

Dept. of Geography, University of California, 405 Hilgard Ave., Los Angeles, California 90095 USA (walter@geog.ucla.edu)

Wynberg, Rachel

c/o Environmental Evaluation Unit, University of Cape Town, Private Bag, Rondebosch 7700, South Africa (wynberg@enviro.uct.ac.za)

Part I

Introduction

1 Landscape Disturbance in Mediterranean-Type Ecosystems: An Overview

P.W. Rundel

1.1 Mediterranean-Climate Regions

Five regions of the world – the Mediterranean Basin, California, central Chile, the Cape Region of South Africa, and Western and South Australia (Fig. 1.1) – share a unique climatic regime with their characteristic conditions of mild wet winters and warm dry summers. This mediterranean-type climatic regime typically has 90% or more of annual precipitation falling in the six cool season months, mild winter conditions with infrequent and relatively mild frosts, and frequent periods of extended summer drought.

The classic vegetation structures in mediterranean-type ecosystems are evergreen shrublands dominated by species with sclerophyllous leaves. These shrublands are termed maquis or garrigue in the Mediterranean Basin (Polunin and Walters 1985), chaparral in California (Rundel and Vankat 1989), matorral in Chile (Rundel 1981a), fynbos in South Africa (Cowling 1992), and kwongan or heathlands in southwestern Australia (Pate and Beard 1984).

Although the mediterranean-type ecosystems are commonly characterized as being dominated by evergreen sclerophyll shrublands, other vegetation structures are also important (Fig. 1.2). Woodlands are widespread in most mediterranean regions, particularly in areas with deeper or nutrient-richer soils, or as riparian woodlands or gallery forests in wetter sites. Both the Mediterranean Basin and California have extensive oak woodlands dominated by the genus *Quercus*. These communities can take the form of closed canopy evergreen woodlands grading into shrublands as in live oak woodlands of southern California and the maquis of Europe, or open savannas of deciduous oaks that are widespread in both regions. Central Chile once had widespread evergreen sclerophyll woodlands (Rundel 1981a), while evergreen eucalypt woodlands termed mallees are widespread in semi-arid areas of Western and South Australia (Specht 1981; Noble and Bradstock 1989). Only the Cape Region of South Africa of all of

Ecological Studies, Vol. 136
Rundel et al. (eds.) Landscape Degradation and Biodiversity in Mediterranean-Type Ecosystems

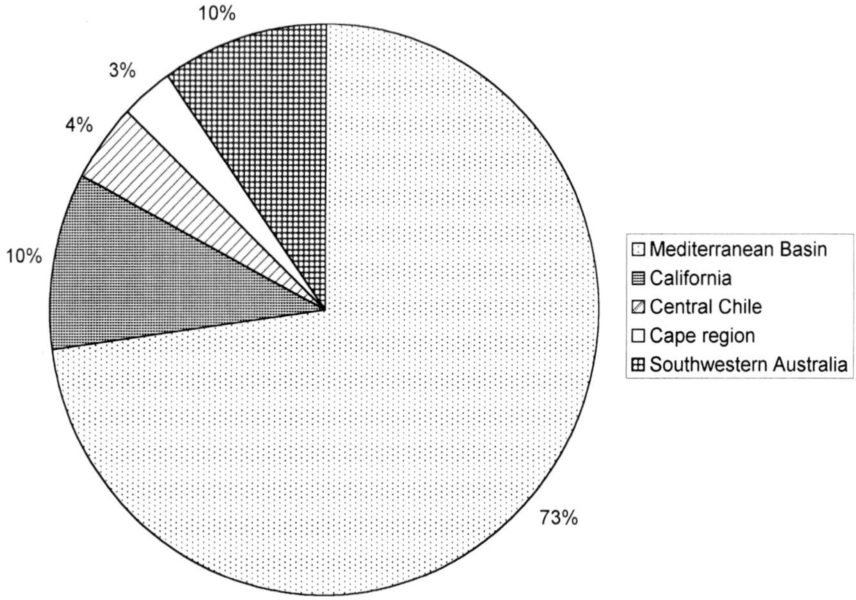

Fig. 1.1. Proportional global distribution of land area of the five mediterranean-type ecosytems. (Data from Cowling et al. 1996)

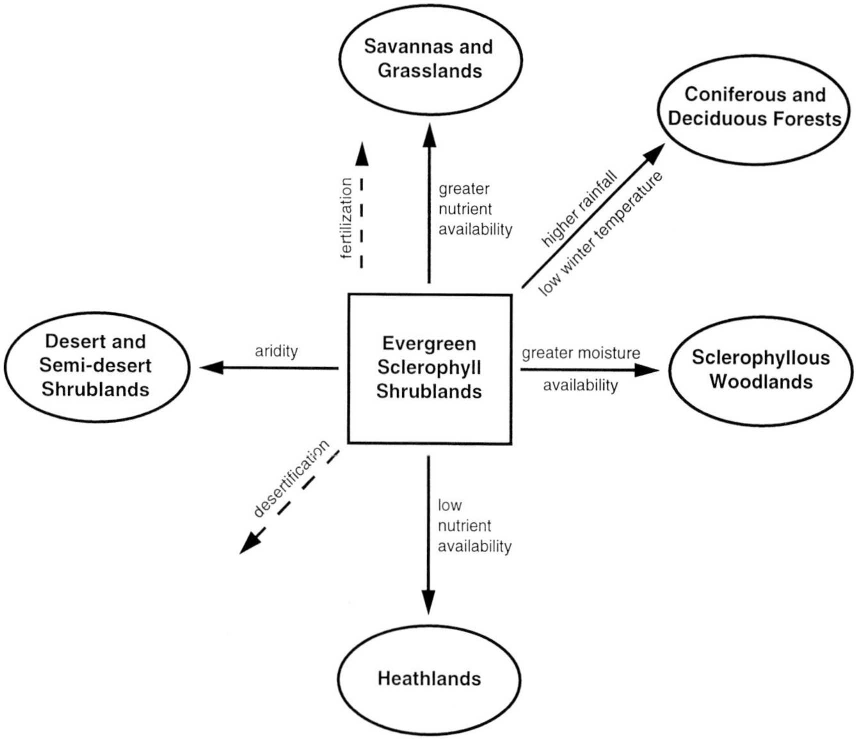

Fig. 1.2. Distributions of mediterranean-climate communities along environmental gradients of moisture availability, temperature, and soil nutrient availability. (Adapted from di Castri 1981)

the mediterranean regions is largely lacking in woodlands, with such communities restricted to scattered relictual stands of low tree diversity along the southern coast (Midgley et al. 1997).

The evergreen shrublands of California and Chile commonly grade off at their arid interior margin and along drier coastal margins to vegetation dominated by drought deciduous shrubs. This community, which may also dominate early successional sequences following disturbance as well as stable islands in arid azonal habitats in evergreen shrublands, is termed coastal sage scrub in California and coastal matorral in Chile (Rundel 1981a). Structurally similar communities with mixed dominance of low evergreen and deciduous shrubs are called phrygana in Greece and the eastern Mediterranean Basin (Polunin and Walters 1985). Deciduous shrubs are largely lacking, however, from similar habitats in the Cape Region of South Africa and southwestern Australia. It is these coastal and semi-arid margins of mediterranean-climate regions that have received much of the most severe human disturbance (Naveh, Chap. 2; Esler et al., Chap. 12).

Including these evergreen shrublands and woodlands, mediterranean-type ecosystems (mediterranean-climate regions sensu stricto) cover a tiny 1.2% of the earth's surface (di Castri 1981). Approximately 73% of this area lies within the Mediterranean Basin (Fig. 1.1) where the presence of the Mediterranean Sea allows this climatic regime to extend far to the east across southern Europe and North Africa to Israel and adjacent parts of the Middle East in Asia. The largest political area of the mediterranean-climatic regime in this region is present in Spain, followed by Turkey, Morocco, and Italy (di Castri 1981). Australia and California have the second largest area of mediterranean shrublands and woodlands with 10% each of the total area. The Cape Region of South Africa forms the smallest total area of the mediterranean-climate region with 3% (Fig. 1.1)

Mediterranean-type ecosystems in a broad sense, i.e. those regions which share elements of mediterranean-climate regimes, include not only these typical shrublands and woodlands, but also an arid margin with desert shrublands and conifer and deciduous forest communities at higher elevations (Fig. 1.2). Mediterranean desert shrublands with annual rainfall levels below 250 mm occur in the winter rainfall areas of the northern Sahara and Arabian deserts, the Mojave Desert and Colorado Desert (the western portion of the Sonoran Desert) in California, the Atacama Desert of Chile, the Succulent Karoo in South Africa, and extensive arid land areas with a winter rainfall regime in Western Australia. High elevation areas in the Mediterranean Basin and in California support conifer forests with associated deciduous tree species, again in mediterranean-climate regions (Rundel et al. 1977; Tomaselli 1981). If mediterranean-climate regimes are broadened to include these arid shrublands and montane areas with similar climatic regimes, then the global land area of such regimes expands to almost 5% (Cowling et al. 1996).

In their contribution to species diversity of vascular plants, mediterranean-climate regions have a significance that far exceeds their relatively small area of coverage. These regions, in the broad sense including winter rainfall deserts and montane areas, include about 48 250 plant species, approximately 20% of the world's total (Cowling et al. 1996). Nowhere outside of lowland tropical rainforests are there ecosystems with higher regional diversities of species. The largest number of these species come from the Mediterranean Basin with an estimated 25 000 species, followed by South Africa with 8550 and southwestern Australia with 8000 species (Table 1.1). The small Cape Floristic region of South Africa is unique in the presence of so many species within a relatively small area, and forms one of the six floristic kingdoms of the world (Takhtajan 1986).

The importance of protecting biodiversity in mediterranean-climate regions is magnified tremendously in comparison to problems being faced in tropical forest ecosystems by the magnitude of dramatic changes to

Table 1.1. Plant species diversity, topographic and climatic diversity, and regional disturbance threats in mediterranean-climate regions of the world. Order of regional disturbance threats roughly follows their significance, beginning with the most critical factors. (Adapted from Cowling et al. 1996)

Region	Area (10^6 km^2)	Native plant species	Topographic heterogeneity	Climatic heterogeneity	Regional disturbance threat
Mediterranean Basin	2.30	25 000	High	Very high	Deforestation, overgrazing, agriculture, urbanization
California	0.32	4 300	High	Very high	Urbanization, agriculture
Central Chile	0.14	2 400	Very high	Very high	Deforestation, overgrazing, agriculture
Cape region, South Africa	0.09	8 550	Moderate	High	Invasive alien plants, agriculture, urbanization
Southwestern Australia	0.31	8 000	Low	Moderate	Agriculture, deforestation, introduced pathogens

mediterranean landscapes that have occurred or are being wrought today by human actions (Groves, Chap. 3). No ecosystems in the world have been altered to the extent of those in mediterranean-climate regions.

1.2 Natural Disturbance Regimes

The natural disturbance regimes in the five mediterranean-climate regions of the world have many similarities in environmental conditions of summer drought, fire, tectonic stability, and flood/erosion cycles. However, the frequency and intensity of natural disturbance regimes differ in a number of distinct ways between regions. Although all five regions experience characteristic summer drought, the magnitude of this drought is particularly severe in California, Chile, and much of the subarid portion of the Mediterranean Basin where 6–8 months or more may pass without measurable rainfall (Rundel 1995). Extreme drought such as this is rare in South Africa and southwestern Australia where summer months frequently have light showers.

The high woody biomass and dense cover of mediterranean-climate shrublands cause them to be highly flammable, particularly under dry summer conditions with low humidity. Fire is thus an important disturbance regime today in all mediterranean-type ecosystems, although the frequency of natural fires probably differed greatly within and between regions (Mooney and Conrad 1977; Rundel 1981b, 1983; Trabaud and Prodon 1993; Oechel and Moreno 1994). The resilience of mediterranean-type shrublands to fire disturbance is a function of the morphological, ecophysiological, and phenological adaptations to postfire regeneration of these stands through resprouting and fire-stimulated reseeding (Keeley, Chap. 14; Vazquez-Yanes and Orozco-Segovia, Chap. 15; Arianoutsou, Chap. 16; Davis, Chap. 17).

The frequency of natural fires appears to be quite different between regions. In South Africa, for example, fynbos vegetation in the Cape region commonly burns at intervals of 10–15 years (van Wilgen et al. 1992), while in the Mediterranean Basin and California natural frequencies are thought to be 30–50 years or more (Rundel and Vankat 1989; Trabaud and Prodon 1993). It is not unusual to encounter chaparral stands that have not experienced fires in more than a century. Although fires are significant natural disturbance events in four of the five mediterranean-type ecosystems, central Chile is an exception to this pattern. The geographic isolation of central Chile, protected from summer storms with lightning by the high Andean Cordillera, has little evidence of fire as an important ecological disturbance regime in the evolution of the native flora. Anthropogenic fires, however, have become common in recent decades.

There are other strong environmental differences in landscape history and stability that make the Cape Region of South Africa and southwestern Australia distinct from the other three mediterranean-climate regions. These two areas lie in geologically ancient and stable landscapes, resulting in highly leached and nutrient-poor soils. In contrast, earthquakes, volcanic activity, orogenic uplift, and other dynamic processes create natural disturbance regimes in California, Chile, and the Mediterranean Basin that are absent in South Africa and Australia. Unlike South Africa and southwestern Australia, the younger landscapes of these three regions have experienced tremendous changes in climate regime and landscape structure in Quaternary and even Holocene times and these changes have had profound impacts on community structure and speciation (Cowling et al. 1996). The remarkable patterns of speciation in fire-sensitive shrub lineages in the Cape Region of South Africa and southwestern Australia have most likely resulted from a combination of relatively mild and stable Quaternary climatic conditions coupled with high fire frequencies in these nutrient-poor habitats (Cowling et al. 1996).

1.3 Anthropogenic Disturbance Regimes

Histories of human impact, beyond differences in natural environmental disturbance regimes, separate the five mediterranean-climate regions of the world. The Mediterranean Basin was the site of early human cultural advances, and indigenous agriculture and animal husbandry have been practiced there for more than 10000 years (Naveh and Dan 1973; Le Houérou 1981; Naveh 1991). Within the other four mediterranean-climate regions, relatively small populations of hunter-gatherers were the dominant human influences until only a few centuries ago. Chile was first colonized by the Spanish in 1542, and the Dutch settled in South Africa in 1652. Settlement of California did not take place until 1769, two and a half centuries after discovery by the Spanish, and population with European ancestry was very small until the discovery of gold in 1848. It was only in 1827 that the British colonized southwestern Australia.

It is not surprising with the cultural differences between colonizing nations, that the different levels of population growth and economic development, and the nature of the natural environment in each region, have combined to produce a complex of both similar and contrasting disturbance regimes in the five mediterranean-climate regions. Changes in fire frequencies and intensities, grazing pressure, urbanization, agricultural expansion, deforestation, and the introduction of exotic species and pathogens are all aspects of regional disturbance to natural vegetation in one or more mediterranean-climate regions (Table 1.1).

Invasions of exotic species into mediterranean-climate regions have become a problem of increasing environmental focus and concern (Mooney and Drake 1986; di Castri et al. 1990; Groves and di Castri 1991). While invasive species commonly share a number of life history attributes which allow rapid reproduction and colonization, an evolutionary history of adaptation to disturbance regimes associated with human activities is also characteristic. The origin of many invasive plant species in the Mediterranean Basin points to a genetic pre-adaptation to disturbance regimes associated with agriculture, grazing, and fire. Critical questions which are now being addressed with respect to invasive species are: (1) which species make good invaders? (Reichard and Hamilton 1997; Rejmanek and Richardson 1996); (2) which communities are most easily invaded? (Perrins et al. 1992; Planty-Tabacchi et al. 1996); and (3) what will be the rate of spread of invading species? (Andow et al. 1993; Shigesada and Kawasaki 1996).

1.4 Regional Disturbance Regimes

1.4.1 Mediterranean Basin

Anthropogenic disturbances such as deforestation, grazing, agricultural development, and fire management have been factors in influencing community structure and diversity for at least 10000 years in the Mediterranean Basin as agricultural civilizations developed and spread in this region (Aschmann 1973; Naveh and Dan 1973). Agricultural development within highly developed political systems of kingdoms or city-states had became widespread in the eastern Mediterranean and Middle East by 2000 BC, and rapidly spread westward to the Atlantic. Although problems of deforestation and erosion date back to Greek and Roman times, many land-use practices were relatively sustainable. Many of the early forms of agropastoral use of Mediterranean woodlands and shrublands continued without significant modification from classic times up until the middle of the twentieth century (Caravello and Giacomin 1993). Changing patterns of land-use and disturbance in the past half century, however, have brought dramatic changes in the structure and diversity of these communities (Naveh, Chap. 2).

Extensive oak forests and woodlands of the Iberian Peninsula dominated by *Quercus ilex* and *Q. suber*, known as dehesas and montados, have played a central role in agroforestry systems of the western Mediterranean Basin for centuries, combining livestock grazing with harvesting of cork and firewood (Pinto-Correia 1993). Coppiced woodlands of *Quercus ilex* and *Q. pubescens* in southern France and Italy have likewise been harvested for centuries for firewood and charcoal. Over the past four decades, however, dramatic alterations have occurred in the landscapes of these ecosystems. In the dehesas and montados, ecological deterioration has been particularly evident due to a change in traditional management with increasing deforestation and clearing to promote intensive agriculture and the extensive development of plantations of *Pinus* and *Eucalyptus*. Related changes have occurred with oak forests and maquis communities in southern France and Italy. Additionally, widespread landscape abandonment associated with increasing urbanization has led to uneven management policies, rapidly expanding frequencies and extents of fire as a disturbance factor, and loss of both landscape and biodiversity (Naveh and Lieberman 1993; Naveh, Chap. 2; Etienne et al., Chap. 7; Papanastasis and Kazaklis, Chap. 8; Samways, Chap. 18).

The stability and biodiversity of Mediterranean Basin ecosystems are strongly tied to the nature of both grazing and fire regimes. As plant communities have developed with moderate grazing pressure as a natural selective regime, plant species have evolved reproductive strategies linked to this disturbance. There has also been adaptation over evolutionary time to fire as a natural environmental factor, and reproduction in many species is strongly related to fire frequency and intensity (Trabaud and Prodon 1993). The frequency and magnitude of fires in the Mediterranean Basin have doubled in the past two decades, with the consequence of decreased forest vigor and sharp alterations in forest structure and soil stability (Kuzucuoglu 1989; Naveh 1990).

What is the relationship between landscape disturbance and biodiversity? Studies of maquis forests and shrublands in northern Israel and southern France have found that high floristic and structural diversity of mediterranean shrublands and woodlands are associated with moderate grazing pressures which maintain open habitats through regular disturbance (Naveh and Whittaker 1979; Naveh 1982; Kaplan 1992). Both floristic and structural diversity have been found to be reduced significantly where an intermediate disturbance regime has been altered through reduced or increased intensity disturbance, land abandonment, or protection from fire. Sites that had been moderately grazed and coppiced in the past were found to have a woody plant community with good structural diversity and a rich floristic diversity contributed by many small annuals, geophytes, and herbaceous perennials. Animal species richness and relative abundance of birds, reptiles, rodents, and isopods showed similar trends with moderate disturbance (Warburg et al. 1978; Farina, 1989; Chap. 20). Small-scale heterogeneity created by variable soil depth and micro-relief, availability of moisture and nutrients, rock outcrops, tree cover, litter cover, and shade all contributed to this increased diversity.

Desertification, a process with linked physical, biological, and socio-economic origins, presents another significant problem of landscape disturbance in the Mediterranean Basin (Fantechi and Margaris 1986; Perez-Trejo 1994). Environmental problems associated with desertification include reduced vegetation cover, degradation in soil structure, reduced infiltration capacity, and salinization, all of which work together in concert to promote secondary effects of increased severity of erosion and changes in hydrologic processes related to ground water recharge and surface flow. Obvious feedbacks to animal population structure and diversity occur as vegetation structure and diversity are altered. Changing patterns of land use, fragmentation of natural plant communities, increased rates of soil erosion, high fire frequencies and intensities, and reduced water resource availability have all intensified desertification in recent decades, with associated reductions in natural biodiversity (Meeuss et al. 1990; Naveh and Lieberman 1993; Perez-Trejo 1994). Socio-economic and political issues frequently act to confound these problems (Naveh, Chap. 2; di Castri, Chap. 23).

1.4.2 California

Populations of native Americans, who became widespread in California only about 6000–10000 years ago, are known to have had a significant impact on natural vegetation. Indigenous tribes of hunter-gatherers harvested many native species for food, fiber, medicine, or building materials, and made wide use of fire to either drive game or clear areas to promote grazing by large animals (Anderson 1993).

Although only small numbers of people of European ancestry had settled in California before the discovery of gold in 1848, large herds of cattle and sheep were already present at this time. This grazing pressure over extensive areas of California grasslands was an important component of converting these plant communities from dominance by native perennials to exotic annuals from the Mediterranean Basin. As grazing pressure began to be reduced, and sheep herding became less economically viable in the 1880s, there was a rapid expansion of agricultural lands in the San Joaquin Valley and southern California. Expanding urbanization and clearance for new agricultural lands began to rapidly occur along the California coast at this time, and has continued to the present (Walter, Chap. 6). Such changes have impacted not only plant populations, but have been very significant for animal populations as well (Quinn, Chap. 20).

Disturbance to natural ecosystems in California over the past two centuries has been highly variable, with some ecosystems remaining relatively pristine while the structure and diversity of others have been changed dramatically by human activities. Most dramatic in change have been grasslands and oak woodlands of the state. These communities became altered from systems largely dominated by native perennial grasses to ones in which introduced European annual grasses and forbs provide almost all of the herbaceous plant cover (Heady 1977). The sources of these invaders are reasonably well documented, with both deliberate and accidental introductions involved (Mooney and Drake 1986; Rejmanek et al. 1991). These species appear to have genetic traits of flexible life-history characteristics, high reproductive effort, low root to shoot ratios, and large seed banks which pre-adapt them to habitats with agriculture, strong grazing pressure, and other forms of human disturbance (Jackson 1985). It is interesting to note that edaphically stressful soils in California such as serpentine-derived or heavy clay soils retain a dominance of native annuals and herbaceous perennials, indicating that the widespread introduced annuals are not competitive in these habitats. The plant species diversity of these edaphic habitats is commonly higher than that of grasslands dominated by introduced species.

Alterations in natural fire frequency and intensity have been a major aspect of changing community structure in California chaparral and woodlands under human influence. These changes began with native Americans who utilized fires in managing plant succession, regrowth, and animal population patterns (Blackburn and Anderson 1993). These early fire management practices were changed dramatically at the end of the nineteenth century as new American ideas of resource management led to fire suppression policies that actively fought fires. The result of this policy, which has only begun to be altered in recent decades, has been a reduction in fire frequency but an increase in intensity of chaparral fires (Minnich 1983; Rundel and Vankat 1989).

1.4.3 Central Chile

Little is known about the possible environmental impacts of indigenous populations of hunter-gatherers who lived in central Chile for about 10 000 years before the Spanish settlement in 1542. With colonization came major changes associated with agriculture and grazing and an economic development that allowed increasing levels of urbanization. With the absence of montane forest vegetation in central Chile, sclerophyll woodlands were rapidly felled for construction material and matorral vegetation was cleared and burned for charcoal production. Agricultural expansion and the cultural use of goats by rural populations combined with this vegetation clearance to dramatically reduce woody cover over large areas (Fuentes and Hajek 1979; Balduzzi et al. 1982; Aronson et al., Chap. 9). This impact, which has continued up until recent decades, is particularly apparent in the semi-arid transition region where overgrazing has led to extensive devegetation and desertification over broad areas (Bahre 1979; Etienne et al. 1986). Much of the vegetation of the central valley of Chile today is dominated by a savanna of *Acacia caven* (Ovalle et al. 1990, 1996; Aronson et al., Chap. 9), with an herbaceous cover that is largely exotic annual grasses and forbs from the Mediterranean Basin (Gulmon 1977). Many rare species of Chilean plants have been seriously threatened by such impacts, with geophytes as a notable example (Hoffmann et al., Chap. 13).

Matorral communites are common in the coastal range and foothills of the Andes, but the structure and diversity of these ecosystems have also been negatively impacted by human actions (Rundel 1981a). Patterns of anthropogenic changes in community structure and succession in both matorral and acacia woodland communities of central Chile have been described in detail by Balduzzi et al. (1982). In comparison to shrublands in California and the Mediterranean Basin, matorral communities are often open with clumped shrub distributions on all but the most mesic sites. It has been suggested that this open structure has been maintained in part by the impact of European rabbits that were introduced to Chile approximately 50 years ago and are now ubiquitous (Fuentes et al. 1983, 1986). The recovery of matorral vegetation following disturbance is very slow. Frequent fires, rabbit grazing, limited seed dispersal, and summer drought stress are important ecological factors limiting the restoration of natural vegetation cover. Changing plant cover, with productivity responding to sharp changes in year-to-year rainfall, has dramatic impacts on animal populations (Jaksic and Feinsinger, Chap. 22).

With the rapid growth of the Chilean economy in the past decade and the centralization of this economy in Santiago, significant changes in disturbance regimes have taken place. As more and more of the rural population moves to the city, there has been extensive landscape abandonment and reduction in disturbance from overgrazing, fuel wood, and charcoal production. The consequent revegetation of these areas, however, has left them open to anthropogenic fires whose frequency and average size have increased rapidly in recent years as continuous vegetation cover returns over areas formerly too open to carry a fire. The lack of woody species with obligate postfire reseeding strategies, and the absence of specialized fire-following annuals point to the conclusion that fire was not an important disturbance regime in the evolution of this flora. Nevertheless, matorral shrubs do resprout rapidly after fire.

More than in any other mediterranean-climate region, landscape clearance for the establishment of tree plantations (*Pinus radiata* and *Eucalyptus globulus*) has been a major factor in landscape alteration in central Chile. More than 65 000 ha of plantations were planted annually during the 1980s, and that rate has nearly doubled in this decade (Aronson et al., Chap. 9). Much of the area planted lies in the foothills of the coastal ranges in what had previously been some of the least disturbed natural ecosystems of hygrophilous and sclerophyll forest. Additionally, vineyard plantations have increased dramatically in extent in recent years in Chile and California, often at the expense of native woodlands.

1.4.4 The Cape Region of South Africa

Human settlement history in the Cape Region of South Africa extends back as much as 150 000 years, and there is clear evidence for active hunter-gatherers of the Later Stone Age in this region 21 000 years ago (Deacon 1992). It is likely that herding of sheep and cattle by Bushmen tribes over at least 2000 years led to significant impacts on the fynbos ecosystems through changes in fire frequency and accelerated rates of erosion. The development and growth of a European population occurred slowly, as much of this region had soils too poor to support typical cereal and vegetable crops. Serious impacts of human disturbance began with the reduction in large animal biomass and diversity with hunting, and the cutting of the relict Afromontane forest patches along the south coast (Deacon 1992).

The remarkable species diversity of the Cape Floristic region within a relatively small spatial area presents unusual problems of human impacts on species diversity (Cowling 1992; Davis and Wynberg, Chap. 4; Cowling and McDonald, Chap. 10). While urbanization, agricultural clearance of areas with renosterveld communities, and tree plantations have had significant impacts on landscapes of the Cape Region, the striking element in the disturbance of fynbos landscapes has been the invasion of alien trees and shrubs (Richardson et al., Chap. 11). The rapid spread and establishment of dominance by such species in relatively undisturbed natural communities are unprecedented in any other mediterranean-type ecosystem (Richardson et al. 1992). The most dramatic changes have occurred with the establishment of Australian acacias along watercourses, and *Hakea sericea* and pine species (*P. pinaster* and *P. radiata*) into mountain fynbos. Changes in fundamental ecosystem processes have occurred with these invasions, including the acceleration of riverbank erosion, reductions in stream flow through increased evapotranspiration, changes in nitrogen cycling dynamics, and altered fire frequencies (Richardson et al. 1992; Stock et al. 1995; Le Maitre et al. 1996). Although undisturbed fynbos ecosystems have been widely invaded by alien plants, these communities do not seem to be unusually favorable for alien animals. Alien birds have been well studied and show correlations with the degree of disturbance to the habitat (Glyphis et al. 1986; Armstrong et al. 1994).

1.4.5 Southwestern Australia

Agricultural development of wheatlands and associated sheep grazing over the past century and a half has had a tremendous influence in southwestern Australia in fragmenting kwongan and mallee habitats within a matrix of agricultural lands. The resultant small fragments of kwongan and mallee vegetation within the wheat belt are generally too small to maintain viable populations of plants and animals, leading to rapid declines in native animal diversity and slower reductions in plant species diversity (Fox and Fox 1986). Only about 7% of original native vegetation remains within the wheat belt area, although this percentage varies greatly within the region, with the oldest settled areas retaining less vegetation (Hobbs, Chap. 5). Before modern settlement, this region consisted of a mosaic of vegetation types, including various woodlands, shrublands, and mallee types (Pate and Beard 1984).

Deforestation has been a major landscape change in native eucalypt forests of the wetter areas of southwestern Australia. Deforested areas are commonly replanted with either native or exotic species, but with mixed success as problems are encountered with both rising water tables and introduced pathogens (Hobbs 1992). The replacement of deep-rooted native trees with secondary vegetation of shallow-rooted herbaceous or semi-woody species has led to reduced community levels of evapotranspiration and thereby upward movement of saline paleosoil profiles. Resultant salinization of surface waters and agricultural land has impacted nearly 3% of the agricultural lands of Western Australia today (Peck 1978). As this progression from landscape clearance to agricultural development to salinization continues to occur, it seems clear that current levels of agricultural production in this important region are not sustainable (Hobbs, Chap. 5).

The establishment of an apparently exotic fungal pathogen, *Phytophora cinnamomi*, following disturbance has had a significant impact on forest and heathland communities. Dramatic effects on the floristic structure of woody species have occurred in areas where this pathogen has become established (Dell and Malajczuk 1989). Die-back of native species infected by *Phytophora* is particularly significant in the Proteaceae, a family which naturally forms a major element of kwongan and woodland cover in southwestern Australia (Hobbs 1992).

Exotic species introductions have also had significant impacts on biological diversity in southwestern Australia. Problems have frequently resulted from the deliberate introductions of plants, mammals, and fish to Western Australia for agriculture, horticulture, or sport without due consideration for the potential threats of invasiveness. Exotic species introductions of animals have also had dramatic effects on the community structure of shrublands and woodlands in southwestern Australia. The most significant of these introductions was that of the European rabbit (*Oryctolagus cuniculus*) which had become established in widespread feral populations by 1859 (Fox and Fox 1986). Although biological controls sharply reduced these populations in the 1950s, rabbits nevertheless had a dramatic effect on plant community structure through a century of their selective grazing and browsing. Such grazing opened up many areas to invasion by exotic plant species.

Introductions of exotic carnivores such as foxes, cats, and ferrets in Australia have taken a heavy toll on native mammals, ground- and hollow-nesting birds, lizards, and insects (Hobbs et al. 1993; Fox, Chap. 19). While habitat fragmentation has traditionally been considered to be the primary cause of declines in Australian faunal populations, the commencement of rapid fragmentation coincided with the arrival of rabbits and foxes. It is likely that these invading species, and the measures used to control them by the human population, were responsible for much of the faunal decline, especially for mammals (Hobbs et al. 1993).

1.5 Resiliency of Mediterranean-Type Ecosystems to Disturbance

There has been an increasing level of interest in mediterranean-climate regions in issues of ecological restoration. Such restoration is normally designed to intentionally alter a degraded system with the goal of reestablishing the structure, function, and dynamics of an historic ecosystem (Fig. 1.3). The approaches used in such projects may take many forms (Bradshaw and Chadwick 1980; Cairns 1986; Jordan et al. 1990). One such form is replacement, in which a degraded ecosystem is replaced with a

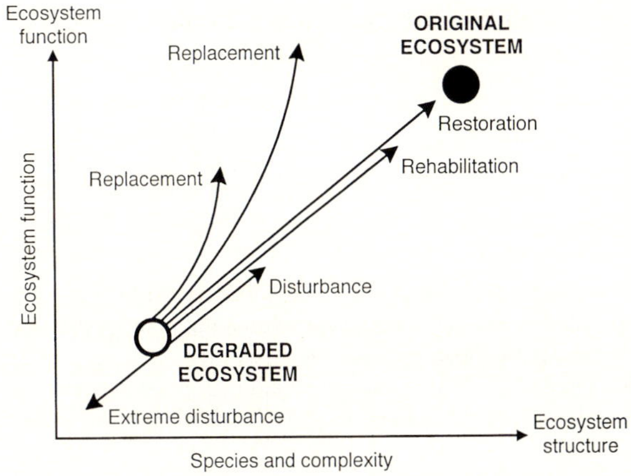

Fig. 1.3. The relationship between species diversity and ecosystem function, measured as productivity, in natural and degraded ecosystems. (Adapted from Bradshaw 1984)

more productive ecosystem, but not the original type. This often occurs, for example, when highly degraded forest or shrubland communities are altered to become grazing lands. In rehabilitation, a degraded ecosystem is restored to the extent that at least there is a return of some of the original ecosystem functions and dynamics, if not the original species. Finally, ecological restoration in the full sense involves the re-establishment of natural community biodiversity as well as function. A critical question in both the philosophical basis of restoration ecology and the successful implementation of restoration projects remains the extent to which our understanding of biodiversity and ecological function can allow us to reassemble a functioning ecosystem from its component parts (Diamond 1990). Despite a major focus of international interest in the relationship between biodiversity and ecosystem function (Schulze and Mooney 1995; Tilman et al. 1997; Chapin et al. 1998), there is much still to be learned about this central topic. Mediterranean-climate ecosystems provide important models for searching for new paradigms relating biodiversity with stability and resilience (Davis and Richardson 1995).

Given the nature of strong and continuing human actions impacting mediterranean-climate landscapes throughout the world, how resilient are these ecosystems to such impacts and alterations in natural disturbance regimes? In past discussions of this theme (Dell et al. 1986), it was concluded that detailed studies of ecological, demographic, and ecophysiological aspects were all critical elements in developing a better understanding of resilience in mediterranean-climate regions. This volume presents an attempt to advance our understanding of this theme through a variety of approaches, including landscape-level assessments and ecological studies, as well as an exploration of themes of biodiversity and disturbance for both plant and animal populations in all five mediterranean-climate regions of the world.

References

Anderson K (1993) Native Californians as ancient and contemporary cultivators. In: Blackburn TC, Anderson K (eds) Before the wilderness: environmental management by native Californians. Ballena Press, Menlo Park, pp 151–174

Andow DA, Kareiva PM, Levin SA, Okubo A (1993) Spread of invading organisms: patterns of spread. In: Kim K, McPheron BA (eds) Evolution of insect pests: patterns of variation. John Wiley, New York, pp 219–242

Armstrong AJ, Van Hensbergen HJ (1994) Comparison of avifaunas in *Pinus radiata* habitats and indigenous riparian habitat at Jonkershoek, Stellenbosch. S Afr J Wildl Res 24: 48–55

Aschmann H (1973) Man's impact on the several regions with mediterranean climates. In: di Castri F, Mooney HA (eds) Mediterranean-type ecosystems: origin and structure. Springer, Berlin Heidelberg New York, pp 363–371

Bahre C (1979) Destruction of the vegetation in northern central Chile. Univ Calif Publ Geogr 23:1–117

Balduzzi A, Tomaselli R, Serey I, Villaseñor R (1982) Degradation of the mediterranean type of vegetation in central Chile. Ecol Medit 8:223–240

Blackburn TC, Anderson K (eds) (1993) Before the wilderness: environmental management by native Californians. Ballena Press, Menlo Park

Bradshaw AD (1984) Ecological principles and land reclamation practice. Landscape Plann 11:35–48

Bradshaw AD, Chadwick MJ (1980) The restoration of land. Blackwell Scientific, Oxford

Cairns J (1986) Restoration, reclamation, and regeneration of degraded and destroyed ecosystems. In: Soulé M (ed) Conservation biology: the science of scarcity and diversity. Sinauer Associates, Sunderland, Massachusetts, pp 153–181

Caravello GU, Giacomin F (1993) Landscape ecology aspects in a territory centuriated in Roman times. Landscape Urban Plann 24:77–85

Chapin FS III, Sala OE, Burke IC, Grime JP, Hooper DU, Lauenroth WK, Lombard A, Mooney HA, Mosier AR, Naeem S, et al. (1998) Ecosystem consequences of changing biodiversity. Bioscience 48:45–52

Cowling RM (ed) (1992) The ecology of fynbos: nutrients, fire and diversity. Oxford University Press, Cape Town

Cowling RM, Rundel PW, Lamont BB, Arroyo MK, Arianoutsou M (1996) Plant diversity in mediterranean-climate regions. Trends Ecol Evol 11:352–360

Davis GW, Richardson DM, eds (1995) Mediterranean-type ecosystems: the function of biodiversity. Springer, Berlin Heidelberg New York

Deacon HJ (1992) Human settlement. In: Cowling RM (ed) The ecology of fynbos: nutrients, fire and diversity. Oxford University Press, Cape Town, pp 260–270

Dell B, Malajczuk N (1989) Jarrah dieback – a disease caused by *Phytophora cinnamomi*. In: Dell B, Havel JJ, Malajczuk N (eds) The jarrah forest. A complex mediterranean ecosystem. Kluwer, Dordrecht, pp 67–87

Dell B, Hopkins AJM, Lamont BB (eds) (1986) Resilience in mediterranean-type ecosystems. Dr W Junk, Dordrecht di Castri F (1981) Mediterranean-type shrublands of the world. In: di Castri F, Goodall DW, Specht RL (eds) Mediterranean-type shrublands. Elsevier, Amsterdam, pp 1–52

di Castri F, Goodall DW, Specht RL (eds) (1981) Mediterranean-type shrublands. Elsevier, Amsterdam

di Castri F, Hansen AJ, Debussche M (eds) (1990) Biological invasions in Europe and the Mediterranean Basin. Kluwer, Dordrecht

Diamond JM (1990) Reflections on the goals and on the relationship between theory and practice. In: Jordan WR, Gilpin ME, Aber JD (eds) Restoration ecology: a synthetic approach to ecological research. Cambridge University Press, Cambridge, pp 329–336

Etienne M, Faúndez L, Valdés J (1986) Evaluación de la tasa de desertificación en la zona árida de Chile central. Ambiente y Desarrollo 2:105–112

Fantechi R, Margaris N (eds) (1986) Desertification in Europe. Reidel, Dordrecht

Farina A (1989) Recent changes of the mosaic patterns in a montane landscape (north Italy) and consequences on vertebrate fauna. In: Baudry J, Bunce RGH (eds) Land abandonment and its role in conservation. Proc Zaragossa/Spain Seminar 10–12, Dec 1989. Options Mediterraneennes (Ser A) 15:121–134

Fox BJ, Fox MD (1986) Resilience of animal and plant communities to human disturbance. In: Dell B, Hopkins AJM, Lamont BB (eds) Resilience in mediterranean-type ecosystems. Dr W Junk, Dordrecht, pp 39–64

Fuentes ER, Hajek ER (1979) Patterns of modification in relation to agricultural practice in central Chile. Environ Conserv 6:265–271

Fuentes ER, Hoffmann AJ, Poiani A, Alliende MC (1986) Vegetation change in large clearings: patterns in the Chilean matorral. Oecologia 68:358–366

Fuentes ER, Jaksic F, Simonetti J (1983) European rabbits vs. native mammals: effects on shrub seedlings. Oecologia 58:411–414

Glyphis JP, Milton SJ, Siegfried WR (1986) Dispersal of *Acacia cyclops* by birds. Oecologia 48:138–141

Groves RH, di Castri F (eds) (1991) Biogeography of mediterranean invasions. Cambridge University Press, Cambridge

Gulmon SL (1977) A comparative study of grasslands of California and Chile. Flora 166:261–278

Heady HF (1977) Valley grassland. In: Barbour MG, Major J (eds) Terrestrial vegetation of California. John Wiley, New York, pp 491–514

Hobbs RJ (ed) (1992) Biodiversity of mediterranean ecosystems in Australia. Surrey Beatty & Sons, Chipping Norton

Hobbs RJ, Saunders DA, Arnold GW (1993) Integrated landscape ecology: a Western Australian perspective. Biol Conserv 64: 231–238

Jackson LE (1985) Ecological origins of California's mediterranean grasses. J Biogeogr 12: 349–361

Jordan WR, Gilpin ME, Aber JD (eds) (1990) Restoration ecology: a synthetic approach to ecological research. Cambridge University Press, Cambridge

Kaplan D (1992) Responses of mediterranean grassland plants to gazelle grazing. In: Thanos CA (ed) Plant-animal interactions in mediterranean-type ecosystems. University of Athens, Athens, pp 75–79

Kuzucuoglu C (1989) Fires in the Mediterranean region. Blue Plan. Ecology 72:371–412

Le Houérou HN (1981) Impact of man and his animals on mediterranean vegetation. In: di Castri F, Goodall DW, Specht RL (eds) Mediterranean-type shrublands. Elsevier, Amsterdam, pp 479–521

Le Maitre DC, van Wilgen BW, Chapman RA, McKelly DH (1996) Invasive plants and water resources in the Western Cape Province, South Africa: modeling the consequences of a lack of management. J Appl Ecol 33:161–172

Manders PT, Richardson DM (1992) Colonization of Cape fynbos communities by forest species. For Ecol Manag 48:277–293
Meeus J, Wijermans MP, Vroom MJ (1990) Agricultural landscapes in Europe and their transformation. Landscape Urban Plann 18:289–335
Midgely JJ, Cowling RM, Seydack AHW, van Wyk GF (1997) Forest. In: Cowling RM, Richardson DM, Pierce SM (eds) Vegetation of southern Africa. Cambridge University Press, Cambridge, pp 278–299
Minnich RA (1983) Fire mosaics in southern California and northern Baja California. Science 219:1287–1294
Mooney HA, Conrad CE (tech co-ord) (1977) Symposium on the environmental consequences of fire and fuel management in mediterranean ecosystems. USDA For Serv Gen Tech Rep WO-3
Mooney HA, Drake JA (eds) (1986) Ecology of biological invasions of North America and Hawaii. Springer, Berlin Heidelberg New York
Naveh Z (1982) Mediterranean landscape evolution and degradation as multivariate biofunctions: theoretical and practical implications. Landscape Urban Plann 9:125–146
Naveh Z (1990) Fire in the Mediterranean: a landscape perspective. In: Goldhammer JG, Jenkins MJ (eds) Fire in ecosystem dynamics. SPB Academic Publ, The Hague, pp 401–434
Naveh Z (1991) Mediterranean uplands as anthropogenic perturbation dependent systems and their dynamic conservation management. In: Ravera OA (ed) Terrestrial and aquatic ecosystems, perturbation and recovery. Ellis Horwood, New York, pp 544–556
Naveh Z, Dan J (1973) The human degradation of mediterranean landscapes in Israel. In: di Castri F, Mooney HA (eds) Mediterranean-type ecosystems: origin and structure. Springer, Berlin Heidelberg New York, pp 373–390
Naveh Z, Lieberman AS (1993) Landscape ecology - theory and applications, 2nd edn. Springer, Berlin Heidelberg New York
Naveh Z, Whittaker RH (1979) Structural and floristic diversity of shrublands and woodlands in northern Israel and other mediterranean areas. Vegetatio 41:171–190
Noble JC, Bradstock RA (1989) Mediterranean landscapes in Australia: mallee ecosystems and their management. CSIRO, East Melbourne
Oechel WC, Moreno JM (eds) (1994) The role of fire in mediterranean-type ecosystems. Springer, Berlin Heidelberg New York
Ovalle C, Aronson J, Del Pozo A, Avendaño J (1990) The espinal: agroforestry systems of the mediterranean-type climate region of Chile. Agrofor Syst 10:213–239
Ovalle C, Avendaño J, Del Pozo A, Aronson J (1996) Land occupation patterns and vegetation structure of the anthropogenic savannas (espinales) of central Chile. For Ecol Manag 86:129–139
Pate JS, Beard JS (1984) Kwongan: plant life of the sand plains. University of Western Australia Press, Nedlands
Peck AJ (1978) Salinization of non-irrigated soils and associated streams: a review. Aust J Soil Res 16:157–168
Perez-Trejo F (1994) Desertification and land degradation in the European Mediterranean. Office for Official Publications of the European Communities, Luxembourg
Perrins J, Williamson M, Fitter A (1992) A survey of differing views of weed classification: implications for regulation of introductions. Biol Conserv 60:47–56
Pinto-Correia T (1993) Threatened landscapes in Alentejo, Portugal: the *montado* and other agro-silvo-pastoral systems. Landscape Urban Plann 24:43–48
Planty-Tabacchi A, Tabacchi E, Naiman RJ, Deferrari C, Dcamps H (1996) Invasibility of species-rich communities in riparian zones. Conserv Biol 10:598–607
Polunin O, Walters M (1985) A guide to the vegetation of Britain and Europe. Oxford University Press, Oxford

Reichard SH, Hamilton K (1997) Predicting invasions of woody plants introduced into North America. Conserv Biol 11:193–203

Rejmanek M, Richardson DM (1996) What attributes make some plant species more invasive? Ecology 77:1655–1661

Rejmanek M, Thomsen CD, Peters ID (1991) Invasive vascular plants in California. In: Groves RH, di Castri F (eds) Biogeography of mediterranean invasions. Cambridge University Press, Cambridge, pp 81–101

Richardson DM, Cowling RM, Le Maitre DC (1990) Assessing the risk of invasive success in *Pinus* and *Banksia* in South African mountain fynbos. J Veg Sci 1:629–642

Richardson DM, MacDonald IAW, Holmes PM, Cowling RM (1992) Plant and animal invasions. In: Cowling RM (ed) The ecology of fynbos: nutrients, fire and diversity. Oxford University Press, Cape Town, pp 271–308

Roy J, Navas ML, Sonié L (1991) Invasion by annual brome grasses: a case study challenging the homocline approach to invasions. In: Groves RH, di Castri F (eds) Biogeography of mediterranean invasions. Cambridge University Press, Cambridge, pp 207–224

Rundel PW (1981a) The matorral zone of central Chile. In: di Castri F, Goodall DW, Specht RL (eds) Mediterranean-type shrublands. Elsevier, Amsterdam, pp 175–201

Rundel PW (1981b) Fire as an ecological factor. In: Lange OL, Nobel PS, Osmond CB, Ziegler H (eds) Physiological plant ecology I: response to the physical environment. Springer, Berlin Heidelberg New York, pp 501–538

Rundel PW (1983) Impact of fire on nutrient cycles in mediterranean-type ecosystems, with reference to chaparral. In: Kruger FJ, Mitchell DT, Jarvis JUM (eds) Mediterranean-type ecosystems: the role of nutrients. Springer, Berlin Heidelberg New York, pp 192–207

Rundel PW (1995) Adaptive significance of some morphological and physiological characteristics in mediterranean plants: facts and fallacies. In: Roy J, Aronson J, di Castri F (eds) Time scales of biological responses to water constraints: the case of mediterranean biota. SPB Academic Publ, Amsterdam, pp 119–139

Rundel PW, Vankat JL (1989) Chaparral communities and ecosystems. In: Keeley S (ed) The California chaparral; paradigms reexamined. Los Angeles County Museum of Natural History, Los Angeles, pp 127–139

Rundel PW, Parsons DJ, Gordon DT (1977) Montane and subalpine vegetation of the Sierra Nevada and Cascade Ranges. In: Barbour MG, Major J (eds) Terrestrial vegetation of California. John Wiley, New York, pp 559–599

Schulze E-D, Mooney HA (eds) (1993) Biodiversity and ecosystem function. Springer-Verlag, Berlin Heidelberg New York

Shigesada N, Kawasaki K (1996) Biological invasions: theory and practice. Oxford University Press, Oxford

Specht RL (1981) Mallee ecosystems in southern Australia. In: di Castri F, Goodall DW, Specht RL (eds) Mediterranean-type shrublands. Elsevier, Amsterdam, pp 203–231

Stock WD, Wienand KT, Baker AC (1995) Impacts of invading N_2-fixing *Acacia* species on patterns of nutrient cycling in two Cape ecosystems: evidence from soil incubation studies and ^{15}N natural abundance values. Oecologia 101:375–382

Takhtajan A (1986) Floristic kingdoms of the world. University of California Press, Berkeley

Tilman D, Knops J, Wedin D, Reich P, Ritchie M, Siemann E (1997) The influence of functional diversity and composition on ecosystem processes. Science 277:1300–1302

Tomaselli R (1981) Relations with other ecosystems: temperate forest, mediterranean coniferous forest, savannahs, steppes and desert shrublands. In: di Castri F, Goodall DW, Specht RL (eds) Mediterranean-type shrublands. Elsevier, Amsterdam, pp 123–130

Trabaud L, Prodon R (eds) (1993) Fire in mediterranean ecosystems. Commission of European Communities, Brussels van Wilgen BW, Richardson DM, Kruger FJ, van Hensbergen HJ (eds) (1992) Fire in South African mountain fynbos: ecosystem, community and species response at Swartsboskloof. Springer, Berlin Heidelberg New York

Warburg M, Ben-Horin RA, Rankevich D (1978) Rodent species diversity in mesic and xeric habitats in the Mediterranean region of northern Israel. J Arid Environ 1:63–69

2 From Biodiversity to Ecodiversity – Holistic Conservation of the Biological and Cultural Diversity of Mediterranean Landscapes

Z. Naveh

2.1 Introduction

Human disturbance in mediterranean-type ecosystems continues to be a subject of serious concern, and nowhere is such disturbance and degradation of natural ecological processes and biodiversity more evident than in the Mediterranean Basin itself. In this chapter I will point out currently prevailing trends in Mediterranean landscapes in the context of our global environmental crisis. This will lead to a discussion of the close interrelations between biological diversity, ecological heterogeneity, and cultural diversity, or in one word, total landscape *ecodiversity* in the Mediterranean Basin, and to recent advances in landscape ecology in the evaluation of this issue. I will conclude by offering some new approaches and tools for holistic ecodiversity conservation as an integral part of the urgently needed environmental revolution.

2.2 The Neo-Technological Impoverishment of the Open Mediterranean Landscape

2.2.1 Current Trends

Current trends in Mediterranean landscapes can be summarized in one sentence as the "bad and good news". I will start with the bad news: the quality of the environment and of the open landscape around the Mediterranean Basin is further deteriorating at an exponential rate in time and space. In a recent volume on landscape ecology (Naveh and Lieberman 1994) we have described these threatening trends of *neo-technological* landscape degradation as a major part of the global environmental crisis. It is operating in the Mediterranean Basin as a vicious circle of mutually amplifying and destabilizing feedbacks of exponentional growth of popula-

Ecological Studies, Vol. 136
Rundel et al. (eds.) Landscape Degradation and Biodiversity in Mediterranean-Type Ecosystems

tions and their needs for open land, coupled with the land use pressures of rapidly growing tourism. These are combined with synergistic processes of greater intensity of traditional and modern agricultural land uses and urban-industrial expansion on one hand, and depopulation and land abandonment on the other.

If the scenarios of The Blue Plan of the United Nations Environmental Program (UNEP 1988) come true, then the total population around the Mediterranean Basin will grow from about 360 million to more than 550 million in the next 30 years, nearly two-thirds of this in countries south and east of the basin. The main threats to environmental quality are in the coastal regions with the largest urban industrial concentrations and the fastest growing tourism developments. Here, population will increase from about 140 million at present to 195–217 million in 2025, an increase of 45–62%. Its urban populations alone will rise from 82 to 145–170 million. In addition to around 200 existing energy and industrial installations of oil ports, refineries and thermal power plants, many more will be required by 2025, in addition to cement plants, steel works, fertilizer plants, etc. At the same time, the tourist flow to the Mediterranean coast will increase from 95 million today to 220 million tourists in 2025 and may even treble. About 40% of these holiday makers are merely seeking relaxation, sun, sand and the sea, without caring much about natural and cultural values. Therefore, their mass-accommodation close to the beaches is the main socio-economic and cultural force driving further uncontrolled development of the rapidly diminishing open shorelines and their intrinsic biological, scenic and other *soft* landscape values over much larger areas. Thus, for instance, Marcuzzi (1991) has used the disappearance of typical sand dune tenebrionid beetles from the Adriatic coast of Italy as a sensitive bioindicator of the detrimental impacts of mass tourism on coastal dune areas.

The resulting rapid loss of almost all natural coastal sand dune habitats and wetlands and the advanced degrees of air and water pollution in the lowlands are combined with a rapid process of homogenization, fragmentation, denudation and destabilization in the uplands. The loss of the vital protection and carrier functions of the vegetated mountain watersheds have far-reaching and many times even catastrophic implications for the intensively cultivated and/or densely populated valleys and coastal areas. The final result is accelerated and widespread decline of the unique biological, ecological, and cultural ecodiversity, productivity and stability, leading to more frequent and more destructive wildfires, and to accelerated soil erosion and flooding.

Unfortunately, the ecological, social, and economic damage inflicted by such land abuse and unrestrained developments are not included in the narrow and short-sighted market-based cost/benefit calculations which are still guiding the decisions on land uses in most Mediterranean countries.

One of the major causes for this ecological debacle is the rapid and careless shift from diversified and stable traditional agriculture to intensive large-scale, agro-industrial farming. Chiefly in the western Mediterranean, such farming has caused widespread land abandonment. But in all Mediterranean countries it is combined with indiscriminate planting of pine and eucalyptus monocultures creating large stretches of monotonous and highly flammable vegetation. These developments have destabilized and despoiled rich and unique landscapes. In spite of their detrimental ecological impacts and their highly dubious economic justification, such practices have been promoted by local governments and were heavily subsidized by the European Unit (EU). These high technology and high-input agricultural practices are further increasing rates of air, water and soil pollution, soil erosion, salinization, and siltation.

A typical example is Extremadura, one of the poorest regions in southern Spain (CEPA 1992). Here, the EU has allocated great sums from its European Structural Funds, in order "to develop and to bring this region into line with the rest of Europe". In reality, however, these developments have undermined the local economy by destroying thousands of hectares of Mediterranean woodlands, thereby altering traditional agricultural practices and degrading the natural, scenic and cultural resources of this region. In addition to the great investments in the lowlands, huge sums have been spent on large-scale afforestation projects, and on so-called conservation and erosion control projects. For these, heavy machinery and bulldozers are used to clear existing maquis vegetation and build roads and terraces. Instead of controlling water run-off and soil erosion by vegetative stabilization and appropriate engineering, these projects have accelerated processes of landscape degradation by leveling large areas and removing their protective natural vegetation cover. In this way, the upper and richest soil profiles which had taken hundreds of years to build have been destroyed. Severe sedimentation occurs as a result, with more than 30 tons of soil per hectare washed away every year.

Among those threats whose danger has not yet been recognized sufficiently is the alarming rise of photochemical oxidant pollutants and especially of tropospheric ozone oxidant pollutants and acid deposition. Such atmospheric pollutants may be playing a major role in the progressive decline of coniferous forests in the Mediterranean Basin (Ferretti et al. 1995; Milan et al. 1996). As in California, these threats from photochemical air pollutants are not restricted to the coastal urbanized zones close to the emission sources, but spread also farther inland (Naveh et al. 1981). In 1987, 82% of *Pinus pinea* in Sardinia, and 78% of these trees in Tuscany showed evidence of pollution damage. Other woody species, especially evergreen oaks and other broad-leaved maquis trees, were also declining (Bussotti et al. 1991, 1992).

In their thorough study of the transformation of agricultural landscapes in Europe, Meeus et al. (1990) predicted that in the future, the *montados* of the Iberian Peninsula and the Italian *coltura promiscua* landscapes will no longer exists, except in history books or national parks and outdoor museums, because of marginalization, abandonment, deterioration and the futile attempts of reforestation of the mountain slopes.

Such a doomsday scenario seems to approach closer every year, and landscape disturbance is now affecting remote mountainous uplands. The speed and extent of human encroachment can be demonstrated in the western Galilee of Israel (Fig. 2.1) Here, within 14 years, most of the open agro-pastoral (so-called rugged land) has been lost or fragmented by monospecific pine plantations and intensive agriculture, as well as by urban-industrial developments.

As a result of large-scale emigration and agricultural intensification (encouraged by EU policies and subsidized wheat prices) over large areas of these uplands in the western Mediterranean, the rich mosaics of coppiced and managed oak and chestnut forests, mountain grasslands, and the terraced mixed vine/olive and crop cultures are being replaced by intensive monocultures of wheat and grapes, or by olive orchards on non-terraced slopes, and by pine plantations and secondary scrub on abandoned pastures and fields. Since the soils of non-terraced slopes are cultivated now by heavy machinery they are suffering from erosion and losing their fertility.

This breaking down of fine-grained patterns into coarser landscape patterns with impoverished floras and faunas has been documented in the Solano Basin of Tuscany (Vos and Stortelder 1992). If this process of biological, cultural and scenic impoverishment continues unimpeded, then, the four richest out of ten major land units will vanish completely by the year 2035. The remaining land units will be chiefly intensively cultivated, heavily sprayed and fertilized fields, with planted coniferous forests spreading over the whole uncultivated area, and secondary communities on abandoned pastures, thereby creating the same species-poor oak and pine forest types over extensive areas (Fig. 2.2). Many more examples of this closely interwoven biological, ecological and cultural impoverishment have been provided recently in a symposium dealing with land abandonment (Baudry and Bunce 1989), and in conference proceedings dealing with the future of Mediterranean landscapes (Farina and Naveh 1993).

One of the best examples of planning efforts toward the resource management of Mediterranean landscapes comes from an interdisciplinary

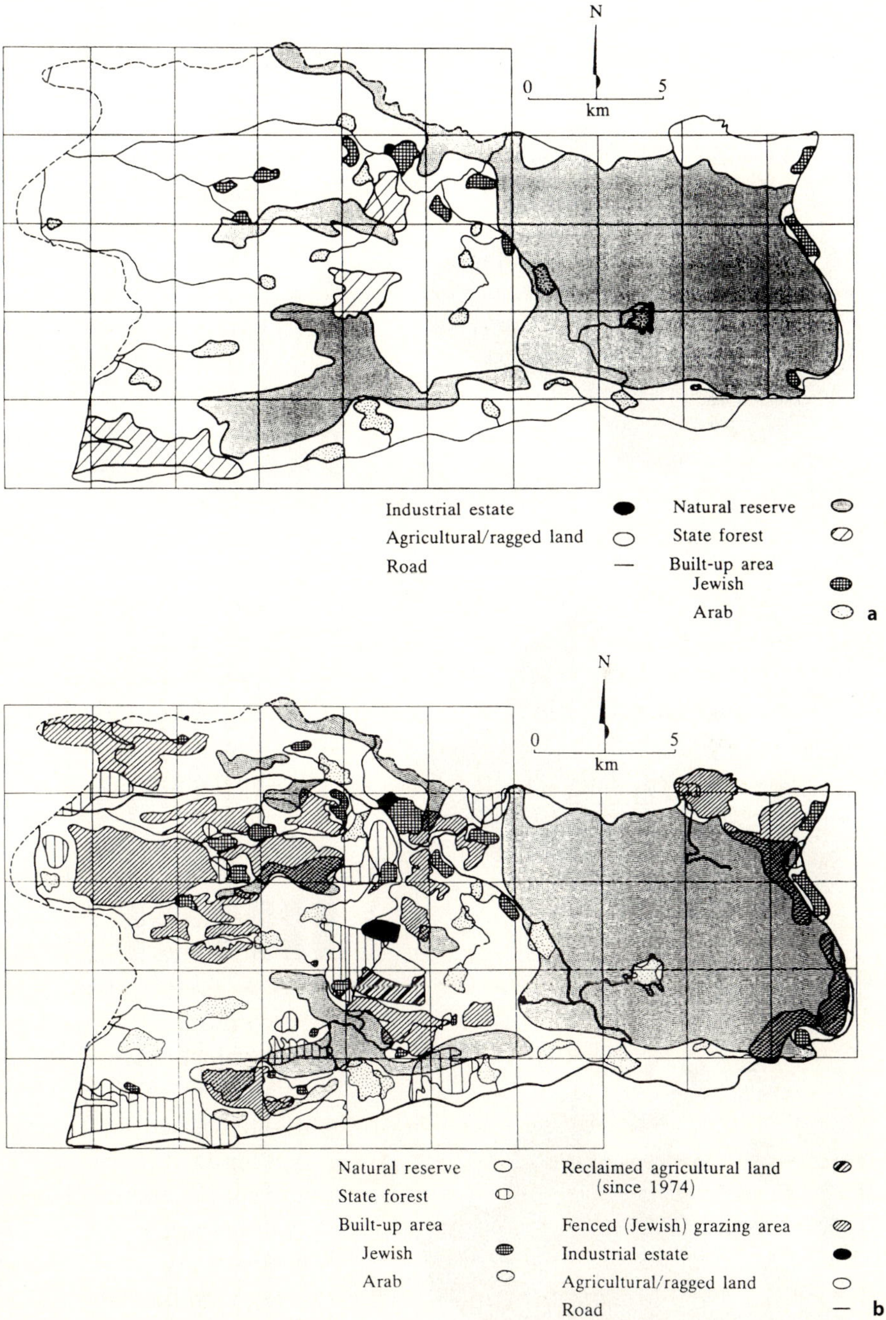

Fig. 2.1. Major land uses in the Yehiam region western Galilee, Israel: *a* 1974; *b* 1988 (Yiftach-El and Law-Yone 1995)

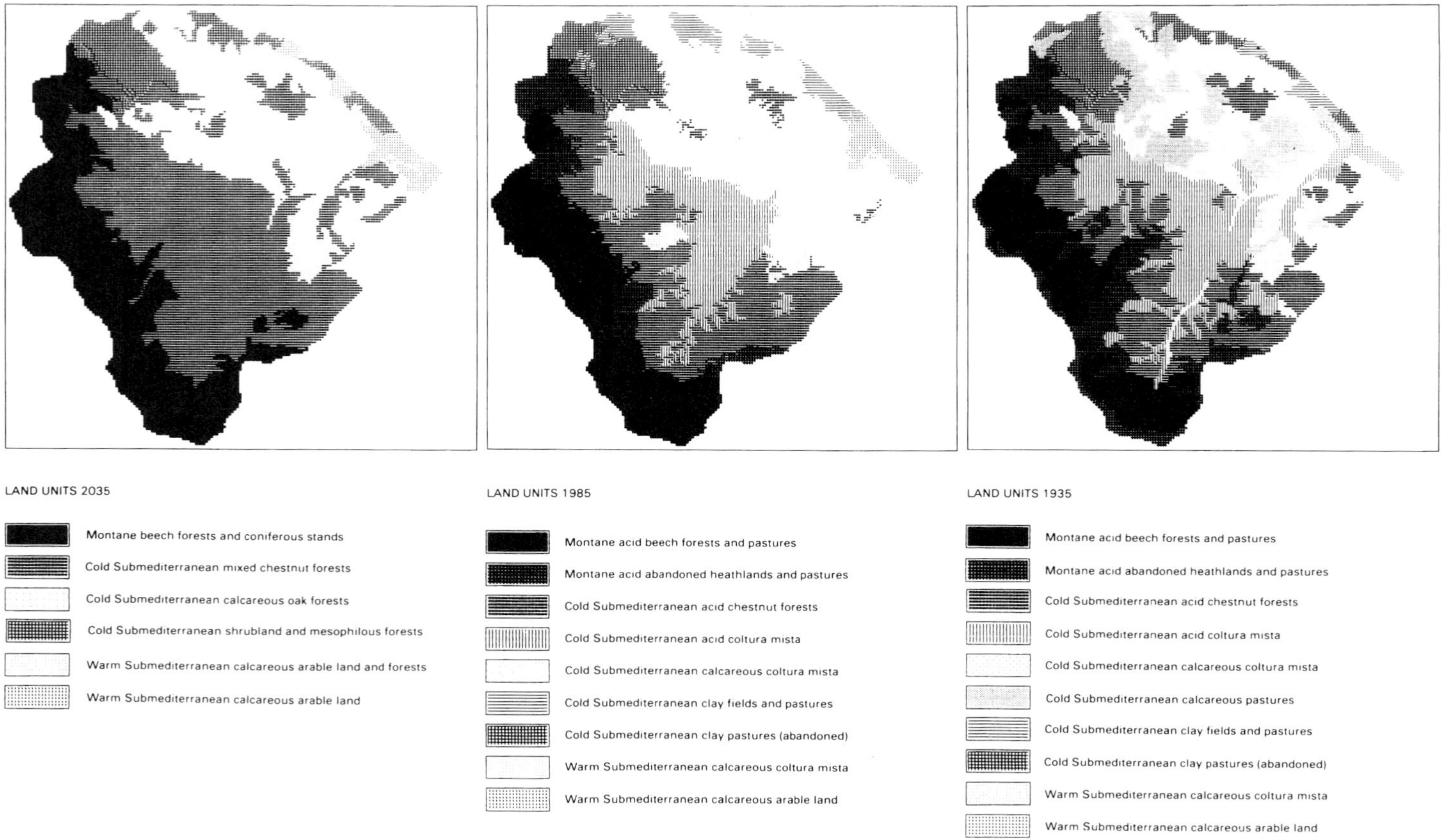

Fig. 2.2. The vanishing occurrence of Tuscan landscape units between 1935 and 2015 (Vos And Stortelder 1992)

study of landscape planning and ecology in western Crete. This study, described in detail in Chapter 7 of this book, was initiated by the Working Group for the Conservation of Landscapes of the IUCN Commission on Environmental Strategies and Planning (WGCL-CESP) as the first case study of *Green Books for Landscape Conservation,* and was funded by the EU EPOCH program (European Program on Climatology and Natural Hazards). The study was carried out by a multinational and interdisciplinary team of landscape ecologists and planners, botanists, physical and cultural geographers, forest and range, and remote sensing specialists.

However, the most severe problems of environmental degradation come from the fastest growing semi-arid countries of the eastern and southern Mediterranean Basin. Here, the major causes for land degradation and desertification are overgrazing and excessive removal of fuelwood. According to FAO studies, in the year 2000 about 140 million people in these countries will suffer a deficit of close to 30 million m^3 of wood to supply their annual needs for cooking and heating. Taking into consideration the low wood production of these forests and woodlands, extreme deforestation of twindling woodlands will likely affect very large areas.

2.2.2 Impacts of Global Change

Predicted regional warming and the destabilization of rainfall and temperature regimes resulting from global change may have severe repercussions on the biological diversity and stability of inland and upland landscapes in the Mediterranean Basin. In addition to the destructive impact of rising sea levels on coastal shores, and the far-reaching effects on hydrological regimes and on agricultural crops, global change could further aggravate the above-mentioned processes of landscape degradation (Jeftic 1993).

In the more humid western Mediterranean ecosystems these predicted climatic changes could further enhance brush encroachment in abandoned and neglected forests and maquis, and thus encourage their floristic impoverishment and increase wildfire hazards. In the subhumid ecosystems such as those of northern Israel, the expected rise in summer temperatures and in evaporation rates will have severe repercussions on the diversity and stability of the most mesic and diverse plant communities (Naveh 1995a).

In the driest semi-arid ecotones of the Mediterranean these climatic stresses would even further aggravate the existing processes of landscape desertification induced by heavy human and livestock pressures. This may cause the advancement of semi-arid and semi-desert formations far into the heart of the Mediterranean biomes. Considering that the desert encroachment into the semi-arid Mediterranean region is presently estimated at more than 2% per annum, this would have truly catastrophic consequences (Le Houerou 1992).

Because of the chaotic and non-linear behavior of atmosphere-biosphere interactions, and their complex and mostly synergistic interactions with other environmental stresses, our present knowledge is not sufficient for any detailed, robust prediction models. Therefore, at this stage the only certain prediction which can be made is that we will face a period of increasing uncertainty in spatial and temporal climatic trends and their ecological effects. We can also anticipate with great certainty that any increase in climatic stress will further aggravate the process of overall landscape degradation (Naveh 1995).

2.3 New Hopes For Mediterranean Landscapes

The good news is that there is now also a rapidly growing public awareness of these perils. There are presently many more research activities and better conservation management policies leading in the right direction; and important international, national and non-governmental initiatives are instilling some hope that this situation can be changed for the better before it is too late in at least in some of these countries.

Among the most outstanding international examples is the Mediterranean Action Plan (MAP) of the United Nations Environmental Program (UNEP), which was approved in 1975 by 16 Mediterranean countries. It called for a series of legally-binding treaties to be drawn up and signed by the Mediterranean governments, the creation of a pollution monitoring and research network, and a socio-economic program to reconcile development priorities with a healthy Mediterranean environment. Up to now, the most important product of this group has been the Blue Plan (UNEP 1988), as the first prospective study on the relationship between the environment and development in the Mediterranean. It has reviewed present trends and also recommended scenarios for sustainable integrated social and economic development of the Mediterranean Basin.

Another promising development is related to the ongoing peace talks in the Middle East which began in 1992, and brought together ministers and top governmental delegates from all Arab countries and Israel. A constructive dialog on environmental issues and their solution at the level of specialists and decision-makers has resulted from these talks. If this and other important international and regional activities are taken seriously by law-makers, planners, politicians and decision-makers it may lead the way for a better future for these landscapes in all participating countries. This will be especially the case if the new EU agricultural policy recognizing the natural, cultural and socio-economic values of marginal lands is put into practice. The same is also true for other important initiatives such as a convention for the protection of Europe's rural landscapes.

The importance of holistic studies on landscape planning was in studies on vanishing Tuscan landscapes (Vos and Stortelder 1992). Based on this report, the regional Tuscan government requested that there be a comprehensive landscape assessment of the ecological impacts of the planned Farva River dam and impoundment. The conclusions of this report led to cancellation of ambitious plans for intensive agricultural development and irrigation of the biologically significant and scenic Grosseto Plains, which would have had disastrous environmental consequences (Pedroli et al. 1988). The regional Tuscan government has supported further ecological landscape studies, as well as conferences, workshops and training courses.

2.4 Biodiversity and Ecodiversity in the Mediterranean Basin

It is now generally recognized among all those who care for the fate of our planet that the most alarming immediate global threat to life on earth is that of biological impoverishment. The rapid reduction of the biotic resources threatens the integrity of the global biosphere landscapes and their life-support systems (Norton 1987; Kim and Weaver 1994a). The severity of this threat can be measured by the predicted per capita shrinkage of all human-modified and used landscapes in relation to population growth. According to the latest report of the Worldwatch Institute (1994), a predicted increase in the world population by 33% by the year 2010 will result in the drop of per capita forest cover by 30%, grazing and pasture land by 22%, cropland by 21%, and irrigated land by 12%. Although representing only a very small portion of all the terrestrial landscapes on earth, the Mediterranean Basin is of special significance because its unique evolutionary and cultural history makes it one of the richest biological regions outside the tropics.

As described elsewhere in detail (Naveh 1984, 1990b; Naveh and Lieberman 1994), we can assume that the final geomorphologic, climatological and biological evolution of Mediterranean landscapes in the Pleistocene coincided with the major phases of Mediterranean human evolution from the Acheulian hunter-gatherer *Homo erectus* to the food-producing neolithic *Homo sapiens*. In this closely physical, biological and cultural environment, both natural and anthropogenic fires became a major driving force in human culture. Human use of fire can be traced back several hundred thousand years in Israel, thanks to findings of stone tools, artifacts, and the charred and fractured bones of 80 different animal species in the fire-swept volcanic soils of the ecotones between semiarid tropical and Mediterranean forests and savannas of the Jordan Valley (Haas 1966; Stekelis 1966). This region was also one of the first cradles of cereal and stock farming about 10000 years ago, and its drought, fire and grazing-resistant annual grasses and legumes were among the first domesticated plants. This process of ecotypic variation and speciation was continued and gradually intensified during the long period of human occupation. Through this process, those Mediterranean uplands which are steep and rocky or have soils which are too poor and/or shallow and for intensive, mechanized agriculture, have become the largest refuge of spontaneously occurring and reproducing plants and animals in Europe, the Near East and North Africa. Because of their longer history of intensive human modification, these habitats have high degrees of structural, floral and faunal diversities compared to mediterranean-type landscapes in Chile and California (Naveh and Whittaker 1979).

The Mediterranean Basin flora is distinguished by its great number of herbaceous plants, including many annuals and exceptionally rich and colorful flowering Asteraceae and geophytes with ornamental value. Among these are many rare and endemic plants. For example, there are more than 1000 different species of flowering plants within a 10km radius from Jerusalem, while Greece contains more than 6000 different species of flowering plants. Much attention is presently devoted by geneticists and plant breeders to important cereals pasture grasses and legumes from the Mediterraean Basin such as Emmer wheat (*Triticum diccocoides*), the progenitor of cultivated wheat. Among woody plants, and especially in the Labiatae, there are many species with great value for pharmaceutical, cosmetic, spice, balsam, and other uses. These are now collected, grown and utilized with increasing intensity for commercial production. A number of deeply rooted, evergreen and hardy shrubs are now widely used for erosion control on steep and denuded slopes.

The genetic diversity on the subspecies level is indicated by the results of a study of 21 plant and animal species from the Mediterranean and steppe region of Israel which revealed high levels of allozyme diversity and DNA polymorphism. This genetic diversity has been hypothesized to correlate with the broad ecological plasticity shown by many of these species (Nevo 1988).

Mediterranean landscapes provide an ecological bridge between Europe, Africa and Asia, and their wetlands serve as vital breeding and feeding stations for hundreds of resident and migrating bird species crossing the Mediterranean Sea between Europe and Africa.

Contrary to a widely held misconception, Mediterranean peoples did not always degrade their habitats and deplete their flora and fauna. They reclaimed these uplands with great skill, utilized their biological productivity and preserved at the same time their organic diversity. This happened when ecological, socio-economic and cultural peaks were reached during biblical and classical times, and much later again during the Renaissance chiefly in the western Mediterranean.

It is not by chance that in these periods landscapes were perceived as the total natural and human living space, combining beauty and utility. In the Bible, the Hebrew word *noff* (etymologically related to *jafeh* meaning beautiful) symbolizes the overall view of King Solomon's Jerusalem and its surrounding green hills. The lofty mountains in which the Greeks erected their temples, symbolized the closeness of God to man in his natural, most beautiful and harmonious surroundings, and the graciously terraced slopes of olive groves and vineyards and interspersed with majestic cypresses and the dark green oak forests of the Tuscan landscape have served as a source of inspiration for many generations of artists and poets.

Di Castri (1981) has clearly described the closely interwoven natural and cultural multidimensional and multiscale heterogeneity of Mediterranean ecosystems and landscapes. This diversity was the result of the closely coupled natural and anthropogenic feedback processes which induced the co-evolution of Mediterranean peoples and their cultural landscapes.

There is a close interdependence between biological and cultural diversity and the threats both by neo-technological landscape degradation and by despoliation. Special efforts were made by a research team studying resource management issues in western Crete to distinguish genuine threats from changes that are part of the inherent cycle of changes, and to link ecological conservation to other branches such as archeological and historical conservation (Grove et al. 1993, 1994). This enabled the recognition of the threats to total landscape ecodiversity by the declining biological, historical, and scenic values. As a result, from the long interplay of a rugged topography with human activities and wild plants and animals, the White Mountain region of Crete continues to be a fine-grained and harmonious Mediterranean landscape, rich in cultural features.

2.5 Mediterranean Landscapes as Perturbation-Dependent Non-Equilibrium Systems

Several prominent American ecologists have proposed a new paradigm in ecology on which conservation and restoration ecology should be based.

This paradigm replaces the metaphor of a balance of nature maintained without human interference with the flow of nature through dynamic change induced by disturbances including humans and their effects (Pickett 1994) or by the recognition of *discordant harmonies* (Botkin 1990).

It was shown more than 20 years ago that the mechanistic and deterministic concepts of a homeostatic stable state equilibrium, presented in classical ecology by the succession-to-climax paradigm, is not suitable for conservation management in the Mediterranean Basin (Naveh 1974; Naveh and Dan 1973). We introduced new insights of systems theory and system dynamics into landscape ecology, especially the new thermodynamic principles of self-organization in non-equilibrium systems (Prigogine 1976; Prigogine and Stengers 1984) and the concepts of homeorhetic flow equilibrium developed by Waddington (1975) as a catastrophe theory of evolution (Naveh 1982; Naveh and Lieberman 1994). Any new paradigm in ecology with relevance for conservation management cannot ignore these and other new insights into systems dynamics.

As has been discussed previously in more detail (Naveh 1987, 1991a), Mediterranean landscapes, as well as most other cultural semi-natural and agro-silvo-pastoral landscapes, behave in thermodynamic terms like dissipative structures. They are maintained and stabilized only by permanently interchanging energy and entropy with their environment. Driven by positive feedbacks of environmental and internal fluctuations, they move to new intermediate states that generate conditions of renewal of higher internal entropy production, while undergoing short- and long-term cyclic fluctuations. Thus they are far from a homeostatic equilibrium stage.

Such cyclical perturbations have been introduced for centuries by rotational grazing, browsing, burning, cutting and coppicing regimes, together with cultivation and other human land uses. These human perturbations were superimposed on the seasonal and annual climatic fluctuations, and their resulting defoliation pressures were incorporated into the landscape, together with these and other natural perturbations at different spatio-temporal scales. They resulted in the establishment of a human-maintained and dynamic long- and short-term flow equilibrium or *homeorhesis* (from the Greek, meaning preserving the flow) between the tree, shrub, herb, and grass layers and the cultural artifacts of forests, woodlands, shrublands and grasslands.

In homeorhesis, the system is not returned to a stable state as in homeostatic climax systems, but moves along the same trajectory of change, as long as these cyclic perturbations are continued with similar intensities and frequencies. In this manner, these perturbation-dependent systems have acquired long-term adaptive resilience and evolutionary metastability. The long-term maintenance of such a homeorhetic flow equilibrium, operating within the great macro- and microsite heterogeneity of such rocky and rough terrain, producing the fine-grained agro-silvo-pastoral land use patterns, apparently played a major role in inducing the unique ecological and cultural ecodiversity of these landscapes.

A typical example for such a human-perturbation dependent landscape and the present neo-technological distortion of homeorhetic flow equilibrium, is the *montado* in southern Portugal and the *dehesa* in southwestern Spain. As described in more detail by Pinto-Correia (1993), this agro-silvo-pastoral system was well adapted to the low edaphic and climatic potentials of these uplands in the complementary utilization of the open evergreen woodlands of oaks (chiefly *Quercus suber*), olives and chestnut trees for cork, fruit, wood and livestock production, and cereal cultivation. Managed extensively for many decades with minimum human inputs, this land use created a fine- grained heterogeneous and attractive landscape with relatively high biodiversity. At the same time it furnished also the necessary *hard* economic outputs for the dispersed rural population. However, today there are highly intensive and extensive land uses that are distorting this human-maintained homeorhetic flow equilibrium and endangering not only the biological and economic productivity, but also the ecodiversity and stability of these unique cultural landscapes (Figs. 2.3–2.4).

The results of our studies of maquis forests and shrublands in northern Israel and southern France (Naveh and Whittaker 1979; Naveh 1982, Schreiber and Naveh, unpubl. data) can serve as a further indication of the disruption of this homeorhetic flow equilibrium. Wherever moderate, traditional defoliation pressures had maintained open, grassy patches for light demanding herbaceous plants, including also highly ornamental geophytes, we found a very high floristic and structural diversity. But where these pressures had ceased, either for the sake of nature protection or as a result of land abandonment, or where they had been intensified, both floristic and structural diversity had been reduced considerably. In most cases, species richness dropped by 75%, namely from more than 100–120 species to less than 30 species per 0.1 ha. Equitability was much lower and the dominance concentration of the most aggressive and shade tolerant trees and shrubs increased significantly. At the same time, structural vegetation diversity was also reduced considerably. Thus, for instance, on Mt. Carmel, the rich herbaceous understory and gap vegetation of semi-open and patchy sclerophyll forests and shrublands contained 93 species, as

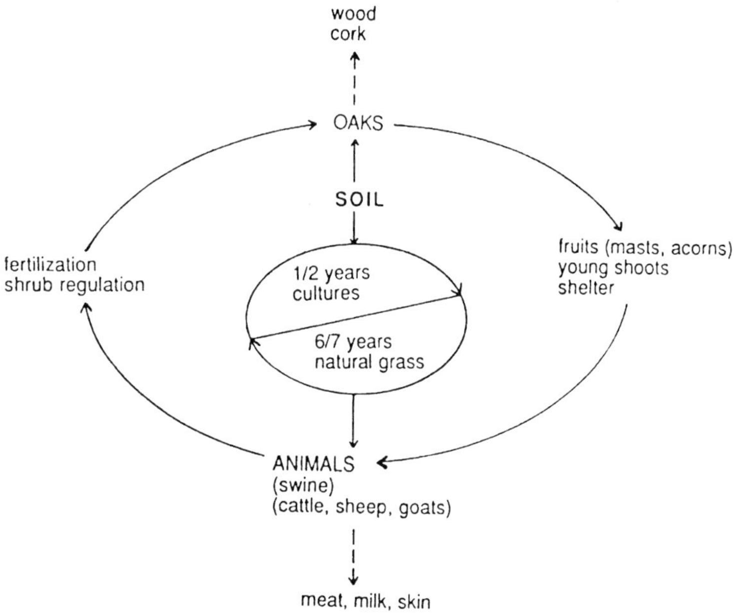

Fig. 2.3. Homeorhetic land use cycles in *montado* landscapes of Southern Portugal. (Pinto-Correia 1993)

opposed to only a few shade tolerant perennial grasses and herbs which could survive under the dense, undisturbed tall shrub canopy, dominated by *Quercus calliprinos*. This is shown in Fig. 2.6, comparing the dominance-diversity curves of the Muhraqa site which at that time had not been disturbed for more than 30 years, and the tall Forty Oaks Grove, which had been grazed and coppiced in the past. Here the woody species are well partitioned between the different strata and species, leading to a much lower dominance concentration and higher structural and floristic diversity contributed to by many small and densely packed annuals (therophytes) and a wealth of flowering geophytes, as well as hemicryptophytes. In simultaneously conducted zoological studies, animal species richness and relative abundance of birds, reptiles, rodents and isopods showed similar trends (Warburg 1977; Warburg et al. 1978 (Fig. 2.5).

In open oak woodlands of Tabor oak (*Quercus ithaburense*), which can be considered as the eastern Mediterranean counterpart to the *montados* and *dehesas*, highest species richness and diversity were reached under

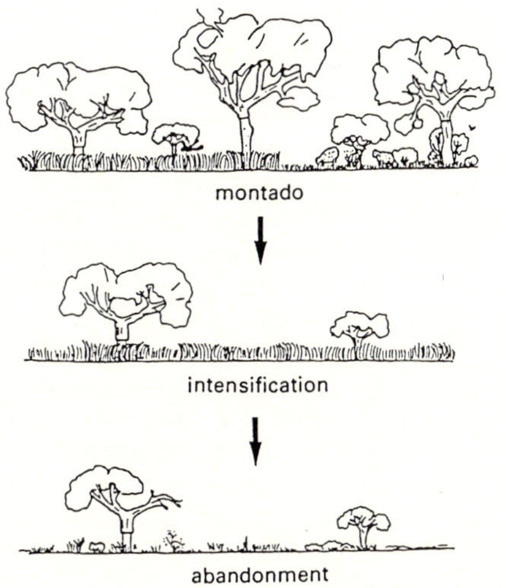

Fig. 2.4a. Degradation of the *montado* by intensified land use (Pinto-Correia 1993). In order to intensify crop production, oak density was reduced, the soil work mechanized, and grazing animals excluded. Trees were severely damaged, soil erosion increased, and crop yields remained too low. The fragile relationships of the *montado* system were broken and the resources deteriorated. The final result was abandonment, since the land no longer had production potential and there was no capacity for regeneration. In a further phase this area will probably be planted with eucalyptus trees for production of wood fiber

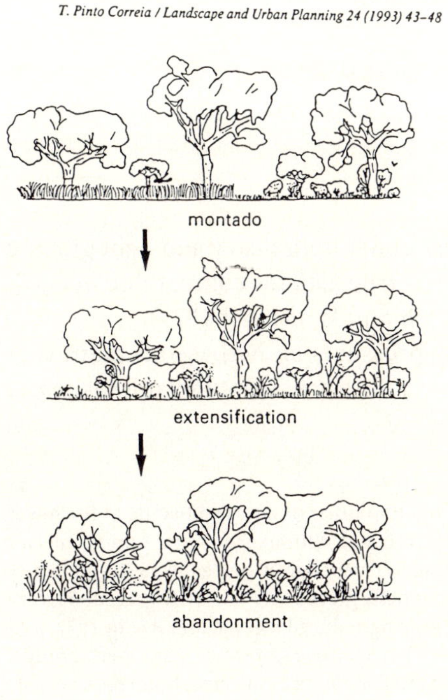

Fig. 2.4b. Degradation of the *montado* by extensification (Pinto-Correia 1993). Due to the end of pig raising as a result of African swine fever and in general the effect of changing socio-economic factors, the soil in the *montado* has been progressively used in a more extensive way. Cultivation and sometimes grazing disappear. The density of the undercover vegetation increases, until it forms a new assemblage of maquis where tree production must also be abandoned. Such shrub formations correspond to a simplification of the landscape, increased risk of fire, and little or no possibility of exploitation

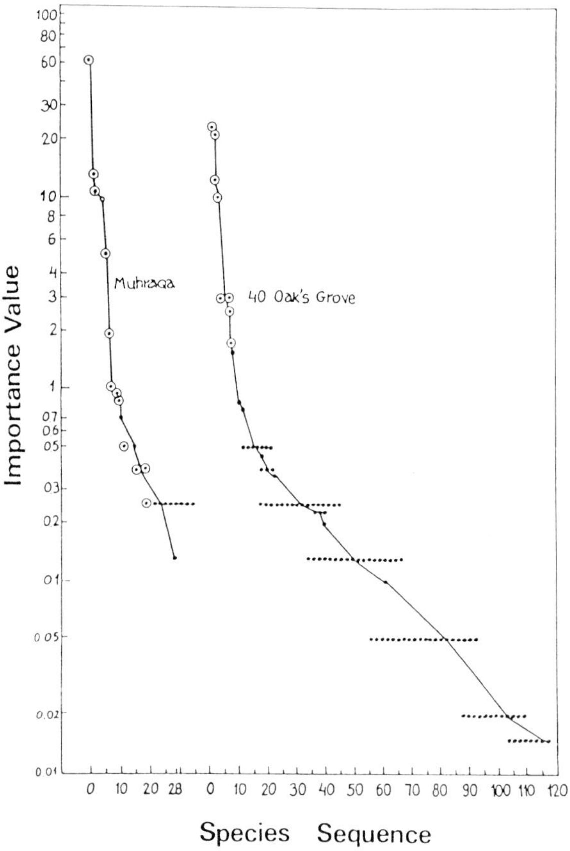

Fig. 2.5. Dominance-diversity curves for all vascular plant species in 1000 m^2 communities of closed and semi-open maquis on Mt. Carmel, Israel (Naveh and Whittaker 1979). Species percentages of mean cover are used as the importance values in a logarithmic scale on the ordinate, and species numbers in the sequence from the most important to the least important on the absicca. *Dots* are herb species and *circled dots* are woody species

moderate grazing pressures and lowest under both heavy and light grazing pressure or complete protection. This is true also for comparable open *Q. dumosa* woodlands in California (Naveh and Whittaker 1979).

The effect of such different grazing pressures on herbaceous plant species diversity is shown in Fig. 2.6a–c. In such complex microsites, a

Fig. 2.6a–c. Species richness and diversity of open woodlands and shrublands in the lower Galilee, Israel, as related to grazing pressure (Naveh and Whittaker 1979): *a* Annual species richness in 1976 and 1977. *KN-3* years protection; *NY* very lightly grazed in 1976 and lightly grazed in 1977; *AA* moderately grazed; *AL* rotationally grazed; *KN+* heavily grazed; *BT* very heavily grazed (spring sampling shortly after grazing reduced *s* in *AA* and *AL*: in 1977, and more intensive grazing increased *s* in *NY* in 1977). b Species richness of major herb groups, 1977. Pasture plots and grazing intensities as in a; *c* Dominance concentration reciprocal (1/c) and equitability (EXP H) of vascular plant species, 1977. Pasture plots and grazing intensities as in *a*

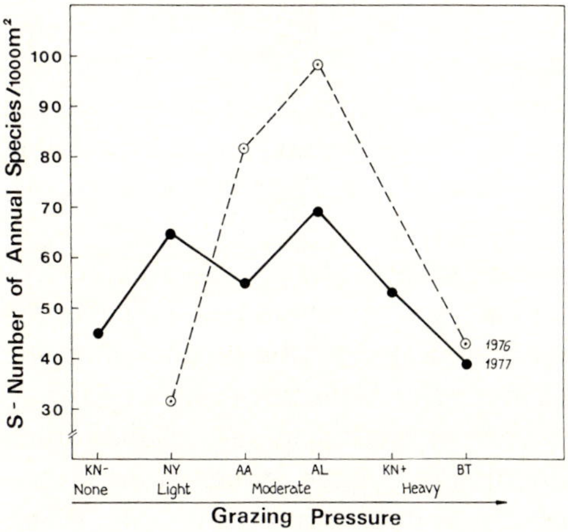

S - Number of Annual Species/1000m²
100
90
80
70
60
50
40
30
1976
1977
KN-
NY
AA
AL
KN+
BT
None
Light
Moderate
Heavy
Grazing Pressure

a

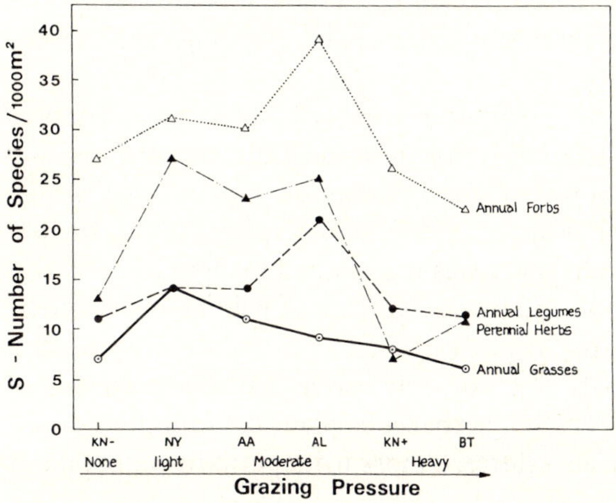

S - Number of Species/1000m²
40
35
30
25
20
15
10
5
0
Annual Forbs
Annual Legumes
Perennial Herbs
Annual Grasses
KN-
NY
AA
AL
KN+
BT
None
light
Moderate
Heavy
Grazing Pressure

b

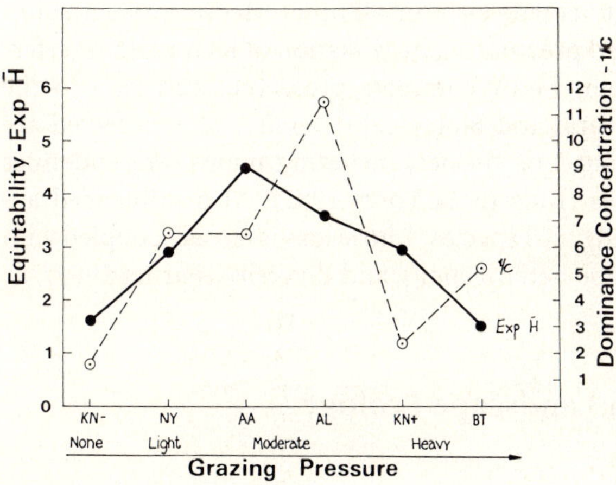

Equitability - Exp H̄
Dominance Concentration - 1/C
1/C
Exp H̄
KN-
NY
AA
AL
KN+
BT
None
Light
Moderate
Heavy
Grazing Pressure

c

dynamic flow equilibrium is maintained between the small-scale heterogeneity created by variable soil depth and micro-relief, availability of moisture and nutrients, rock outcrops, tree cover, litter cover, shade and the abundance of different plant species, as affected by grazing and trampling (Mann 1986; Naveh 1991a; L. Olswig-Whittaker, M. Blumler, and Z. Naveh, unpubl. data).

Terrace walls, rock outcrops and their shallow soil pockets and crevices can serve as refuges for many species and genotypes which are well adapted to these microsite scales. The latter is indicated by the distinct allozyme polymorphism of adjacent populations of *Hordeum spontaneum*, growing either in the open, deeper soil patches or nearby in very shallow rock pockets. For such a rock refuge population, higher drought tolerance is essential to ensure survival and viable seed production (Nevo et al. 1986). In many cases such rock niches are archeological remnants of ancient olive or wine presses, or other cultural artifacts, increasing ecological heterogeneity and biological diversity.

A recent survey of all nature reserves in northern Israel, over an area of about 3000 ha of mixed vegetation (maquis, woodlands, shrublands, and grasslands), corroborated some of our findings on much larger scales. According to Noy-Meir and Kaplan (1991), and Kaplan (1992), highest vascular plant richness and diversity was maintained by light to moderate grazing of domestic livestock, and by free grazing of wild animals, chiefly mountain gazelles and boars. In protected and too lightly grazed sites, species richness and diversity was not only lower, but the frequency of wildfires was also higher. Protected herbaceous plant communities dominated chiefly by tall grasses, also suffered more from periodic outbreaks of rodent populations.

This pattern of diversity is true also for all other Mediterranean countries. As has been described previously, the cessation of all human interference by land abandonment or by complete protection can lead to the impoverishment of structural and biological diversity. Most affected are the light demanding herbaceous species, including many rare endemics and ornamental geophytes (Ruiz de la Torre 1985). Also influenced are many vertebrates, especially bird species, whose loss is closely coupled with the loss of overall landscape heterogeneity and diversity (Farina 1989).

2.6 Ecodiversity and Landscape Ecology

At present, there are strong tendencies to shift some of the major conservation efforts from the protection of rare and endangered species and their natural habitats and ecosystems in restricted nature reserves and parks to the protection of most valuable open semi-natural and rural landscapes. Therefore, the need for developing a sound conceptual and methodological foundation for the conservation of total landscape ecodiversity has also become acute on a global scale. On the international level, these efforts are carried out by IUCN and UNESCO, chiefly through the establishment of Biosphere Reserves and World Heritage Sites. But as has been described recently in detail by Lucas (1992), there are also many other regional and local success stories of protected natural and managed cultural landscapes. Lucas emphasized their important role in the strategies for sustainable living as outlined in the important document *Caring For the Earth* (IUCN/WWF/UNEP 1991), and provided a very useful guide for the selection, implementation, legalization and management of such protected landscapes. In this context, a clear distinction has to be made between:

1. Solar-powered biosphere landscapes or *bio-landscapes*, which are driven mainly by solar energy and its biological conversion and are regulated to greater or lesser extents by natural, bio-physico-chemical information. In addition to the natural and near natural landscapes, these also include the two most important cultural biosphere landscape classes for ecodiversity conservation, namely: (a) semi-natural and agro-silvo-pastoral (or semi-agricultural) landscapes in which their natural, (spontaneously occurring and reproducing) biota has been modified by human land uses; and (b) low-input traditional agricultural landscapes and their agro-ecosystems. The latter are subsidized only to a low degree by fossil energy and chemicals, but are still controlled to a large degree by natural information and processes. To the latter belong the traditional agro-ecosystems as well as planted forests and sown pastures in rural landscapes, and we may also include here the rapidly expanding organic farming systems.
2. Intensively cultivated, fertilized, and chemically-treated *agro-industrial* landscapes which depend on solar energy for their production. They are, however, heavily subsidized by large inputs of low-grade energy of entropy and pollution-producing fuels and chemicals, and are controlled almost entirely by human agro-technology. As such they have no specific conservation values, and may even have detrimental environmental impacts on the remaining open landscapes and their biological and ecological diversity and stability (Naveh 1987).
3. Fossil-fuel powered technosphere landscapes, with highways, bridges and power lines that are entirely artificial, are human-made and controlled. They are driven by the technological conversion of high-grade solar energy into low-grade energy, which produces great amounts of

entropy, pollution, and waste products. The overall environmental impact of these landscapes on our total human ecosystem is indicated by the huge amount of energy that flows annually through great urban-industrialized centers. According to Odum (1993), this flow can be measured in millions of kcal per square meter, as compared to only thousands of kcal in solar powered landscapes.

The exponential and generally uncontrolled expansion, and chiefly adverse role, of the partly and wholly fuel-powered landscape classes is one of the major causes of our global environmental crisis. However, as has been shown in successful examples of rural conservation management in Great Britain (Green 1985) and of other rural landscapes elsewhere (Lucas 1992), there is still hope that the antagonistic relations between these two major cultural landscape classes can be transformed into mutually beneficial, and even symbiotic relations.

The broadening of conservation perspectives from species and natural ecosystems to cultural bio-landscapes and their intrinsic and instrumental values requires the broadening of spatial scales and conceptual and methodological dimensions from the realm of natural sciences to the realm of humanities, dealing with human land uses. The uniqueness of these cultural bio-landscapes as the tangible meeting point between nature and mind cannot be conceived, studied and managed by conventional, mechanistic and mostly reductionist bio-ecological approaches and methods. They require a multidisciplinary landscape-ecology approach to land use, helping to bridge the gaps between bio-ecology and human ecology. These issues have been discussed elsewhere in detail (Naveh 1990a, 1991b, 1994, 1995b). Here I can point only to some of the methodological and practical aspects of ecodiversity conservation.

In recent years landscape ecology has made great strides in the development of integrative and cross-disciplinary methods, enabling workers to identify, analyze and synthesize and, at least partly, to quantify in more holistic ways complex natural and cultural patterns and processes at different temporal and spatial scales. This has been achieved with the help of recent advances in computer hardware and software, remote sensing and satellite images, and Intelligent Geographic Information Systems (IGIS) combining artificial intelligence with geographic information systems. As has been reviewed recently in more detail (Naveh and Lieberman 1994), these have been utilized in landscape ecological studies together with the application of innovative mathematical and system theories, such as information theory, fractal geometry, fuzzy logic, and hierarchy theory.

A first important step for ecodiversity conservation has been made in California by Davis et al. (1990). They proposed the use of comprehensive

GIS models to organize existing biodiversity data and to improve the spatial aspects of their assessment for the US National Biodiversity Center along multilevel spatial scales from the local site to the regional and the biogeographical scale. In each spatial level the specific cultural features of land use and their effect on environmental quality and landscape structure were also taken into consideration. The model shows clearly the complexity of biodiversity assessments for management and conservation purposes. It is meant to facilitate the needed cooperation and coordination across administrative and political borders.

A practical ecodiversity approach has already been applied in Denmark for conservation management of small and locally distributed communities (e.g. natural or artificial lakes and pools, rivers, canals, drainage ditches, hedges, road verges, tree rows, brush thickets, bogs, and marl pits) as natural and cultural biosphere islands in the monotonous agro-industrial steppe (Agger and Brandt 1988).

Some landscape ecologists are also paying attention to ecological heterogeneity along different spatial scales and its relation to biodiversity. Among others they have broadened the spatial scales of the diversity indices from the single species level to the three-dimensional landscape level by transforming the Simpson dominance-diversity index into a *landscape eveness* index, and a *landscape patchiness* index (Romme 1982), and into a *landscape contrast* index (Hoover and Parker 1991).

In Germany, Haber (1990) has broadened the Shannon-Weiner species diversity index into an *ecotope diversity index* to compare the variability between different ecotopes or communities as affected by land use within one landscape unit. Such a *gamma ecotope diversity index* should be used as a guideline for nature conservation and biodiversity conservation in agricultural landscapes through a differentiated land use strategy. Haber proposed to preserve at least 10–15% of the area within each regional natural unit for natural, semi-natural and semi-agricultural ecotopes (communities) of extensively managed pastures and forests.

However, in all these studies the vegetation served as the sole diversity parameter. For systematic measurement of the multidimensional landscape ecodiversity, such strictly biological parameters have to be expanded. They should also incorporate other relevant ecological and cultural parameters at different spatial and perceptional scales, such as micro- and macrosite heterogeneity in soil and topography, as well as cultural artifacts and land use variables.

For this purpose these species diversity indices could be transformed into even more holistic ecodiversity indices following Whittaker's (1965) classification: (1) *alpha ecotope ecodiversity* of biotic, ecological and cultural components within each ecotope; (2) *beta ecotope ecodiversity* be-

if their contents can reach all those who care for, those who live from and those who deal with these landscapes at all levels of decision making. Such a tool could be the Green Books for Landscape Conservation (GBLC), developed by the above-mentioned IUCN-CESP Working Group for Landscape Conservation. For this purpose they should fulfill the following functions:

1. In contrast to the IUCN Red Data Books for endangered plants and animals, which deal with the abstract taxonomic species level, these Green Books should deal with the concrete space/time defined landscape level. Its scales range from the smallest mappable landscape ecotope to regional landscape scales of 200–400 km^2. It is on this level of terrain that decisions on land uses and conservation measures are made, which determine the fate of plant and animal populations, and of all other landscape values and functions. It should also be realized that the threats to such tangible and familiar landscape units have much more meaning and public appeal than threats to species or to spatially vaguely defined and even intangible ecosystems. This will be even more the case if these Green Books present relevant information not only on endangered natural assets but also on all other crucial issues and perils to cultural, socio-economic, historical and scenic landscape assets, comprising the total landscape ecodiversity.
2. In contrast to the species Red Data Books, and to the common type of land studies and surveys, GBLCs should recommend alternative and more sustainable land use practices and conservation strategies, including zoning and protecting, for each specific landscape unit within the regional landscape systems.
3. In order to avoid the "top-down" syndrome of conservation plans prepared by experts and imposed by administrators, efforts should be made for maximum involvement of the local populations from the early planning stages. With the help of such open dialogue and by approaching conservation "from the roots", their fullest comprehension and cooperation should be ensured. They should demonstrate above all, how demands to safeguard intrinsic biological and cultural "soft" landscape values can be reconciled with controlled utilization of "hard" values, which are vital for sustainable socio-economic advancement.
4. To arouse awareness of the dangers to these landscapes, and apprehension about their future, comprehension of the complex problems and their best solutions, and motivation for active involvement in their realization, these Green Books should not be only descriptive but anticipatory. This could be achieved with the help of different optional scenarios on the fate of these landscape units under alternative land use

options. However, although based on advanced scientific methods, this information should be presented to the planner, land manager and user, the political decision maker and the public at large in clear non-technical language with ample illustration by maps, figures and photos.

5. In contrast to the fate of most scientific reports, which are filed away, GBLCs should be used as a practical background document for planning, policy and management guidelines, usable by professionals, politicians and the public. They should serve not only as a first database and dynamic landscape model, but should lay out the foundations for a long-term impact evaluation and forecasting tool.
6. Although containing information on present and anticipated environmental impacts, such Green Books for threatened landscapes differ from the regular environmental impact statements. The latter are merely defensive and passive reactions to planned, urban-industrial or other economic development projects, limited to specific sites and their potential threats. But GBLCs should be active and positive initiatives to change already prevailing undesirable trends and to prevent their continuation with the help of innovative, holistic, long-term and multi-beneficial planning and management strategies on regional landscape scales.
7. Published under the auspices of IUCN in English, as well as by the local authorities in the language(s) of the country and/or the ethnic region, they should be as widely distributed as possible, to reach all those directly or indirectly concerned with these threatened landscapes and to all those who can utilize their contents. With appropriate modifications, they could have relevance for public and school education in conservation, as well as for tourists visiting the region.
8. GBLCs should be used as the major tool for "rescuing" endangered landscapes from Red Lists of Endangered Valuable Landscapes, to be prepared according to the guidelines of IUCN on a worldwide basis in each country. But they should be regarded also as a emergency tool wherever necessary, even before such Red Lists are ready.

The IUCN General Assembly at Buenos Aires in January 1994 adopted a resolution, drafted by the WGCL-CESP recommending actions to be taken for the conservation of threatened landscapes and for the promotion of interdisciplinary approaches between natural and social sciences in the planning and management of threatened landscapes. It urged the undertaking of a pilot study to identify regionally and nationally significant landscapes under threat, to establish means and processes to promote their conservation, and to bring together researchers and practitioners to prepare case studies as a vehicle for developing strategies and methodologies for holistic landscape conservation.

Now, the IUCN is working toward Red Lists for Endangered Landscapes, and the experience gained in the western Crete study will be utilized for the development of a uniform Green Book methodology and for further case studies in Italy, Portugal, Israel, Costa Rica, and other countries, depending on the financial support available for each project.

2.7 Discussion and Conclusions

One of he major conclusions to be drawn from this discussion is that holistic conservation strategies for cultural biosphere landscapes should be guided by the recognition that ecodiversity is the tangible expression of the interplay of biological, ecological and cultural processes and patterns. These are determined by biodiversity, ecological micro- and macrosite heterogeneity, human land uses and cultural artifacts. The last two are either enriching or impoverishing both biodiversity and overall landscape heterogeneity. As demonstrated in Fig. 2.7, these are coupled by positive, mutually amplifying and self-augmenting feedbacks: the greater the eco-

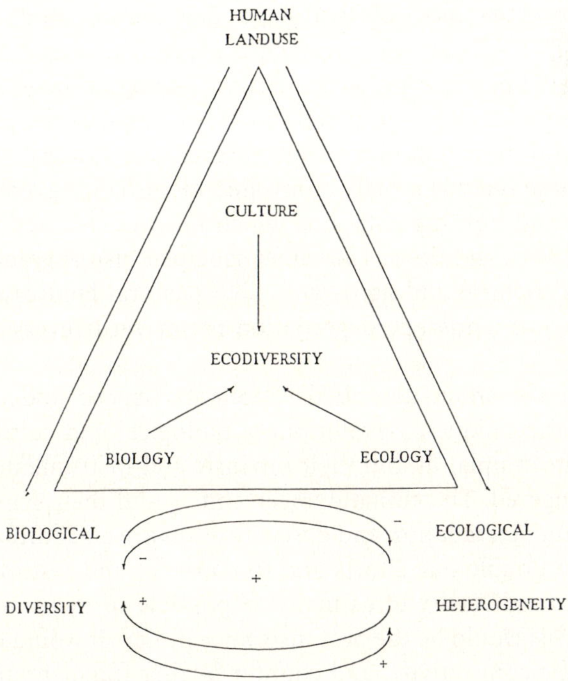

Fig. 2.7. The landscape shaping forces of the biology-ecology-culture triangle determining ecodiversity

logical landscape heterogeneity, the greater the chance for biological diversity and at the same time floristic, faunal and structural vegetation diversity enhances ecological heterogeneity. But the opposite is also true. Landscape homogeneization fragmentation, and despoliation diminishes biodiversity which in turn further reduces landscape heterogeneity.

I have attempted to show that in perturbation-dependent Mediterranean cultural bio-landscapes, total landscape ecodiversity is determined by the maintenance of a dynamic homeorhetic flow equilibrium between these three major forces. In all these Mediterranean uplands, the disruption of this dynamic homeorhetic flow equilibrium is the result of neo-technological landscape degradation. It is characterized by the unfortunate combination of the cessation of human agro-silvo-pastoral activities and disturbance processes, following depopulation on one hand, and on the other hand by the rapid and careless shift from diversified and stable traditional agriculture to large-scale agro-industrial farming, coupled with overgrazing and indiscriminate planting of monocultures of pine and eucalyptus. High landscape ecodiversity and vital biological, ecological, cultural and socio-economic functions and values can only be ensured by the maintenance of a dynamic homeorhetic flow equilibrium on both micro- and macroscales.

Further neo-technological landscape degradation and despoliation cannot be prevented simply by protective measures. Maximum attainable biodiversity, within a fine-grained matrix of ecological heterogeneity patterns, fulfilling all above-mentioned functions, requires the perpetuation, simulation, and/or restoration of all natural and cultural patterns and processes. These include strictly controlled disturbance pressures by cutting, grazing and burning rotations which are well adapted in time and space to local site conditions. The same principles also apparently apply to all other semi-natural and agro- and silvo-pastoral landscapes in which neo-technological landscape degradation is not yet irreversible (Ricklefs et al. 1984).

In view of the above-described exponential speed and extension of this process, the unique co-evolutionary biological and cultural richness of Mediterranean uplands and their intrinsic and instrumental values are gravely endangered. The climatic uncertainties and their synergistic couplings with land degradation and desertification are an additional compelling reason to double our efforts and to conserve and restore the health, integrity and ecodiversity of as many as possible of these most valuable landscapes. This would be the best insurance policy. It would ensure sufficient landscape connectivity, and counter further fragmentation, homogenization, pollution and scenic despoliation. It would provide options for sustainable life-supporting and other vital ecosystem and landscape functions and keystone species, and afford best chances for the survival of rich biotic communities and their further evolution (Naveh 1995a).

However, it has to be realized that the conservation of ecodiversity has to face many strong cultural, demographic, political and socio-economic forces, all driving in the direction of the cumulative loss of biodiversity, ecodiversity and stability, and thereby enhancing neo-technological landscape degradation. Unfortunately, the dominant cultural values and recognized models of nature and landscapes cannot constrain or may even encourage the vicious circle of exponential population growth and unrestrained growth of consumption and acquisition.

It is therefore obvious that the remedies for our environmental crisis should be sought not only in the scientific, technological, political and socio-economic spheres, but also in the spheres of cultural, spiritual and ethical values, and living norms. Biosphere landscapes should not merely be viewed as a source for our materialistic satisfaction, but also as a source of enlightenment and enjoyment, and as has been shown recently in a convincing manner by Roszak (1992) thereby also as a source of mental health.

In a recent publication (Kim and Weaver 1994a), biodiversity and landscapes were approached from a broad interdisciplinary viewpoint that has much relevance to these issues. Much of this publication dealt with the paradoxical conflict that human life depends for survival on biodiversity and landscapes, yet at the same time, modern human activity threatens the survival or existence of these natural systems. The editors defined its roots in human values and the limits of human cognition and pleaded for an "imperative for change" from myopic and hedonistic values toward a new *Green* equilibrium in the relationship between human society and biodiversity. These could be reached through education and a new consensus on local, national and international scales (Weaver and Kim 1994).

In their recent challenging *Beyond the Limits* (Meadows et al. 1992), three world-renowned systems analysts and modelers went even farther. They claimed, rightly, that in order to avoid the collapse of our earth system, we need a far-reaching environmental revolution which should lead to a radical shift from consumption to conservation and from unrestrained quantitative growth to lasting qualitative improvement and development.

This recognition of the urgent need for such a global environmental revolution and the feasibility of its realization, if humanity will accept this challenge in time, is no longer regarded as merely the utopian dream of radical environmentalists and *deep* ecologists. This view is now shared by some of the most prominent economists, managers and political decision makers (Gore 1991; King et al. 1991; Tolba 1992).

As outlined elsewhere (Naveh and Lieberman 1994), this environmental revolution should lead to a new post-industrial symbiosis between human society and nature which will ensure a viable and healthy global ecosphere. This can be realized only, if the unrestrained expansion of technosphere and agro-industrial landscapes will be replaced by their total integration with the natural and cultural biosphere landscapes.

In addition to many other encouraging initiatives, the conservation of the total ecodiversity of our highly valuable and most cherished cultural landscapes can be regarded as one important step toward this cultural revolution. In order to ensure its success, we will have to integrate scientific ecological knowledge with ecological wisdom and ecological ethics. This should help us learning from the past, comprehending the present and envisaging the future of our total human ecosystem landscapes, and translating these insights into practical, large-scale conservation activities in the field.

References

Agger P, Brandt J (1988) Dynamics of small biotopes in Danish agricultural landscapes. Landscape Ecol 3:227–240

Baudry J, Bunce RGH (eds) (1989) Land abandonment and its role in conservation. Options Mediterraneenes Ser A.15

Botkin DB (1990) Discordant harmonies: a new ecology for the twenty-first century. Oxford University Press, New York

Bussotti F, Ferretti M, Cenni E, Gellinni R, Clauser F, Grossoni P, Barbolani E (1991) New type forest damage to mediterranean vegetation in Southern Sardinian forests (Italy). Eur J For Pathol 21:290–300

Bussotti F, Gellini R, Ferretti M, Cenni E, Pietrini R, Sbrilli G (1992) Monitoring in 1989 of Mediterranean tree condition and nutritional status in southern Tuscany, Italy. For Ecol Manage 51:81–93

CEPA (Coordinadora Extremena de Proteccion Ambiental) (1992) Dealing with disparity. European structural funds in southwest Spain. Ecologist 22:91–96

Davis FW, Stoms DM, Estes JE, Scepan J (1990) An information systems approach to the preservation of biodiversity. Int J Geogr Syst 4:55–78

Di Castri F (1981) Mediterranean-type shrublands of the world. In: di Castri F, Goodall DW, Specht RL (eds) In Mediterranean-type shrublands. Elsevier, Amsterdam, pp 1–52

Farina A (1989) Recent changes of the mosaic patterns in a montane landscape (North Italy) and consequences on vertebrate fauna. In: Baudry J, Bunce RGH (eds) Land abandonment and its role in conservation. Proc Zaragossa/Spain Seminar 10–12 December 1989. Options Mediterraneennes Ser A 15:121–134

Farina A, Naveh Z (eds) (1993) Landscape approach to regional panning: The future of Mediterranean landscapes. Landscape Urb Plann 24. Special Issue, July 1993

Ferretti M, Bussotti F, Cozzi A, Cenni E (1995) Forest decline and environmental pollution in Italy. Agr Medit

Gore A (1992) Earth in balance: forging a new common purpose. Earthwatch, London

Green B (1985) Countryside conservation. The protection and management of amenity ecosystems, 2nd ed. Unwin Hyman, London

Grove AT, Ispikoudies J, Kazaklis A, Moody JA, Papanastasis V, Rackham O (1993) Threatened Mediterranean landscapes: west Crete. Research Report to EU. Cambridge University, Department of Geography, Cambridge

Grove AT, Ispikoudies J, Kazaklis A, Moody JA, Papanastasis V, Rackham O (1994) Green Book: Threatened Mediterranean landscapes of western Crete. Adapted and translated into Greek by V. Papanastasis. Integrated Mediterranean Project of Crete, Aristotele University of Thessaloniki, and Mediterranean Agronomic Institute of Chania, Crete

Haas G (1966) On the vertebrate fauna of the Lower Pleistocene site Ubeidiya. Israel Academy of Science and Humanities, Jerusalem, pp 33–68

Haber W (1990) Using landscape ecology in planning and management. In: Zonneveld IS, Forman RTT (eds) Changing landscapes: an ecological perspective. Springer, Berlin, Heidelberg New York, pp 217–232

Hardin G (1968) The tragedy of the commons. Science 162:1243–1248

Hoover SR, Parker AJ (1991) Spatial components of biotic diversity in landscapes of Georgia, USA. Landscape Ecol 5:125–136

IUCN/UNEP/WWF (1991) Caring for the earth. A strategy for sustainable living. IUCN, Gland, Switzerland

Jeftic L (1993) Implications of expected climate change in the Mediterranean region. In: Garber M, Cohen A, Magaritz M (eds) Regional implications of future climate change. Israel Academy of Sciences and Humantities, Jerusalem, pp 278–302

Kaplan D (1992) Responses of Mediterranean grassland plants to gazelle grazing. In: Thanos CA (ed) Proc 6th Int Conf Mediterranean Climate Ecosystems, September 1991. University of Athens, Athens, pp 75–79

Kim KC, Weaver RD (eds) (1994a) Biodiversity and landscapes: a paradox of humanity. Cambridge University Press, New York

Kim KC, Weaver RD (1994b) Biodiversity and humanity: paradox and challenge. In: Kim KC, Weaver RD (eds) Biodiversity and landscapes: a paradox of humanity. Cambridge University Press, New York, pp 3–27

King A, Richter G, Schneider B (1991) The first global revolution: from the problematique to the resolutique, a "blueprint for the twenty-first century" by the Council of the Club of Rome. Pantheon Books, New York

Le Houerou HN (1992) Change in vegetation and land-use by the year 2050: a prospective study. In: Jeftic L, Miliman JD, Sestini G (eds) Climate change and the Mediterranan. Edward Arnold, London, pp 175–232

Lucas PHC (1992) Protected landscapes. A guide for policymakers and planners. Chapman and Hall, London

Marcuzzi G (1991) Man-made heterogeneity in some Italian habitats – beaches of the Puglia region and Matera province. Econ Montana Lin Ecol 4:21–24

Meadows DH, Meadows DL, Randers J (1992) Beyond the limits: global collapse or a sustainable future. Earthscan, London

Meeus J, Wijermans MP, Vroom MJ (1990) Agricultural landscapes in Europe and their transformation. Landscape Urban Plann 18:289–352

Milan M, Salvador R, Mantilla E, Artnano B (1996) Meteorological and photochemical air pollution in southern Europe: experimental results from EC research projects. Atmos Environ 30:1909–1924

Naveh Z (1974) The ecological management of non-arable Mediterranean uplands. J Environ Manage 2:351–471

Naveh Z (1982) Mediterranean landscape evolution and degradation as multivariate biofunctions: theoretical and practical implications. Landscape Plann 9:125–146

Naveh Z (1984) The vegetation of the Carmel and Nahal Sefunim and the evolution of the cultural landscape. In: Ronen A (ed) Sefunim prehistoric sites in Mount Carmel, Israel. BAR Intern Ser 230, Oxford, pp 23–63
Naveh Z (1987) Biocybernetic and thermodynamic perspectives of landscape functions and land use patterns. Landscape Ecol 1:75–85
Naveh Z (1990a) Landscape ecology as a bridge between bio-ecology and human ecology. In: Svobodova H (ed) Cultural aspects of landscape. PUDOC, Wageningen, pp 45–58
Naveh Z (1990b) Ancient man's impact on the Mediterranean landscape in Israel: ecological and evolutionary perspectives. In: Bottema S, Entjes-Nieborg G, Van Zeist W (eds) Man's role in the shaping of the eastern Mediterranean landscape. AA Baalkema, Rotterdam, pp 43–50
Naveh Z (1991a) Mediterranean uplands as anthropogenic perturbation dependent systems and their dynamic conservation management. In: Ravera OA (ed) Terrestrial and aquatic ecosystems, perturbation and recovery. Ellis Horwood, New York, pp 544–556
Naveh Z (1991b) Biodiversity and ecological heterogeneity of Mediterranean uplands. Econ Montana - Lin Ecol 24:47–60
Naveh Z (1995a) Conservation, restoration and research priorities for Mediterranean uplands threatened by global change. In: Moreno JM, Oechel WC (eds) Anticipated effects of a changing global environment on mediterranean-type ecosystems, pp 482–508
Naveh Z (1995b) Interactions of landscapes and cultures. Landscape Urb Plann 32:43–54
Naveh Z, Dan J (1973) The human degradation of Mediterranean landscapes in Israel. In: di Castri F, Mooney HA (eds) Mediterranean-type ecosystems: origin and structure. Springer, Berlin Heidelberg New York, pp 370–390
Naveh Z, Lieberman AS (1994) Landscape ecology - theory and applications. 2nd edn. Springer, Berlin Heidelberg New York
Naveh Z, Whittaker RH (1979) Structural and floristic diversity of shrublands and woodlands in northern Israel and other mediterranean areas. Vegetatio 41:171–190
Naveh Z, Steinberger E, Chaim S (1980) Photochemical air pollutants: a new threat to Mediterranean conifer forests and upland ecosystems. Environ Conserv 7:301–309
Nevo E (1988) Genetic diversity in nature; patterns and theory. Evol Biol 23:217–246
Nevo E, Geiles A, Kaplan D, Golenberg EM, Olsvig-Whittaker L, Naveh Z (1986) Natural selection of allozyme polymorphism: a microsite test revealing ecological genetic differentiation in wild barley. Evolution 40:13–22
Norton BG (1987) Why preserve natural variety? Princeton University Press, New Jersey
Noy-Meir E, Kaplan D (1991) The effect of grazing on the herbaceous Mediterranean vegetation and its implications on the management of nature reserves. Interim Report to the Nature Conservation Authorities, Jerusalem. (in Hebrew)
Odum EP (1993) Ecology and our endangered life-support systems. 2nd edn. Sinauer, Sunderland, Massachusetts
Papanastasis V (1993) Legal status of land tenure and its implication for open landscapes of western Crete. Landscape Urb Plann 24:273–278
Pedroli GMB, Vos W, Dijkstra H, Rossi R (1988) The Farma River barrage effect study. Giunta Regionale Toscana. Marsilio, Firenze, Italy
Pickett STA, Parker VT (1994) Avoiding the old pitfalls: opportunities in a new discipline. Restor Ecol 2:75–79
Pinto-Correia T (1993) Threatened landscapes in Alentejo, Portugal: the *montado* and other agro-silvo-pastoral systems. Landscape Urb Plann 24:43–48
Prigogine I (1976) Order through fluctuation: Self-organization and social systems. In: Jantsch E, Waddington CH (eds) Evolution and consciousness: human systems in transition. Addison Wesley, Reading, pp 93–130
Prigogine I, Stengers I (1984) Order out of chaos. Bantam Books, New York

Ricklefs RR, Naveh Z, Turner RE (1984) Conservation of ecological processes. Commission on Ecology Paper No. 8. IUCN, Gland, Switzerland
Romme WH (1982) Fire and landscape diversity in subalpine forests of Yellowstone National Park. Ecol Monogr 52:199–221
Roszak T (1992) The voice of the earth: an exploration of ecopsychology. Simon and Schuster, New York
Ruiz de la Torre JR (1985) Conservation of plants within their native ecosystems. In: Gomez-Campo C (ed) Plant conservation in the Mediterranean. De Junk, The Hague, pp 197–219
Stekelis M (1966) Archeological excavations at Ubaidiya. Israel Academy of Sciences, Jerusalem, pp 1–32
Tolba M (1992) Saving our planet: challenges and hope. Chapman and Hall, London
UNEP (1988) Mediterranean blue plan regional activity centre: futures of the Mediterranean basin. Executive summary and suggestions for action. United Nations Environmental Programme, Mediterranean Action Plan, Mediterranean Blue Plan Regional Activity Centre, Sophia Antipolis, France
Vos S, Stortelder AHF (1992) Vanishing Tuscan landscapes: landscape ecology of a submediterranean-montane area (Solano Basin), Tuscany, Italy. Pudoc, Wageningen
Waddington CH (1975) A catastrophe theory of evolution: the evolution of an evolutionist. Cornell University Press, Ithaca, New York
Warburg M (1977) Plant and animal species diversity along environmental gradients in the Mediterranean landscape of Israel. Animal species diversity. Research Report 450. Binational Israel-American Science Foundation. Technion, Israel Institute of Technology, Haifa
Warburg M, Ben-Horin RA, Rankevich D (1978) Rodent species diversity in mesic and xeric habitats in the Mediterranean region of northern Israel. J Arid Environ 1:63–69
Weaver RD, Kim KC (1994) Biodiversity and humanity: toward a new paradigm. In: Kim KC, Weaver RD (eds) Biodiversity and landscapes: a paradox of humanity. Cambridge University Press, New York, pp 393–423
Whittaker RH (1965) Dominance and diversity in land plant communities. Taxon 21:213–251
Worldwatch Institute (1994) State of the World 1994. Norton, New York and Earthscan, London
Yiftach-El O, Law-Yone H (1995) Regional policy and minority attitudes in Israel. Environ Plann 27:1261–1301

3 Ecological Indicators of Landscape Degradation

R.H. Groves

3.1 Introduction

Some previous publications on mediterranean-type ecosystems have covered the effects of the major ecological factors, such as fire (Mooney and Conrad 1977), nutrient deficiency (Kruger et al. 1983) and water deficits in summer (di Castri et al. 1988), that interact to influence ecosystem function in the winter-wet, summer-dry climates typical of the five regions of the world having a mediterranean-type climate. Other publications have tackled more general themes such as co-evolution (di Castri and Mooney 1973), plant-herbivore interactions (Arianoutsou and Groves 1994) and resilience (Dell et al. 1986). This present Volume also takes a more general theme - that of landscape degradation.

Landscapes in all five major regions of mediterranean-type climates are degraded to varying extents, irrespective of the different periods of human occupation of those different regions. The word "degradation" has a negative connotation. Seligman and Perevolotsky (1994) define degradation as "a change of state with an implied negative value assessment". Use of the word thus implies a diminution in landscape quality, and especially I submit, in those attributes associated with human occupation and usage of such landscapes. A degraded landscape is less productive, less resilient, less diverse or less beautiful, although usually it is the reduction in economic benefit that is implied. But the latter is too narrow a view of degradation and the processes associated with it; in what follows I shall concentrate on ecological considerations.

Degradation is a continuum of change marked at various points on that continuum by demonstrable and measurable changes in landscape or ecosystem attributes. It is the reverse of restoration or rehabilitation and students of landscape degradation may learn a lot from the burgeoning literature on restoration ecology, if only by reversing the direction of the continuum of change. In this contribution I wish to list some of the ecological indicators of early stages of degradation and discuss further only

Ecological Studies, Vol. 136
Rundel et al. (eds.) Landscape Degradation and Biodiversity in Mediterranean-Type Ecosystems

several of them as they apply to the different regions of mediterranean-type climate.

3.2 Ecological Indicators of Degradation

Specific ecological indicators of early degradation include:

1. A reduction in the rate of litter accumulation;
2. An increased rate of movement of surface soil particles;
3. An ingress of light-demanding invasive plants;
4. A decrease in the vegetative cover of perennial in favour of annual plants;
5. A change in the phenology of the vegetation;
6. An increased frequency of fire;
7. An increase in the woody (resprouting) element of the vegetation.

Whilst there may be argument about the significance of some of these indicators, and certainly about their order of importance, in the overall degradation process, I wish to discuss only four of them in what follows.

3.2.1 Soil Particle Movement

In the Mediterranean Basin, cradle of western European civilisation, urbanisation first was centred on the often narrow coastal plains lying between steep hills and mountains and the Mediterranean Sea itself. Agriculture was initially confined to the fertile alluvium associated with the silt plains where river systems entered the sea. Associated human activities included timber gathering and grazing in the hinterland. Whilst the hilly inland regions were naturally subject to some erosion, human activities seem to have increased the rate of erosion of soil particles from the hills to the rivers and the sea. Thirgood (1981) cites a change from a pristine situation where surface runoff would normally comprise about 20% of total rainfall to the degraded state of some eastern Mediterranean mountains where the percentage surface run-off is as high as 70–80%. Thirgood further writes that "Silting has clogged rivers, formed deltas and created malarial marshes that have often led to the demise of cities and the resettlement of populations in the poorer hill country away from the potentially fertile silt plains" (Thirgood 1981, p. 8).

So considerable was the increased rate of soil particle movement in parts of the Mediterranean Basin, that cities such as Ravenna, formerly on the coast, came to be situated inland of the present coastline. Ancient Antioch, a city of 400 000 people in Pauline times, is now under more than 5 metres of detritus from the Taurus and Lebanon ranges (Thirgood 1981).

Whilst certain regions of the Mediterranean Basin provide the most compelling evidence for an increased rate of soil particle movement as an indicator of overall degradation, present records from the San Gabriel mountains behind Los Angeles in southern California (A.M. Gill, pers. comm.), together with historical records of once-navigable Chilean river mouths now no longer so (Thirgood 1981, p. 162), show that the phenomenon remains a primary indicator of landscape degradation in most mediterranean-climate regions.

3.2.2 Changed Phenology (Perennial Towards Annual)

The five regions of mediterranean-type climate are characterised by the presence of sclerophyll shrublands (variously called garrigue, maquis, phrygana, chaparral, matorral, fynbos, heath or kwongan, depending on the region in which they occur). All five areas are also characterised by various types of natural grasslands and it is in these grasslands, with or without a tree stratum, that the changes from indigenous perennial tussock ("bunch") grasses to annual forbs and grasses introduced from the Mediterranean Basin are most evident. The sequence of changes accompanying this shift in life cycle has been well described for both southeastern Australia (Moore 1970) and by Burcham (1982) for California. For the present purposes, however, I wish to discuss the change from perennials to annuals in central Chile that has accompanied the increased dominance of the former matorral by tall shrubs of *Acacia caven*.

Acacia caven is a winter-deciduous species of presumed sub-tropical origin (Aronson 1992). Flowering of *A. caven* occurs in early spring after which leaves develop to survive through the hot dry summer (Fig. 3.1). Under the tree layer is an herbaceous community of annual forbs and grasses, introduced from the Mediterranean within the past 300–400 years since Spanish settlement of Chile. These are often the same species that are found in modified grasslands in California and southern Australia (Gulmon 1977; Jackson 1985). In all three regions, these herbaceous species typically establish following autumnal rain and flower and set seed in spring (Fig. 3.1), thereby avoiding the decreased availability of surface soil moisture in summer.

One of the first floristic indicators of change in mediterranean-climate shrublands is the appearance of annuals, and especially of annual grasses of Mediterranean origin. They are notable because of the relative absence

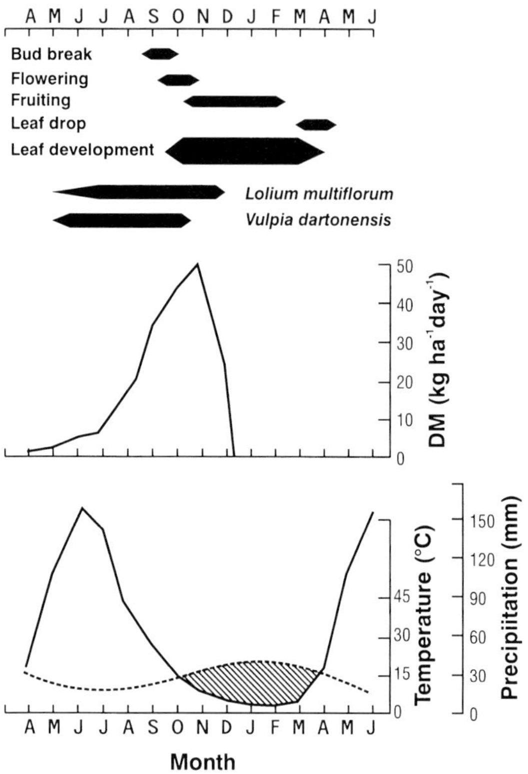

Fig. 3.1. Phenology of *Acacia caven*, and life cycle and growth of two associated annual Mediterranean grasses, in relation to a typical climogram for the subhumid zone of Chile (Ovalle et al. 1990). DM = dry matter production

of perennial grasses from pristine shrubland areas. Because the introduced annual grasses are usually shallow-rooted, their maximum growth occurs in early spring and earlier than most of the shrubs they have replaced. Aronson et al. (1993) cite 30 perennial and 15 annual species in matorral at one particular site in Chile, which changes to 16 perennial and 46 annual species in a mixed community dominated by *A. caven*. At severely degraded sites, the number of perennials is further reduced to 4 and there are 17 annuals (Table 3.1).

The maximum growth of *Acacia caven* trees in summer is reminiscent of the summer growth of the dominant shrubs in Australian sclerophyll communities (Specht and Rayson 1957), a growth pattern which seems anomalous in regions having a winter rainfall regime typical of mediterranean climates.

The change from predominantly perennial to predominantly annual species is accompanied by a change in phenology of the major species and such a change will have significance for food availability for animals as well as for water use at the ecosystem level.

Table 3.1. Vital ecosystem attributes, in late spring, of matorral and the succeeding espinal in various stages of degradation in the subhumid zone of Chile. (Aronson et al. 1993)

Stages of ecosystem degradation				
Vital attribute	Matorral	Esp 1	Esp 2	Esp 3
No. annuals	15	46	27	17
No. perennials	30	16	8	4
Total plant cover (%)	170	95	70	10–30
Soil organic matter (%)	3–5	2.7	1.0	0.2
Length of water availability (months)	8	7	6.5	5.5

3.2.3 Changed Hydrology

One result of a change in the phenology of disturbed plant communities will be a change in the hydrologic regime. An indication of this change may be the general invasion of mediterranean-type shrublands along rivers or drainage channels by introduced tree species, such as of *Acacia dealbata* and *A. melanoxylon* in both South Africa and Chile, of *Acacia saligna* in South Africa, of *Robinia pseudoacacia* in Chile and *Fraxinus ornus* in southern France (Ovalle et al. 1990; Lepart and Debussche 1991). Aronson et al. (1993) show that the change from matorral to a degraded community referred to as espinal in Chile is accompanied by a progressive reduction in maximum soil water reserves from 100–120 mm available for matorral to only 30 mm available for degraded espinal (Table 3.1). The length of the period of water availability was reduced from 8 to 5.5 months for the same communities.

An increase in the number of invasive species may be accompanied by a changed hydrology and especially by a changed relationship between rainfall and runoff. Le Maitre et al. (1996) quantified this relationship for catchments in the mediterranean-climate region of South Africa and showed that riverine invasion of the fynbos after fire by *Acacia saligna* could result in reduced runoff and about 30% less yield of water for nearby urban Cape Town.

A further indicator of a changed hydrologic regime is the prevalence of salinisation of land in several regions of mediterranean-type climate, of which southwest Western Australia provides a contemporary example.

3.2.4 Increased Fragmentation of Landscape

Etienne et al. (Chap. 6, this volume) show the progressive fragmentation of plant communities that go to make up a typical southern French landscape, centred on the township of Puéchabon, near Montpellier. This was a landscape already considerably modified by prehistoric peoples as well as by the Greeks and Romans who cleared land and cut down trees. With the development of abbeys in the region in the twelfth and thirteenth centuries, further landscape modification ensued, especially the removal of wood for timber, glassworks and charcoal production. A diverse agricultural pattern, combined with a variety of forestry pursuits and animal production, further fragmented the landscape for much of the nineteenth and early twentieth centuries. Currently, the region is characterised by a declining rural population, almost no animal husbandry but an increase in seasonal hunting and woodgathering for adjoining urban populations. An increasing number of absentee property owners who reside in the area for only one month in the summer is a major societal change in the region.

Such landscape fragmentation is also a feature of southwest Western Australia where cereal cropping and sheep grazing have created in less than 200 years a complex pattern of uses of land, with few remaining areas of the natural shrublands that once characterised this region (Hobbs and Saunders 1991). Fragmentation has been accompanied by population increases of some animal species and decreases in others.

3.3 Concluding Discussion

In this brief overview of degradation in mediterranean-climate ecosystems I have chosen to concentrate on only a few of the many indicators of change in these lands. Whilst such changes may well characterise relatively undisturbed systems, I contend that the rate of change in them has increased as a result of landscape degradation induced by human settlement. Some of these changes in ecological attributes will create changes for the insects, the birds and the small mammals that depend on the plant communities for their survival. Changes in these groups may well appear that also reflect a change in landscape attributes, and especially the degree of fragmentation of the changed landscape. Sometimes these indicators will be subtle and difficult to determine. Sometimes they may be determined only retrospectively.

The challenge for students and research scientists interested in change in mediterranean-type ecosystems is to document such changes and to avert wherever possible further change if it is believed to be degradative. This will involve political and societal action as well as scientific research. If the quality of life for people occupying mediterranean-type landscapes is to be maintained, and even improved, it is incumbent on us all to alert land management authorities to the most likely consequences of those changes and how they may affect our fellow humans. And I do not refer only to economic benefit!

Finally, let us not overlook the fact that some of the predominant factors in mediterranean ecology, such as fire and fuel management, nutrient availability and soil water regimes, will interact to determine the rate of change in landscape attributes. The resilience of the systems undergoing change will be similarly influential. In this way, I hope that the contributions that follow will reflect the significant progress made in our understanding of mediterranean-type ecosystems since the first volume specifically on the subject of mediterranean ecology was published over 20 years ago as a result of a previous conference, also held in Chile (di Castri and Mooney 1973).

Acknowledgments. I wish to thank James Aronson (CEFE/CNRS, Montpellier) for giving so freely of his time to talk about landscape change in mediterranean-climate areas, and especially to educate me about Chilean ecology, before I had the chance to see the country for myself.

References

Arianoutsou M, Groves RH (eds) (1994) Plant-animal interactions in mediterranean-type ecosystems. Kluwer, Dordrecht

Aronson J (1992) Evolutionary biology of *Acacia caven* (Leguminosae, Mimosoideae). Interspecific variations in fruits and seeds. Ann Mo Bot Gard 79:558–569

Aronson J, Floret C, Le Floc'h E, Ovalle C, Pontanier R (1993) Restoration and rehabilitation of degraded ecosystems in arid and semi-arid lands. II. Case studies in southern Tunisia, central Chile and northern Cameroon. Restoration Ecol 1:168–187

Burcham LT (1982) California rangeland, 2nd edn. Center for Archaeological Research, Davis, Publ No 7, University of California, Davis

Dell B, Hopkins AJM, Lamont BB (eds) (1986) Resilience in mediterranean-type ecosystems. Dr W Junk, The Hague

di Castri F, Mooney HA (eds) (1973) Mediterranean-type ecosystems: origin and structure. Springer, Berlin Heidelberg New York

di Castri F, Floret C, Rambal S, Roy J (eds) (1988) Time scales and water stress. Proc 5th Internatl Conf Medn Ecosystems, Montpellier. IUBS, Paris

Gulmon SL (1977) A comparative study of the grassland of California and Chile. Flora 166:261–278

Hobbs RJ, Saunders DA (1991) Re-integrating fragmented landscapes – a preliminary framework for the Western Australian wheatbelt. J Envir Manage 33:161–167

Jackson LE (1985) Ecological origins of California's mediterranean grasses. J Biogeogr 12:349–361

Kruger FJ, Mitchell DT, Jarvis JUM (eds) (1983) Mediterranean-type ecosystems: the role of nutrients. Springer, Berlin Heidelberg New York

Le Maitre DC, van Wilgen BW, Chapman RA, McKelly DH (1996) Invasive plants and water resources in the Western Cape Province, South Africa: modelling the consequences of a lack of management. J Appl Ecol 33:161–172

Lepart J, Debussche M (1991) Invasion processes as related to succession and disturbance. In: Groves RH, di Castri F (eds) Biogeography of mediterranean invasions. Cambridge University Press, Cambridge, pp 159–177

Mooney HA, Conrad CE (eds) (1977) Proceedings of a symposium on the environmental consequences of fire and fuel management in mediterranean ecosystems. USDA Gen Tech Rep WO-3, Washington, D.C.

Moore RM (1970) South-eastern temperate woodlands and grasslands. In: Moore RM (ed) Australian grasslands. ANU Press, Canberra, pp 169–190

Ovalle C, Aronson J, del Pozo A, Avendano J (1990) The espinal: agroforestry systems of the mediterranean-type climate region of Chile. Agrofor Syst 10:213–239

Seligman NG, Perevolotsky A (1994) Has intensive grazing by domestic livestock degraded Mediterranean Basin rangelands? In: Arianoutsou M, Groves RH (eds) Plant-animal interactions in mediterranean-type ecosystems. Kluwer, Dordrecht, pp 93–103

Specht RL, Rayson P (1957) Dark Island heath (Ninety-Mile Plain, South Australia). I. Definition of the ecosystem. Aust J Bot 5:52–85

Thirgood JV (1981) Man and the Mediterranean forest. A history of resource depletion. Academic Press, London

Part II

Land Use Conflicts in Mediterranean-Type Ecosystems

4 Land Use Conflicts in the Western Cape Region of South Africa

G. DAVIS and R. WYNBERG

4.1 The Gap Between Ecology and Environmental Policy-Making

The body of scientific knowledge generated by ecology and other related disciplines appears to be severely limited in the contribution that it can make to real-world environmental management. One of the basic constraints is that ecologists do not usually have the means at their disposal to convince environmental policy- and decision-makers that the biophysical basis of human life is both finite and fragile. This difficulty of institutionalising baseline ecological knowledge is now recognised as a global problem (Boyle and Carpenter 1994). It is easy to understand that the political arena requires policy-makers to be assertive and unambiguous, both to convince their constituencies of their sound leadership, as well as to make policies implementable. This confidence, however, is anathema to most scientists, whose professional view of life often comprises intersecting clouds of uncertainty, tied together by hypothetical flowcharts. Nevertheless, a quantitative model of ecosystem function, based on scientifically sound data, is an essential component of any effective long-term environmental management plan. For ecologists to play a meaningful role in the process, they need to ask themselves: (1) How can scientifically generated data describing the functioning of ecosystems best be integrated into land management information bases?; and (2) How can scientific rigour be made an intrinsic part of the environmental policy-making machinery?

The natural ecosystems of the mediterranean-climate region of South Africa have been well studied over the past two decades (Richardson et al. 1995). However, these studies have rarely had the capacity or incentive to address the complexity added by consideration of dense human populations and their environmental impacts. A predictive understanding of the interactions between ecosystem function and human demands will become increasingly important for sustainable management of environmental

Ecological Studies, Vol. 136
Rundel et al. (eds.) Landscape Degradation and Biodiversity in Mediterranean-Type Ecosystems

resources – from the conservation of biodiversity, through the provision of clean water, to sustainable agricultural production. In this Chapter we provide an overview of the types of land use found in South African mediterranean-type ecosystems (MTEs), and describe situations in the region where there is a conflicting need for land as a resource. Our goal is to explore, in the context of political evolution in South Africa in the 1990s, opportunities for translating existing scientific knowledge into practical and implementable solutions for environmental management, and to consider some of the choices (and their possible repercussions) that need to be made by policy-makers.

4.2 Characteristics of South Africa's MTEs

A mediterranean-climate is one broadly characterised by winter rainfall and summer drought – the "olive climate" (Aschmann 1973) – but an exact definition is elusive (Hobbs et al. 1995). South African MTEs are confined to the southern and south-western portions of the country, and are contained largely in the Western Cape political province (Fig. 4.1). Here the folded mountain ranges, primarily sandstone in composition, with nutrient-poor quartzitic soils, lie in a broad arc parallel to the coast. They are fringed by a coastal plain, which in places is covered by shale-derived soil, providing relatively good soil for agricultural purposes. The natural vegetation is predominantly a scelerophyllous and species-rich shrubland, containing approximately 8500 higher plant species in an area of 90 000 km^2 (Richardson et al. 1995). It is called fynbos ("fine bush", and pronounced "fane bos") because of the prevalence of small-leaved plants, and is typified by the diverse plant families of the Ericaceae, Proteaceae and Restionaceae. Most plant species in the fynbos vegetation are well adapted to the dry summers and nutrient poor soils. Fire too, plays an important role in most fynbos systems. It is a process to which most plant species are adapted in terms of the cycling of scarce nutrients, and the regeneration of vegetation cover in the post-fire environment.

Human impacts in the region, however, are significant. Most of the land with soil suitable for agriculture has already been transformed for the production of wheat, deciduous and citrus fruit crops, and dairy herding. The mountain regions, on the other hand, are reasonably well preserved. They are also extremely important as water catchments, and, in the face of increasing human impacts on natural systems, significant as reserve areas for the region's exceptionally high plant species diversity. In the metropolitan region of Cape Town, rapid urbanisation is the direct cause of much

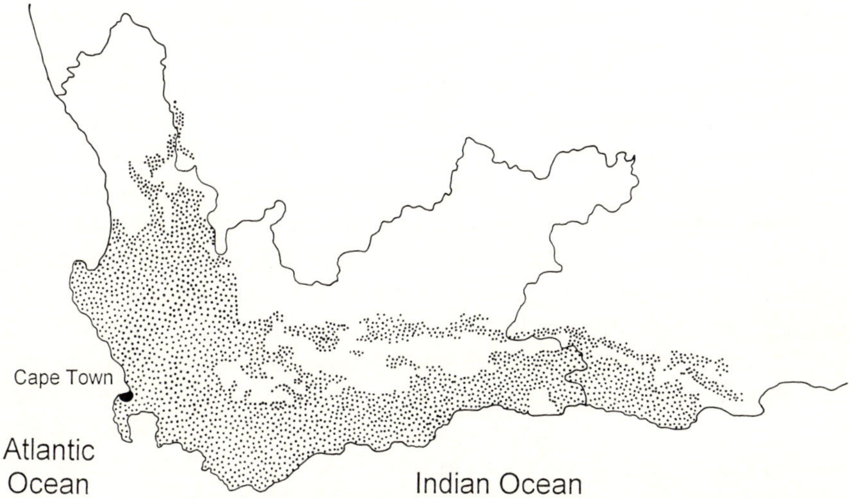

Fig. 4.1. A map of the Western Cape, South Africa. The *shaded* portion represents fynbos vegetation, while the solid line is the political boundary of the Western Cape region

degradation in the lowland areas. People impoverished by the iniquitous policies of the apartheid era, and desperate for shelter and work, habitually set up informal shack settlements on the city edge. The greater city, now estimated to have a population of over 3 million (Bekker 1995) also leaves "footprints" into the rural and undisturbed mountain landscapes, through its increasing demands for water, food, energy and housing.

4.3 Historical Land Use Patterns in MTEs of South Africa

Human occupation of the Mediterranean-climate region at the southern tip of Africa probably goes back to the early post-Pleistocene, when San hunter-gatherers exploited natural resources at a very low intensity, in what must have been an entirely sustainable fashion (Hanekom and Liebenberg 1994). Later, between 2000 and 1500 years ago, Khoi-khoi nomadic pastoralists moved into the region down the west coast. Because of environmental limitations, and the steep rainfall gradient between the coast and the mountains, seasonal migration was necessary to sustain livestock. The arrival of these pastoralists, and the presumed interference of the grazing and browsing of domestic stock with hunter-gatherer access to wild resources, possibly represents the first example of a fundamental conflict over the use of natural resources in South African MTEs.

At the south-western tip of the continent, the first sedentary occupation of the region occurred in 1652, when the Dutch East India Company set up a refreshment station at the site now occupied by Cape Town. Almost immediately, regulations governing the use of natural resources were imposed by the colonists (Rabie and Fuggle 1992). Therewith conflict over land use entered a new phase, characterised by hostilities between European settlers and African communities over access to natural resources such as land, wildlife, water and plants (see, for example, Koch et al. 1990; Carruthers 1995). Such conflicts were further aggravated by the laws and practices of apartheid, introduced from 1948 onwards. Black people were removed from land which they had tended, crowded into "homelands", and often denied access to natural resources upon which they depended, such as thatching, medicinal plants, wild food plants, building materials and fuelwood. In so doing, a fundamental change was brought about in the way in which people related to nature.

The erosion of the traditional land ethic which governed the relationship of indigenous peoples to the environment, and the control which people lost over their land, has had profound implications for land use management in South Africa. Unlike the Mediterranean Basin, and parts of Africa, where political and agricultural development passed through progressive phases of sophistication, the MTEs of South Africa embody an historical land use shift which moved abruptly from nomadic subsistence, to the European-style land management of sedentary agriculture and industry practised today. In contemporary South Africa, the notion of sustainable land use practice in MTEs does not rest, as it does in the Mediterranean Basin, on a long association of humans with the intrinsic production potential of the land. Both crops and techniques of cultivation were imported from Europe and applied to the undeveloped landscapes of the new colony.

4.4 The Special Case of South Africa in the 1990s – MTEs and the Reconstruction and Development Programme

Contemporary land use practices in South Africa are rooted in a colonial context that underpins many land use conflicts in South African MTEs. The first South African government representative of all the country's inhabitants was formed in 1994. This government of national unity recognised as one of its priorities a balancing of the scales between those with, and those without land. Several legal and institutional mechanisms were initiated to effect this, most notably through a comprehensive land reform programme which includes: restitution of land to those who were forcibly removed from their land; the redistribution of land to the disadvantaged and the poor; and land tenure reform to improve the tenure security of all South Africans (Department of Land Affairs 1996). The overall development framework guiding such initiatives is the Reconstruction and Development Programme (RDP) – one of the primary tools adopted by the Government of National Unity for rebuilding the country (African National Congress 1994). Its five inter-related programmes are directed at meeting basic needs; developing the country's human resources; building the economy; democratising the state and society; and establishing structures and plans for implementing such proposals.

Such objectives are clearly critical in light of the country's dire development needs, including an unemployment rate of nearly 50% (Development Bank of Southern Africa 1995). But to a large extent they are only achievable through integrating environmental considerations into the development agenda. The provision of clean water, healthy working and living environments, electricity, primary health care (including traditional medicine systems), and food security, are all fundamental environmental issues which have, until recently, been sorely neglected by ecologists and environmental scientists in South Africa. A substantial research community exists, concerned with issues of ecological process and biodiversity protection, but such research now stands accused in the new democratic South Africa of benefiting only those with access to luxury use of the environment. MTE ecologists in South Africa therefore face a two-fold challenge: (1) to adapt their skills to issues of immediate human concern; and (2) to provide cogent transdisciplinary arguments for continued support of ecological research, and the protection of unspoiled systems and threatened biota. One of the most powerful ways in which to do this is to develop full-cost environmental accounting systems which reflect the true economic value of South Africa's natural resources, and to foster initiatives which demonstrate both conservation and development benefits. South Africa's new Constitution provides an appropriate framework, by proclaiming in its Bill of Rights that:

> Everyone has the right (a) to an environment that is not harmful to their health or well-being; and (b) to have the environment protected, for the benefit of present and future generations, through reasonable legislative and other measures that – (i) prevent pollution and ecological degradation; (ii) promote conservation; and (iii) secure ecologically sustainable development and use of natural resources while promoting justifiable economic and social development.
>
> *Constitutional Assembly 1996*

4.5 Areas of Land Use Conflict in the Fynbos Region

Management of natural resources is still very much an evolving process, even in the so-called developed world. In South Africa, whose economy has distinct "first-" and "third-world" elements, environmental management of land use must address not only the balance between economic development and environmental sustainability, but it must also accommodate a history of political inequity, and the restoration of basic human rights. These rights in many cases include issues of land tenure and access. Historical denial of other rights, such as access to education and formal political process, further complicates resolution of conflicts because unlanded stakeholders are also often disadvantaged in their access to effective bargaining tools. Ultimately, however, all conflicts revolve around the axis of conservation versus development. Conservation, on the one hand, is concerned with limits to development, and is aimed at the preservation of species and habitats – either ethically in terms of humanity's assumed custodianship of Nature, or as a means of holding open future land use options. Development, on the other hand, can range from being low profit, low-impact and sustainable, to making a major contribution to the gross national product by effectively mining the resource. Table 4.1 describes in general terms where conflicts may arise when different types of land use are considered. These zones of interaction between perspectives are considered in more detail below, taking into account some of the specific attributes of the mediterranean-type ecosystems of the Western Cape.

4.5.1 Conservation: Sustainable Land Use and Perceived Future Options

Conservation became a politically contentious issue during the liberation struggle in South Africa, especially in the decade leading up to the recognised start of the democratic process in the early 1990s. Conservation, because it was associated with national parks, conservation areas, recreational landscapes, and even knowledge which was generally inaccessible to the politically disenfranchised sectors of the community, was broadly perceived as elitist and supporting luxury use of the environment. Political normalisation in South Africa, helped on by international pressures and incentives such as the UN environmental conventions on biodiversity and desertification, is allowing conservation to be redefined in a human context. Whereas previously conservation was based on a vague notion of preserving the natural heritage on an ethical and aesthetic basis, threatened species and habitats are now being afforded economic value for indirect and non-consumptive use by tourism and recreation. Even the

Table 4.1. A contingency table of conflicts between different types of land use in the Western Cape (see text for details)

	Conservation: maintaining future options	Ecosystem services	Recreation and tourism	Agricultural production	Housing
Ecosystem services: esp. water production	Compatible, except for dams in areas of high biodiversity				
Recreation and tourism	Partly compatible for ecotourism and wilderness recreation, but not for *laissez faire* commercial development	As for conservation			
Agricultural production: from small-scale to agro-industry	Can be compatible on a small scale for conservation and sustainable use of biodiversity	Compatible with careful management and monitoring	Can be compatible with ecotourism with regard to historical and functional landscapes		
Housing: growing cities and the urban interface	Generally not compatible but good design can reduce the impacts	Not generally compatible; runoff, infiltration, and groundwater systems usually adversely affected	Not generally compatible, except for historical villages and cityscapes	Especially conflicting at the urban interface	
Industry: manufacturing, oil refining, nuclear power	Generally not compatible, but servitudes can be important conservation areas	Pollution is the biggest problem	Not compatible	Especially conflicting at the urban interface, where pollution despoils crops, and where land prices preclude economic viability of farming	Not compatible except for small businesses and light industries in the informal sector

psychological and spiritual upliftment that almost all human cultures associate with nature will probably be given bargaining chips as the resources become more scarce, and disadvantaged communities become more involved in the total political process.

The spectacularly high number of plant species in the fynbos region is a diversity that rivals even tropical rainfall regions (Cowling et al. 1992). Moreover, some 70% of these species are endemic to the region (Bond and Goldblatt 1984), an additional factor that makes the region of particular conservation importance on an international scale. The vegetation comprises different structural types, largely associated with different soil types, each of which has different conservation status. Mountain fynbos vegetation, which occurs on the poorest soils and in the most inaccessible regions, is still 89% natural, and about 53% protected (Rebelo 1992). Coastal renosterveld on the richer, shale-based soils, however, has been transformed, primarily by agriculture, such that only 15% of the natural vegetation remains, with only 0.3% of it formally conserved. Other coastal vegetation, such as that along the west coast of the province, is severely threatened by holiday resort developments (Sowman 1991), while sensitive wetlands on the lowland Cape Flats area to the east of metropolitan Cape Town containing several narrowly endemic plant species, are severely threatened by settlements that are part of the massive influx of people to the urban environment.

Conservation strategies will probably always conflict with the strong need for humans to use land as productively as possible for their own purposes, especially those activities that have limited forward planning. Until the parameters of sustainability can be presented convincingly to the engineers of development and economic growth, this situation will probably persist, especially in the Western Cape, where extreme poverty cannot be overcome without extensive redistribution of resources.

4.5.2 Ecosystem Services: The Ecological Relationship Between Humans and Their Environment

Perhaps the most significant functional demand on pristine land in the Western Cape is for the catchment of water for use by the growing urban population. Quantitative scientific studies have contributed to the development of sophisticated water yield models that predict considerable reduction in the available runoff from catchment areas that have been invaded by woody alien plants that are far more profligate in their use of water than are their indigenous counterparts – they might account for a reduction of up to 80% in runoff depending on the season and intensity of the rain, and the density of the infestation (Burgers et al. 1995). Mountain land in its natural state therefore has considerable economic value as a supplier of ecological services. Because of its inaccessibility and nutrient poor soils, there are seldom major conflicts over its use for catchment and conservation. An exception to this may be the proposed siting of water storage dams, where due to the high degree of plant diversity and narrow endemism in the vegetation, a flooded valley can threaten several species with extinction.

Apart from the provision of water, ecosystem services are often difficult to define, and more difficult to defend against the strong imperatives of development, especially when it is intended to correct gross social injustices. Another instance of a potential water-related ecosystem service being affected by land use for development, is the contamination of groundwater. There are, for instance, at least two aquifers that could be affected by urban growth in the vicinity of Cape Town (see Sect. 4.5.5 below).

4.5.3 Recreation and Tourism: Low Impact Land Use with Opportunity for Income Generation?

Ecotourism is a commercial activity that is theoretically able to convert the intrinsic properties of natural landscapes into an influx of foreign exchange. It is another sector which can be described as being dependent to a great extent on undisturbed ecosystems. It has been estimated that tourism, if all multiplier effects are taken into account, may contribute as much as 10% to employment and the gross regional product (Bridgeman et al. 1992).

Conflict, however, does arise between the desirability of conserving natural landscapes, and the infrastructure that modern tourism demands. Often understated by proponents of fynbos ecotourism as an environmentally friendly panacea for economic woes, are the direct and indirect impacts of hosting large numbers of migrants and temporary residents, and the infrastructural needs for harvesting the foreign exchange. For instance, Table Mountain, the imposing flat topped mountain that towers more than 1000 m behind the city of Cape Town, would not be a tourist destination without its aerial cableway and other facilities that make the experience comfortable and spectacular; nor would the renowned Cape Point Nature Reserve at the tip of the Cape Peninsula (sometimes misrepresented as the southern most tip of Africa), with its spectacular vistas and pristine landscapes, be as saleable without a restaurant, curio store and toilet facilities. In general, the tourist demand for parking-lots, coffee-shops, hotels, paved walkways, balustrades, public telephones and the countless other trappings of modern life, all conflict significantly with the supposed consistency of ecotourism with nature conservation goals.

4.5.4 Agriculture: Are Transformed Systems Sustainable?

Early in the region's colonial history, agriculture acted to sequester the better soils of the region (see "Conservation" in Sect. 4.5.1 above). However, access to farming land in the Western Cape has been restricted almost entirely to white farmers of European descent, with black and coloured residents of the rural areas being regarded by successive governments primarily as a source of cheap labour. These disenfranchised sectors of the community, as a result, have been largely dependent on employer farmers, or the State, for accommodation, schooling and health care. The first democratically constituted government was elected in 1994, and legislation for effecting corrective measures of land restitution and redistribution was passed soon after (see Sect. 4.3 above).

Agriculture is by far the most important primary economic sector in the Western Cape at present, representing 6.1% of the Gross Regional Product (Thomas 1995). Most of the agricultural land is currently dedicated to the production of grain (7756 km^2), fruit (485 km^2), and wine-grapes (916 km^2) (Bridgman et al. 1992). Under the new democratic dispensation, established commercial farmers will probably continue to apply the larger scale techniques of highly mechanised monocultural farming, while newly established small-scale farmers will out of necessity have to use more labour intensive methods, with possibly a far greater mix of crops and livestock. Although conflicts may occur over different production methods, this could possibly provide some latitude for environmentally progressive planning.

The changing economic climate is another influence on the way in which land is used for agricultural production. For instance, market liberalisation and South Africa's membership of the World Trade Organization, together with associated agreements such as GATT, have pushed up the market price of local wheat way beyond the landed cost of imported grain. These economic drivers will likely cause dryland agriculture in the Western Cape to contract, and the intensive production of deciduous fruits and wine to expand. This in turn may have major implications for the management of fynbos ecosystems for water supply, a resource which is by no means in abundant supply considering the demands of a growing urban population (see Sect. 4.5.5 below).

The more extensive modes of production, although they cannot usually turn large profits in a market economy, may be regarded as investments in low impact management for sustainable land use. This includes production opportunities like low intensity small livestock and game farming, or the harvesting of natural products such as plants with decorative, aromatic or medicinal value. Clearly this second option for land use is less in conflict with conservation requirements, or the foundation of an extensive ecotourism industry, but as a contributor to economic growth, its direct contribution is small. Nevertheless, such low impact and labour intensive operations can provide an important economic buffer for rural communities with limited access to resources for technological development.

4.5.5 The Urban and Peri-Urban Environment as a Zone of Land Degradation

Probably the most important influences that humans have on land-degradation are those associated with civilisation's ultimate product – the city. Although considerable attention is now usually paid to urban planning in recognition of this fact, economic, historical, and social factors tend to frustrate the effectiveness of human generated design. This, at least, appears to be the case for Cape Town, the mediterranean-climate megalopolis of South Africa, and the only large city in this climatic zone.

Since the site was settled in the mid-17th century as a refreshment station on the sea-route between Europe and the Far East, the greater city has grown to over 3 million people. Like in any modern city, land is in demand for several uses, the main ones being housing, business, industry and recreation. Environmentally, these are exceptionally challenging demands for Cape Town planners. The reasons for this are mainly the constraints placed on urban development by the environment (Fig. 4.2).

Development in a southerly direction down the Cape Peninsula is limited by the mountainous terrain, the high conservation status of the vegetation (there are over 2000 higher plant species in an area of just 470 km^2; Richardson et al. 1995), and the recognised need to preserve landscapes as a resource for tourism. Development to the north is restricted by the presence of the Koeberg nuclear power station, a contributor to the national energy grid that was commissioned in the early 1980s, and which precludes high density settlements within a radius of 20 km. To the east and north-east urban development can only occur at the cost of transforming the agriculturally productive lands that are the chief production sites of wheat and wine. The Cape Flats region, a low lying area of calcareous duneland 30 km south-east of the city centre, was designated for extensive housing schemes in the early 1980s. Large township complexes, designed according to the apartheid policy of central government, and consistent

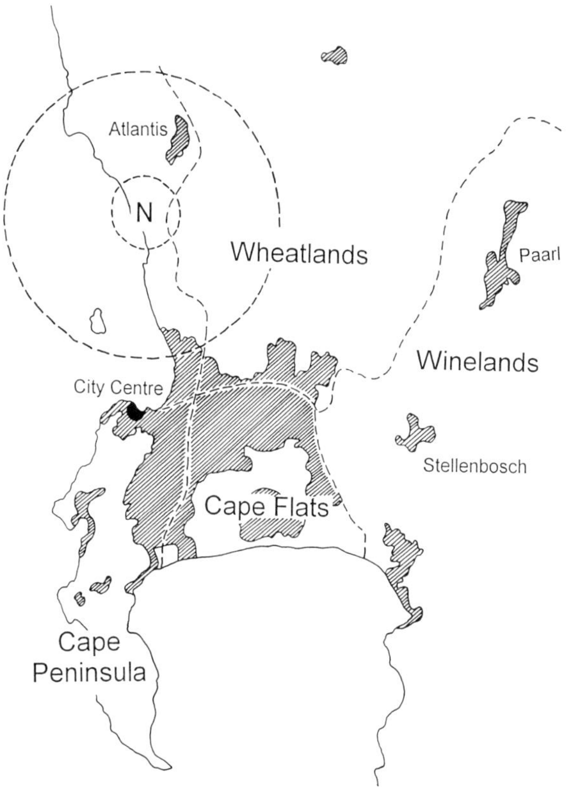

Fig. 4.2. A map of the Cape Town metropolitan area showing features mentioned in the text. The Cape Flats is a duneland area overlying the False Bay aquifer; N = the Koeberg nuclear power station with a 5 km exclusion radius, and a 20 km radius inside which settlement is restricted

with the Group Areas Act (Platzky and Walker 1985), were built to house all black residents in the greater Cape Town area. Apart from the destruction of seasonal wetland habitat, and the possible loss of an unknown number of endemic plant species, the sandy substrate of much of the designated area is highly mobile during the windy summer season, and waterlogged during the winter, creating unpleasant and often unhealthy living conditions. As mentioned above, domestic pollution is also threatening the potability of water in the False Bay aquifer. Other informal housing settlements on the western edge of the Cape Flats are based on the well drained quartzitic sand deposits at Philippi. As housing sites, these are in conflict with some of the better vegetable growing lands in the Cape Town area. A third stakeholder in this land is the glass industry. The sand deposits in this area are of high quality, and supply 50% of the best industrial sand to the South African glass-making industry (Gasson 1993).

At the other end of the economic spectrum to the squatter settlements, are the middle-class suburbs built on the fertile granitic soils of the Constantia Valley in the eastern Cape Peninsula. These lands were important production areas for the Cape grape and wine industry until the 1970s, when suburban land became more valuable as real estate for up-market domestic housing than for agricultural production. The same pattern is currently repeating itself to some extent in the winelands east and north east of the metropolitan area. Between 1969 and 1981, cultivated land in the metropolitan area was lost to urban sprawl at a rate of one hectare per day (Bridgman et al. 1992).

A main focus of the government's Reconstruction and Development Programme is to build an economy that can provide release for people caught in the cycle of poverty arising from underdevelopment and the imposed disadvantages of the apartheid era. Industrial reconstruction is obviously a crucial part of this recovery process, and one which must be accommodated in the metropolitan development strategy. Like housing, however, industrial development is almost always in direct conflict with other forms of land use, especially conservation, management for ecosystem services, and tourism.

Not only is it at the city margins that the primary conflict between development and conservation occurs. Within the city, open spaces form important "green lungs", providing humans with physical and psychological relief from the pressures of urban life. Once again, real estate value in a changing economy can allow such land to come under intense development pressure. Many such pieces of land are under weak statutory protection, and are continuously being converted to housing and business premises. Important pieces of "incidental" conservation land of this type lie within horse race courses, military areas, powerline easements, urban commonages and road reserves. The Kenilworth race course on the western fringe of the Cape Flats, for instance, was noted as serving as a refuge for 17 rare and endangered "red data book" plant species (McDowell 1989).

4.6 Conclusion: Conflict and Land Degradation

Competition for tenure of land is perhaps one of the oldest and most basic areas of conflict in the history of human relations. In South Africa, the series of human migrations into the mediterranean-climate region over the past 10 000 years has caused considerable conflict as the arriving cultures imported new land use practices: from hunting and gathering, through pastoralism and agriculture, to manufacturing and business. Often this conflict has resulted in antisocial behaviour and physical violence. A 1990s version of such territorial conflict is the violent confrontation that erupts over the invasion of open spaces by immigrant squatters in the peri-urban areas, sometimes between metropolitan authorities and informal residents, but also sometimes between the squatter groups themselves as they compete for control of their living space.

Although conflict over land use can often be destructive, the underlying process of dialogue between different users of the resource is an extremely useful one. Correctly managed, and with sufficient democratic structures in place to avoid extreme polarisation, the various requirements of land users can contribute significantly to establishing an efficient management process. Problems arise, however, when some stakeholders are insufficiently empowered to have their contributions recognised by others. A growing lobby in this political process is emerging from the synergy between the many players in the arena: from the practitioners of ecology, economics and sociology, to the community-based organisations that have a firm political base, but limited access to the theoretical basis for environmental management. The emerging profile of credible environmental forums for exploring development options is an optimistic sign in the Western Cape.

References

ANC (1994) The reconstruction and development programme: a policy framework. Umanyano, Johannesburg

Aschmann H (1973) Distribution and peculiarity of Mediterranean ecosystems. In: di Castri F, Mooney HA (eds) Mediterranean-type ecosystems: Origin and structure. Springer, Berlin Heidelberg New York, pp 11–19

Bekker S (1995) The Western Cape: taking the high road. Indicator 12(3):58–60

Bond P, Goldblatt P (1984) Plants of the Cape flora – a descriptive catalogue. J S Afr Bot Suppl 13:1–455

Boyle TP, Carpenter RA (eds) (1994) Institutionalization of ecological knowledge: an international perspective. Ecol Int 21:1–103

Bridgman DHM, Palmer I, Thomas WH (1992) South Africa's leading edge? A guide to the Western Cape economy. Association for the Promotion of Economic Growth in the Western Cape (WESGRO), Cape Town

Burgers CJ, Marais C, Bekker SJ (1995) The importance of mountain catchments for maintaining the water resources of the western Cape Province and the need for optimal management. In: Boucher C, Marais C (eds) Managing fynbos catchments for water. Foundation for Research Development, Programme Report number 24, Pretoria, pp 99–123

Carruthers J (1995) The Kruger National Park. A social and political history. University of Natal Press, Pietermaritzburg

Constitutional Assembly (1996) Constitution of the Republic of South Africa. As adopted by the Constitutional Assembly on 8 May 1996

Cowling RM, Holmes PM, Rebelo AG (1992) Plant diversity and endemism. In: Cowling RM (ed) The ecology of fynbos: nutrients, fire and diversity. Oxford University Press, Cape Town, pp 62–112

Department of Land Affairs (1996) Green paper on a new land policy. Government Printer, Pretoria

Development Bank of Southern Africa (1995) South Africa's nine provinces: a human development profile. Development Information Paper 28. Development Information Group, Centre for Policy, Information and Evaluation, Pretoria

Eckert J (1995) Rural Western Cape: harvesting growth? Indicator 12(3):64–66

Gasson B (1993) The environmentally viable city: a performance concept and a case study. Proc 30th World Conf of the Int Federation of Landscape Architects, Cape Town

Hanekom D, Liebenberg L (1994) Utilisation of National Parks with special reference to the costs and benefits to communities. Bull Grassland Soc S Afr 5(2):25–36

Hatcher B (1994) The reconstruction and development programme: an environmental perspective. Bull S Afr Inst Ecol 13(2):6–9

Hobbs RJ, Richardson RM , Davis GW (1995) Mediterranean-type ecosystems: opportunities and constraints for studying the function of biodiversity. In: Davis GW, Richardson DM (eds) Mediterranean-type ecosystem: functions of biodiversity. Springer, Berlin Heidleberg New York, pp 1–42

Koch E, Cooper D, Coetzee H (1990) Water, waste and wildlife: the politics of ecology in South Africa. Penguin, London

McDowell C (1989) Conservation and horse racing: the unseen connection. Veld Flora 75(2):36–39

Platzky L, Walker C (1985) The surplus people: forced removals in South Africa. Ravan, Johannesburg

Rabie MA, Fuggle RF (1992) The rise of environmental concern. In: Fuggle RF, Rabie MA (eds) Environmental management in South Africa. Juta, Cape Town, pp 11–25

Rebelo AG (1992) Preservation of biotic diversity. In: Cowling RM (ed) The ecology of fynbos: nutrients, fire and diversity. Oxford University Press, Cape Town, pp 309–344

Richardson DM, Cowling RM, Bond WJ, Stock WD, Davis GW (1995) Links between biodiversity and ecosystem function: evidence from the Cape Floristic Region. In: Davis GW, Richardson DM (eds) Mediterranean-type ecosystem: functions of biodiversity. Springer, Berlin Heidleberg New York, pp 285–333

Sowman M (1991) Impact of resort development in the coastal zone, Veld Flora 77:120–123

Thomas W (1995) The Western Cape: maintaining a leading edge. Indicator 12(3):61–63

5 Impacts of Land Use on Biodiversity in Southwestern Australia

R.J. Hobbs

5.1 Introduction

The ecosystems of southwestern Australia developed in isolation from other parts of Australia and the world, and this, together with an array of other factors has resulted in unique and diverse biotic assemblages and a high degree of floristic endemism (Hopper 1979, 1992). These have been subject to human modification for many thousands of years, following the colonization of the area by aboriginal peoples. Evidence is accumulating that these people had considerably more influence on Australian ecosystems than was previously thought, and their activities may have resulted in massive changes in vegetation and the extinction of megafauna (Flannery 1994; Kohen 1995).

It seems likely, therefore, that many of the ecosystems of southwestern Australia were already significantly influenced by human activities when the first European explorers traveled the coasts in the 15th century. The establishment of a British settlement on the Swan River in 1829 saw the commencement of another wave of change in southwestern Australian ecosystems. It is the consequences of this wave of changes which I wish to explore in this chapter.

5.2 Current Status of the Biota

The state of Western Australia comprises one third of the Australian continent, and yet contains under 2 million people. Most of these people live in the south western corner of the state, and a high proportion live in the Perth metropolitan area. Despite the low population density and concentration of people in the major city, human activities have had profound impacts on most southwestern Australian ecosystems.

Ecological Studies, Vol. 136
Rundel et al. (eds.) Landscape Degradation and Biodiversity in Mediterranean-Type Ecosystems

The southwest botanical province ($0.3 \times 10^6 km^2$) is a recognized hotspot for floristic biodiversity (Groombridge 1992; Hopper et al. 1996), with an estimated 8000 species and approximately 75% endemism. This area also has the highest concentrations of declared rare flora (which includes a variety of designations of endangerment). For Western Australia overall, out of a total of around 12000 vascular plant species, 52 are thought to have gone extinct, while 232 are considered threatened, and a further 1200 are of uncertain status due to lack of sufficient information (Hobbs and Mooney 1998). Many of these threatened species are in the agricultural area (Hopper et al. 1990). The flora is still under active research, and estimates of total species numbers are approximate. On current estimates, approximately 8.6% of the total flora is introduced, a figure which is lower than for Australia as a whole.

A report by the government of Western Australia (1992) recognized 11 plant communities under imminent threat of major structural and floristic change, and a further 17 communities which are geographically restricted and/or presently or potentially subject to endangering processes. The main threatening processes affecting a range of communities thought to be in imminent danger of major structural and floristic change are illustrated in Fig. 5.1. The most frequently cited threatening processes were land clearing, secondary salinisation and introduced disease. These processes are also important threats to many of the species currently listed as endangered.

No extinctions have been recorded for Western Australian birds, reptiles, amphibia or freshwater fishes. However, Australia has one of the worst extinction rates for mammals in the world, but the extinction rate

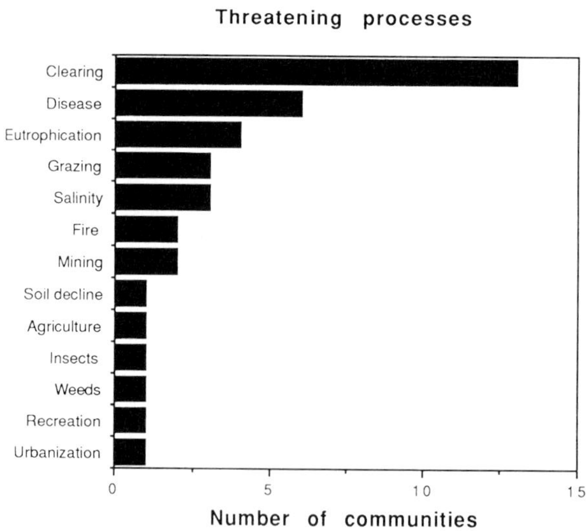

Fig. 5.1. Major threatening processes recognised for plant communities in imminent danger of major structural and florisitic change in the short to medium term and/or which are geographically restricted or subject to endangering processes. Data from Government of Western Australia (1992)

masks a much more widespread problem of range contractions (Strahan 1983; Burbidge and McKenzie 1989; Kennedy 1990). Mammals which were formerly widely distributed across the continent are now restricted to small refugia or offshore islands. Of the remaining mammal species, 21% are considered threatened. It follows from this that many populations of these mammals have gone extinct across the continent, and this is certainly true for the agricultural area. Kitchener et al. (1980) estimated that only 12 of the 43 species of native mammal recorded from the wheatbelt at European settlement were still common. There is increasing evidence that these population extinctions may have had system-level repercussions, particularly where the species involved had significant impacts on their environments (Noble 1993), although it is now difficult to assess these impacts accurately due to lack of baseline data.

Introduced predators, particularly foxes and cats are thought to be one of the major causal factors leading to the complete or near extinction of many Australian marsupials (Burbidge and McKenzie 1989; Friend 1990). Indeed, Western Australian forests have provided the last refuge for a number of species, presumably because fox numbers remained lower than in other areas. Predator control is now practiced in many areas, with obvious success indicated by increases in abundances of native mammals (Kinnear et al. 1988; Friend 1990). Poisoning of introduced predators using 1080 (sodium monofluoroacetate) poison is possible because the resident mammal populations have developed a tolerance to the poison through coevolution with plants containing it (Twigg and King 1991). A recent large scale program of fox eradication entitled "Western Shield" promises to allow recovery or reintroduction of endangered mammals in many forest and non-forest areas.

Invasive plants are another major threat to many ecosystems across Australia (Humphries et al. 1991; Humphries 1993). Individual invaders, such as bridal creeper (*Myrsiphyllum asparagoides*) have the potential to crowd out native species and alter vegetation composition and structure. Herbaceous species, especially grasses, also have the potential to alter fire regimes by changing the structure and availability of fuel (D'Antonio and Vitousek 1992). In the region of 1500 species are currently naturalized across Australia, with 848 species recorded from Western Australia (Keighery 1991). At least 220 species are recognized as noxious weeds across Australia (Parsons and Cuthbertson 1992), and many are problems in native ecosystems. In addition to existing problems, there are likely to be many more species which become problems in the future (Hobbs 1993b). These include, for instance, pine species planted in native forest areas which have been found to be invasive elsewhere in the world (Richardson et al. 1994).

The relatively high levels of invasion by mammals, plants and fish in Western Australia undoubtedly result mainly from the deliberate introduction of these organisms for agriculture, horticulture or sport. Recent indications suggest that new species of plants continue to be introduced (Keighery 1995), and there are recent examples of introduced species rapidly extending their range (Dodd and Moore 1993). Australia continues to import plant species without due consideration for the potential threats of invasiveness. For most animal groups, the rates of introduction are now probably slower, and most of the introductions have already occurred. Nevertheless, Hutchison and Armstrong (1993) detail the history of recent invasion of a southwestern Australian river system by *Perca fluviatillis*. There is still considerable potential for inadvertent introduction of non-native species, for instance as contaminants of agricultural produce, in ballast water or carried by various forms of transport (Carlton and Geller 1993; Hedgpeth 1993).

5.3 Types of Modification

The southwest of Australia is a heterogeneous area in terms of climate, landforms, soils, vegetation and land uses (Hobbs et al. 1995). The area can be subdivided into several botanical districts, major catchments and climatic zones: a combination of these results in 84 "natural resource zones" (Environmental Protection Authority 1994). Here I discuss three major zones: the coastal plain, the forest region and the agricultural region. The history of development of each of these regions is detailed in Jarvis (1986).

Hobbs and Hopkins (1990) recognized four major categories of land use in Australia, based on the degree of modification of vegetation (Table 5.1). Each type of use exists in each of the areas examined, but the extents differ. Clearly, complete removal of vegetation through urbanization or industrial development is mostly restricted to the coastal plain, but even within the urban area, native ecosystems are retained in a more or less modified state within small bushland areas. Mining has occurred in all three zones, but is now mostly confined to the coastal plain and forest areas. Vegetation replacement by agriculture has occurred in all three zones, as has modification by a variety of activities. Conservation areas also exist in all three zones, and these are usually imbedded in the modified production matrix. An important feature, therefore, is the intermixing of different land uses in each area, and the potential conflicts that arise from this.

Table 5.1. Major land use categories, based on degree of modification of the natural vegetation (modified from Hobbs and Hopkins 1990)

Category	Examples
1. REMOVAL	
	-urban development
	-mining
	-transport
	-industrial development
2. REPLACEMENT (with intensively managed systems)	
	-agriculture
	-horticulture
	-plantation forestry
3. UTILIZATION (exploitation of native vegetation)	
	-non-plantation forestry
	-pastoralism
	-recreation
4. CONSERVATION (no deliberate modification)	
	-conservation reserves
	-vacant crown land (uncommitted government-owned land)
	-water catchment

5.4 The Coastal Plain

The coastal plain is the area most affected by urban and industrial development. The metropolitan area of Perth is one of the fastest growing urban areas in Australia, and urban development has resulted in considerable reduction in areas of native vegetation and degradation of remaining areas (Fig. 5.2). Planning of new urban developments has rarely incorporated considerations of biodiversity conservation, and existing bush areas are frequently put under threat of further development. Until recently, there were relatively few data with which to assess the conservation value of areas of native bush prior to development, but recent botanical surveys have provided a database which can facilitate this process. Indeed, some of the major ecosystems of the coastal plain are still relatively poorly researched. For instance, the *Banksia* woodlands of the Swan Coastal Plain are still poorly understood, and yet are increasingly under threat from development and degradation (Burbidge 1989; Hopper 1989).

Areas of bush remaining in the urban area are under considerable threat from a variety of factors. Their small size, isolation and human use means that they have to be actively managed (Scheltema and Harris 1995). Par-

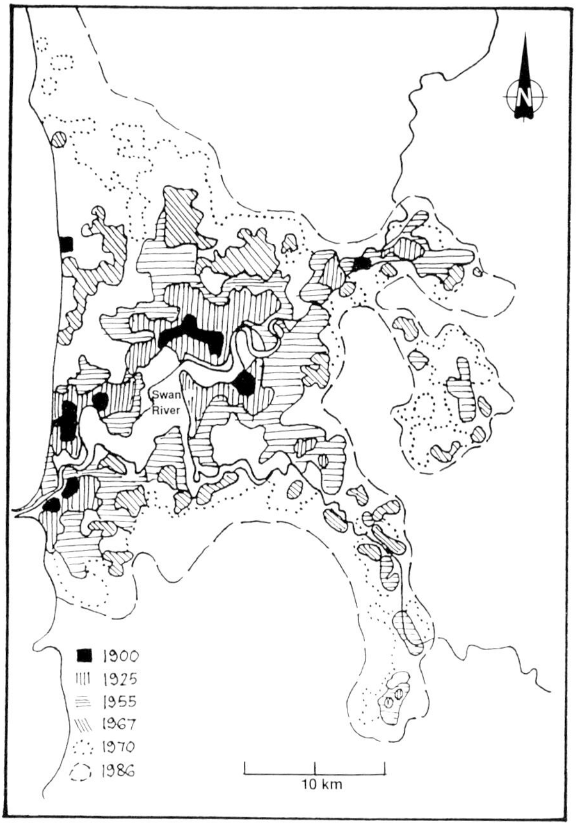

Fig. 5.2. Expansion of urban Perth between 1900 and 1986. (Adapted from Hiller 1987)

ticular threats come from altered fire regimes, weed invasion and, particularly in the case of wetlands, altered hydrology and eutrophication. In many bush areas, fire regimes have been altered dramatically, and arson is a major cause of bushfires (Dixon et al. 1995; Pigott and Loneragan 1995).

The urban population is becoming more aware of the values of urban bushland and open space, and is putting local government and developers under increasing pressure to maintain remaining bush areas. Further to this, there is increasing participation in management issues, with local "Friends" groups undertaking management and restoration activities.

5.4.1 Perth's Offshore Islands: Vignettes of Ecosystem Modification

While ecosystem modifications can be generalized to some extent, it is important to recognize regional and local variations in the importance of different factors. This is true in all the regions considered here, and can be illustrated by examining the three largest islands off the coast of Perth.

The three islands provide a microcosm of the changes in native ecosystems brought about by human activities. The three islands, Rottnest (1600 ha), Garden (1200 ha) and Carnac (16 ha) all have the same general geology and soils, but differ in their history of human use. The following accounts are derived from Rippey and Rowland (1995) and Pen and Green (1983). Rottnest Island is a popular tourist destination and holiday resort. Prior to European settlement, it was, from early accounts, covered by a vegetation mosaic containing significant amounts of woodland dominated by *Callitris preissii* and *Melaleuca lanceolata*. Its use from 1838 as a prison settlement led to cutting of timber for firewood, clearing, and frequent fires used by both Aboriginal and European settlers during hunting. Increased incidence of fire led to a decline in the fire sensitive *Callitris* and *Melaleuca*, which were replaced by the more fire-tolerant *Acacia rostellifera*. By the early 1900s this covered two thirds of the island. In the 1920s the quokka *Setonix brachyuris* (a small wallaby) was declared a protected species and numbers increased dramatically thereafter. High quokka populations prevented post-fire regeneration of tree species and a series of extensive fires greatly reduced the area of woodland which was replaced by low scrub dominated by *Acanthocarpus preissii*. Woodland cover continued to decline dramatically into the 1980s (Fig. 5.3).

About 45% of the 191 plant species currently found on Rottnest are non-native (Rippey and Rowland 1995). The non-native *Trachyandra divaricata* has become widespread on Rottnest in degraded areas, but is now probably preventing much of the island from blowing away. Reforestation programs have been underway for some time, and much of the woodland now present on the island is a result of these activities. Past plantings have included eucalypts such as *E. gomphocephala* and *E. platypus*. Evidence exists for the existence of *E. gomphocephala* woodland in prehistoric times (Churchill 1959), but the appropriateness of revegetating with this and other eucalypt species has been questioned in recent times, Notably, the revegetated areas do not support the same levels of bird diversity as *Callitris* and *Melaleuca* woodland (Saunders and de Rebeira 1993). Current revegetation efforts involve replanting mostly *Callitris* and *Melaleuca*.

At the time of settlement, Garden Island was covered with *Callitris preissii* and *Melaleuca lanceolata* woodland similar to that on Rottnest. The island was taken over by the Department of Defence in 1915, habitation increased and vegetation clearance and modification ensued. In the 1950s, *Acacia rostellifera* scrub was the most abundant vegetation type, but woodland was still abundant. The construction of a naval base commenced in 1969. The provision of water and the establishment of gardens increased the food supply for the resident Tammar wallabies *Macropus eugenii*. Al-

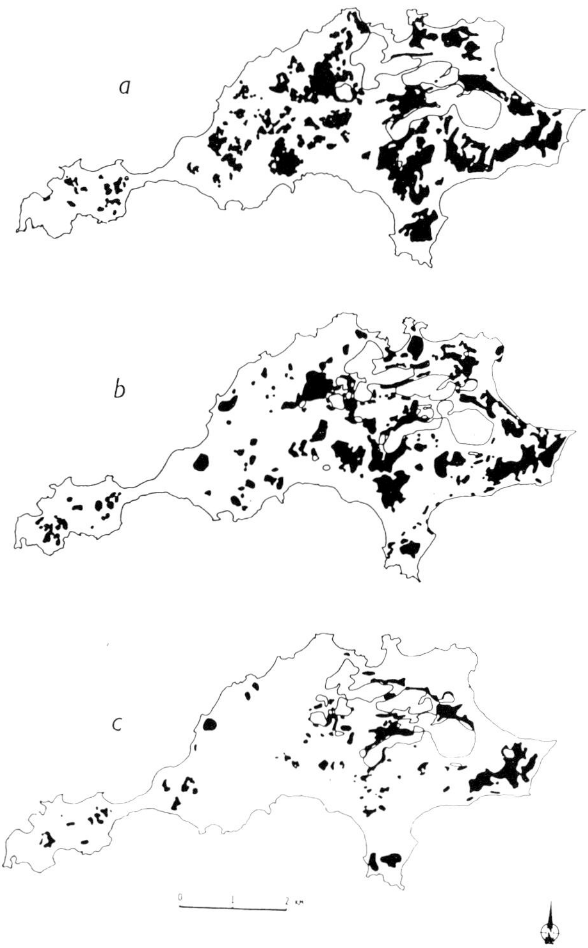

Fig. 5.3a–c. Open forest/woodland vegetation distribution (*shaded*) on Rottnest Island in (a) 1941, (b) 1955 and (c) 1982 (redrawn from Pen and Green 1983)

though unrecorded, numbers of tammars are thought to have increased dramatically, and these now prevent regeneration of *Acacia rostellifera* by grazing on seedlings. Centers of acacia thickets are dying out and being replaced by *Acanthocarpus preissii.* Garden Island does not have the same tourist pressures as Rottnest, and did not experience regular burning, and hence still has extensive woodland and thicket cover. 46% of the 174 plant species found on the island are non-native. The non-native arum lily *Zantedeschia aethiopica* has invaded large areas of the island over the last 30 years, and an extensive (and expensive) eradication program is underway.

Carnac Island (16 ha) is relatively undisturbed and little visited by tourists, partly because of the large number of tiger snakes *Notechis scutatus* on the island. However, the island vegetation was greatly influenced by rabbits, which were probably introduced to the island in the 1820s (Rippey and Rowland 1995). These persisted until they were eradicated in 1969. The rabbits kept the vegetation very open, but since their eradication, dense vegetation cover has developed, with tall *Acacia rostellifera* scrub covering the central portion of the island. Of the 97 plant species on the island, 57% are non-native.

The three islands illustrate the range of impacts imposed on the coastal plain and more generally on southwestern Australian ecosystems. Vegetation clearance, harvesting, modification of fire and grazing regimes and introduction of non-native organisms have impacted most ecosystems in the area to some extent.

5.5 The Forests

The forests of Western Australia, in common with those in other parts of the country have been the focus of considerable political debate. More controversy surrounds the human use of forests than any other ecosystem in Australia. The history of human use of forests has been well documented by Dargavel (1995), who discusses the exploitation of timber resources and changing societal expectations concerning forest use and conservation.

Natural forest dynamics are determined largely by three factors which predominate in the area: i.e. a mediterranean-type climate, with its well-defined summer drought; soils with low nutrient status; and the incidence of fire and other disturbances (Kruger et al. 1983; Dell et al. 1986; Dell et al. 1989; Davis and Richardson 1995). Forest systems of Western Australia are also impacted by a number of human activities. The most important of these are forest management practices, in particular timber harvesting and fuel reduction burning, and the impacts of invasive species, including the introduced pathogen, *Phytopthora cinammomi.*

5.5.1 Phytopthora

The introduced fungus *Phytopthora cinammomi* is undoubtedly a major factor influencing the forest ecosystem in Western Australia and in other parts of the country (Dell and Malajczuk 1989). A recent symposium has highlighted the impacts of the disease on a variety of forest components (Withers et al. 1994). Depending on the severity of attack, *Phytopthora* causes the loss of overstorey and understorey plant species. A large number of plant species have been shown to be susceptible to the fungus, and it is likely that others are impacted by the changes in microclimate and local moisture relations caused by loss of overstorey cover. This fungus, and other diseases, are also important in other non-forest communities, including the floristically rich kwongan heath (Wills 1993; Wills and Keighery 1994).

5.5.2 Forest Management Practices

Opinions vary as to the extent to which current forest management practices affect biodiversity and ecosystem processes. It has recently been argued that timber operations and prevailing prescribed burning practices have little impact on the overall forest system (Abbott and Christensen 1994). However, this may be partially due to the lack of monitoring and research into such potential impacts (Calver et al. 1996). In most parts of Australia, little research work has been conducted into the long-term impacts of timber or burning operations (Williams and Gill 1995). The impact of disturbance-causing activities concentrates on individual species, and little long-term monitoring has been implemented (Lands and Forest Commission 1994). In 1977 the Senate Standing Committe on the Environment stated that "The extreme lack of knowledge on the biological sphere . . . is hampering responsible decision making". In 1993, the Resource Assessment Commission (1993) still had to conclude that "the level of information on impacts appears insufficient for most current uses".

Few long-term studies have been conducted on impacts on biodiversity or system processes. For instance, Schofield et al. (1989) stated that "No long term detailed studies on the effects of different silvicultural systems on the hydrology of the jarrah forest have been carried out". Studies that have been conducted do indicate an impact of timber harvesting. Increases in streamflow (91–182%) were reported by Borg et al. (1987) following heavy logging, and Bari et al. (1994) have shown marked increases in streamflow and groundwater discharge following clear felling. These impacts were highest immediately after logging, but persisted for at least 8 years. These findings concur with experimental work carried out elsewhere (Bormann and Likens 1981).

Impacts of logging on other system components are equally poorly documented, although there are indications from forests elsewhere that impacts on soil properties can be substantial (Rab 1994). Work by Mawson and Long (1994) has suggested that current logging practices significantly alter habitat suitability for some bird species, although the validity of the methods used has been questioned by Burrows et al. (1995). Similar discussions on the impacts of logging have occurred in Victoria, where Attiwill

(1994a,b) has implied that logging is in many ways equivalent to natural forest disturbance, a conclusion contested by Lindenmayer (1995) (see also subsequent response by Attiwill 1995).

The jarrah forest is currently subjected to widespread short-rotation fuel reduction burning, which has evolved to reduce the risk of destructive wildfires. Controversy surrounds the questions of whether such a burning regime is effective and whether it has adverse impacts on the forest ecosystem (McGrath 1985; Tingay 1985; Underwood et al. 1985). It has been claimed that the current regime mimics the regime prevailing prior to European settlement (Hallam 1975; Burrows et al. 1995), although the evidence for this is not compelling, and indeed, basic knowledge on Aboriginal fire regimes is scant (Williams and Gill 1995). Burrows et al. (1995) stated that fire scars on *Eucalyptus marginata* trees indicate the occurrence of moderate to severe fires in the forest occurred with a mean interval of 81 years. They then used historical accounts of aboriginal burning and lightning records to conclude that these severe fires must have been accompanied by low intensity fires every 2–5 years. The question remains as to whether such a regime of low intensity fires prevailed over the whole forest or was restricted to areas most frequented by aborigines. There is some evidence to suggest that such regimes did not prevail everywhere (Lamont and Downes 1979).

Abbott and Christensen (1994) suggest that the current fire regime (and logging activities) are "... a minor, irregular and relatively insignificant perturbation ...". On the other hand, McCaw and Burrows (1989) concede that "While many studies have examined the effects of one, or occasionally several fires, on plant and animal communities in the forest, the basis for predicting longer term effects of different fire regimes is limited". Indeed, the impact of one fire may be minimal, although even this conclusion is open to question (e.g. Majer and Abbott 1989). Again, evidence from Victoria points to a potentially detrimental impact of fuel reduction burning (Hamilton et al. 1991), although the conclusions of this study again have been questioned (McCaw 1993). Nevertheless, it is the overall fire regime (i.e. frequency, intensity, season, size etc.) which shapes the vegetation in the long-term, and such long-term data on vegetation changes under current fire management practices are not available (Williams and Gill 1995).

Analyses such as that by Abbott and Christensen (1994) look at relatively short time scales and suggest no significant ecosystem changes. For instance, extensive logging has been carried out mainly within the last century (Dargavel 1995), and the present fuel reduction fire regime has been in place for less than 40 years – a short time span in terms of forest dynamics. These time frames are probably too short for an adequate assessment of the long-term sustainability of current management practices. Longer term impacts could be significant, but will not be noticed if relevant monitoring systems are not in place. Even if impacts are detected, it could be some considerable time before a policy or management response is implemented. A clear example of the types of lag involved in responding to problems is the salinization of agricultural land caused by past land clearance, discussed below. This problem was first documented in the 1920s (Wood 1924), but is only now being acted on at the policy level.

The degree of debate over the importance of changes to forest ecosystems arising from management practices indicates the difficulty in reaching conclusions on the issue. The problem is further compounded by potential non-linearities in response, and complex interactions between system components (Hobbs 1996). It is clear that *all* management has *some* impact on the ecosystem (even if the management is to do nothing). The important question is whether the level of impact is acceptable or not. The acceptability or otherwise of any particular impact will change as society's expectations and priorities change. Increasingly, society as a whole is demanding more careful use of forest resources and recognizing the value of forests from perspectives other than timber production. In particular, revenue from tourism is increasing rapidly, and most people want to see vertical rather than horizontal forests. While some progress is being made in developing management systems which cater for multiple uses of forests, it would be fair to say that considerable conflict still exists. Resolution of these conflicts requires increased scientific data and understanding but also, and perhaps more importantly, more effective communication and mutual respect between different viewpoints.

5.6 Agricultural Areas

The main agricultural area, or wheatbelt, was settled by Europeans during the first half of the 19th Century, but rapid development commenced only during this century. The area suitable for agriculture is determined by the distribution of suitable soils on one hand, and adequate rainfall on the other. The area was extensively cleared of native vegetation during this century, with a rapid rise in the rate of clearing after 1945 (Saunders et al. 1985). Over the wheatbelt as a whole, approximately 7% of native vegetation remains, although this percentage varies greatly within the region, with the oldest settled areas retaining less vegetation. Prior to clearing the area consisted of a mosaic of vegetation types, including various woodland, shrubland and mallee (multi-stemmed eucalypt) types. These vegetation types were cleared to differing extents, depending on the suitability of the underlying soils for agriculture. Woodlands, in particular, were heavily cleared, with some types being all but eliminated from the region (Beard and Sprenger 1984).

The native vegetation left in the region is present as a large number of mostly very small remnants. A few large areas are reserved as national parks or nature reserves, and these account for a large proportion of the native vegetation remaining (Wallace and Moore 1987). Within the area administered by the Department of Conservation and Land Management as the wheatbelt region, there are over 600 nature reserves, which range in size from 0.4 to 309000 ha. The three largest reserves account for 64% of the reserved area, and the median size of reserves is 114 ha (Wallace and Moore 1987). In addition to the areas designated as nature reserves or set aside for other reasons (e.g. water reserves, townsite reserves) are a great many remnants on private land. In some areas, these privately owned remnants constitute a large proportion of the remaining native vegetation. For instance, in the 1680 km^2 area studied in detail by us in the central wheatbelt, 77% of remnant vegetation is on private land, and is mostly in small patches, often less than 10 ha (Arnold and Weeldenburg 1991; Saunders et al. 1993a). In addition to this, vegetation has been retained along road verges, which often contain significant proportions of the vegetation remaining in the area. Road verges vary in width and condition and provide some degree of linkage between remnant areas (Hobbs et al. 1993a).

The wheatbelt landscape has thus been rapidly transformed from a continuous mosaic of predominantly perennial vegetation types to a landscape dominated by agriculture based on annual crops and pastures, within which small fragments of the original vegetation have been retained. This transformation has had profound implications for the native biota, through habitat loss and isolation. Importantly, however, the process of transformation to agriculture has also fundamentally altered landscape processes, which in turn are now having important impacts on both the native and agricultural sections of the landscape.

5.6.1 Impacts on Biota

An area of 1.56×10^5 km^2 of land has been cleared for agricultural production in this area, resulting in significant reductions in many vegetation types (Saunders and Hobbs 1992). Of the 44 vegetation types recognized in the area (Beard and Sprenger 1984), many have been significantly reduced in extent, predominantly due to agricultural development, and nine have less than 10% of their original extent remaining. Two restricted types are believed to have been almost completely eliminated. In addition to habitat loss due to clearing, many vegetation types are threatened by degradation of various sorts, as discussed previously and in Fig. 5.1.

By reducing available habitats and by driving populations extinct, land clearing, fragmentation and degradation are producing a legacy of species that no longer retain populations that are viable in the long term and hence may be doomed to extinction. These same processes are significantly impacting large areas of production and conservation lands in southwestern Australia (Hobbs et al. 1993b; Withers et al. 1994; George et al. 1995). The processes leading to endangerment and extinction are hence also leading to declining agricultural production.

As noted previously, mammal extinctions have been a feature of the development of the wheatbelt. Despite there being no species extinctions recorded for the other faunal groups, evidence on birds suggests significant range contractions for many species. Over half the bird species recorded from the agricultural area in the southwest of the state have declined in distribution or abundance since the start of this century (Saunders and Ingram 1995). This has resulted in population extinctions at local and regional scales (Saunders 1989, 1990, 1993). This work indicates that different groups have been impacted in different ways, with resident passerines being particularly affected (Fig. 5.4). Saunders' work also highlights the fact that a suite of native species have increased their ranges and abundance dramatically in response to habitat changes. For instance, bird species formerly found mainly in the arid zone are now abundant in the agricultural area. An important example is the galah (*Cacatua roseicapella*), which has invaded the Western Australian agricultural area from the arid zone, is implicated in woodland decline (due to its habit of ringbarking trees) and competes with native species such as Carnaby's cockatoo *Calyptorhynchus funereus* (Saunders 1990; Saunders and Ingram 1995).

5.6.2 Impacts of Non-Native Species

Domestic stock, predominantly sheep, affect remnant areas in several important ways by removing understorey vegetation, preventing regeneration of dominant species and altering soil structure (Scougall et al. 1993; Norton et al. 1994; Yates, Norton and Hobbs, in prep.). Movement of domestic stock into remnants significantly increases nutrient input, since stock feed in the paddocks and defecate in the remnants (Scougall et al. 1993). They thus represent an important influence on remnants which arises from the surrounding matrix.

Another particularly important influence on the native biota in remnants is the influx of feral and invading animals. Foxes (*Vulpes vulpes*), cats (*Felis*

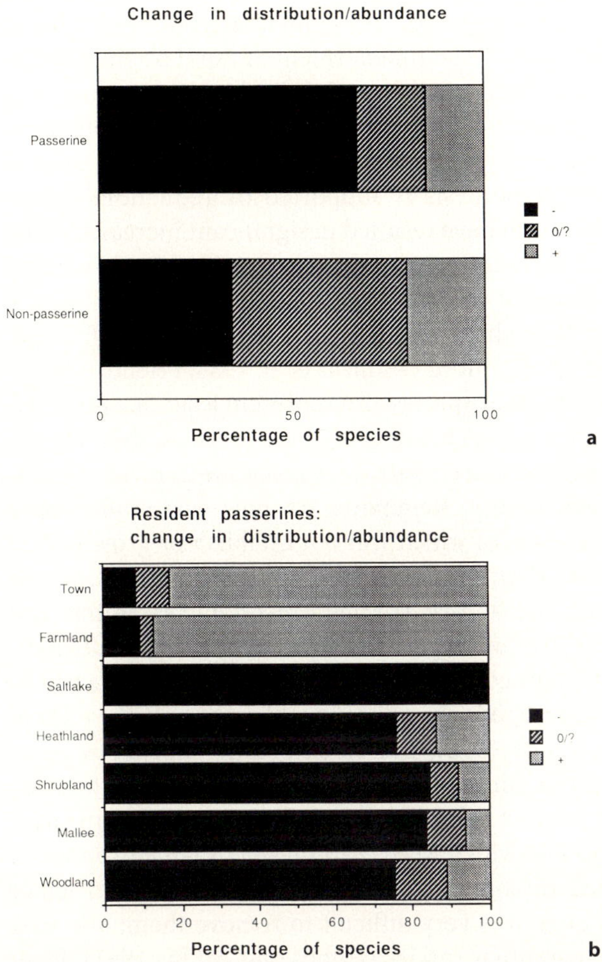

Fig. 5.4a,b. Changes in range and/or abundance of birds in the wheatbelt of Western Australia between the periods 1900–1937 and 1987–1990, as estimated from historical records and volunteer recordings. *a* Total passerine and non-passerine species. *b* Resident passerines categorised in relation to habitat. - Decline in range and/or abundance; 0/? no change or insufficient data to assess change; + increase in range and/or abundance. Data from Saunders and Ingram (1995)

catus) and rabbits (*Oryctolagus cuniculus*) are particularly important in the context of the Western Australian wheatbelt. Although rabbits were introduced to eastern Australia in the middle of the last century, they reached southwestern Australia only in the early part of this century (Stodart and Parer 1988). Fox and cat invasions probably occurred at about the same time. Rabbits reached plague proportions and led to large-scale habitat modification, while foxes and cats predated heavily on native mammals, ground- and hollow-nesting birds, lizards and insects (Hobbs et al. 1993).

While most emphasis has been placed on fragmentation as the cause of declines in faunal populations, the commencement of rapid fragmentation coincided with the arrival of rabbits and foxes. It is likely that these invading species, and the measures used to control them by the human population, were responsible for much of the faunal (especially mammal) decline (Hobbs et al. 1993). This hypothesis is supported by reductions in fox numbers through bating, which have resulted in significant increases in the populations of such native mammals as the black footed rock wallaby (*Petrogale lateralis*), numbat (*Myrmecobius fasciatus*) and woylie (*Bettongia penicillata*), either where populations of these species remained, or where they have been reintroduced (Kinnear et al. 1988; Friend 1990).

As elsewhere, non-native plant species also represent a significant biotic flow into remnant areas from the surrounding matrix. The native vegetation is mostly composed of long-lived perennial species with many adaptations to low nutrient availability. Remnants are now surrounded by a matrix of annual species, most of which will not establish or grow well in the unmodified soils within remnants. However, if the soil within remnants is disturbed or enriched, the annual species can establish (Hobbs and Atkins 1988b; Hobbs 1991). Some vegetation types, especially shrublands, are remarkably resistant to weed invasion, whereas others, such as woodlands are more susceptible. Weeds are often a problem only at the edge of remnants (Hester and Hobbs 1992), but stock grazing can allow weed invasion throughout entire remnants (Scougall et al. 1993). The intactness of the vegetation cover and the degree of soil disturbance are primary determinants of the degree to which weeds can invade. A large proportion of remnants are invaded to some extent (Saunders et al. 1993a). Once weeds have invaded an area, it is very difficult to remove them, and they prevent regeneration of perennial species (Hobbs and Atkins 1991). Plant invasions thus represent one of the most important threats to the maintenance of native vegetation in remnant areas.

5.6.3 Ecosystem Impacts

The switch from continuous cover of native vegetation to small fragments in a predominantly altered matrix brings with it many important consequences for the remnant vegetation, over and above the biogeographic effects of reduced area and increased isolation. These have been reviewed by Saunders et al. (1991), and stem from changes in physical, chemical and biotic flows in the altered landscape. Also relevant are changes in the disturbance regime and the spread of disturbance through the fragmented landscape (Hobbs 1987; Baker 1992), with resulting changes in ecosystem dynamics (e.g. Yates et al. 1994).

Hobbs and Saunders (1993) discussed the changes in ecosystem processes which have occurred in the wheatbelt landscape. Profound changes in nutrient, energy and water fluxes result from the replacement of perennial vegetation with predominantly annual crops and pastures.

In terms of nutrients, the landscape changes from one in which considerable internal nutrient redistribution occurs, but little net loss is evident, to one in which large inputs of fertilizer occur, and large losses of nutrients result from wind and water erosion and harvesting (Hobbs 1993c). In the fragmented wheatbelt landscape, remnant native vegetation on low nutrient soils are set in an agricultural matrix with elevated nutrient levels and increased mobility of topsoil. Loss of topsoil by wind and water erosion, especially during episodic storm events, is a major problem over much of the wheatbelt and can cause significant long-term yield reductions (Nulsen 1993). The impacts of this soil movement on remnant vegetation may also be profound, since water movement and wind bring nutrient-rich soil into remnant areas and hence alter the nutrient regime. This has important follow-on implications in terms of weed invasion, which is promoted by nutrient increase (Hobbs and Atkins 1988a).

Changed energy fluxes and wind regimes result from the change from tall evergreen vegetation to the alternating short annual vegetation and bare soil associated with agriculture (Hobbs 1993c). As well as influencing local microclimates, these changes may have profound regional scale implications. Rainfall patterns in southwestern Australia have changed over the past century, such that annual amounts have declined in the agricultural areas, but increased in the adjacent inland areas (Pittock 1988). Smith et al. (1992) have recently suggested that these changes could be associated with the effects of changing albedo on cloud formation. Evidence for this comes from a phenomenon frequently observed in satellite images for the area whereby clouds are found only above the large areas of native vegetation remaining in the wheatbelt and to the east of the line between cleared and uncleared land.

Perhaps the most important alteration to the wheatbelt landscape resulting from the extensive fragmentation of the native vegetation has been the effect on hydrology. In a mediterranean landscape with a pronounced summer drought, water availability is an important determinant of plant and animal distribution, growth and survival. Within uncleared vegetation, plant architecture, soil structure and soil biotic activity act to channel and redistribute rainfall, and most is utilized and transpired or evaporated, with little runoff or transfer to the water table (Nulsen et al. 1986; McFarlane et al. 1992, 1993). Removal of native vegetation and replacement with annual crops and pastures results in considerably less efficient use of rainfall, increased runoff and greater inputs to the water table. This has led to rising water tables, with rises of 50 cm $year^{-1}$ or more recorded in some areas (George et al. 1995). Beneath much of the wheatbelt are considerable quantities of salt stored at depth, resulting from a long history of input in ocean-derived rainfall coupled with poor drainage (Hingston and Galaitis 1976). As water tables rise, the stored salt is mobilized, and wherever the water table reaches the surface, salinization occurs. Salinization and water-logging have resulted in the loss of large areas of productive farmland, and pose a continuing threat to a significant proportion of the wheatbelt (Nulsen 1993; McFarlane et al. 1993).

Rising water tables also threaten the fragments of native vegetation. While native vegetation within remnants still uses water more efficiently than the surrounding crops and pastures, the overriding influence of the agricultural matrix means that water tables are rising even under large remnants (Salama et al. 1994). Many low lying areas have already been affected, and most of the previously freshwater lakes have become saline (Froend et al. 1987; Froend and McComb 1991; Hobbs et al. 1993b). In some areas the water table is rising so quickly that remnant vegetation will be destroyed within the next 5–10 years, while in lower rainfall areas, the timescale is longer but the problem is no less insidious.

5.7 Management Responses

The catalogue of threats to native ecosystems and landscape processes presented above may appear to present an insurmountable challenge, especially for the agricultural areas. However, the importance of the unique biota of southwestern Australia and the importance of the production enterprises demand that solutions to the problems be found.

In the forest region, solutions to current land use conflicts lie in more open assessment of the impacts of management practices and an increased willingness of both government agencies and conservation groups to negotiate mutually acceptable outcomes. It is still open for debate whether such outcomes are either possible or likely. Here, and in the urban areas, the potential for conflict between conservation and development is likely to continue.

In the agricultural regions there are fewer conflicts of interest, but much more intractable biophysical problems. The problems faced by conservation and agriculture stem from the same basic cause – i.e. the extensive removal of the native vegetation – and hence common solutions to the problems should be available (Hobbs and Saunders 1991; Hobbs et al. 1993a). From an agricultural viewpoint, measures are needed which reverse the current trends of degradation and loss of productive land. A variety of approaches to this are possible, but all involve replacing the current annual-based agriculture with a more diverse system that includes perennial and deep-rooted pastures and extensive revegetation (Lefroy et al. 1993a,b). From a conservation viewpoint, the requirement is for a reduction in the deleterious effects of matrix-derived influences on remnants and an increase in the area and connectivity of habitat.

There is currently a major push by government agencies and an increasing interest from the farming community for revegetation programs to reduce land degradation. While it seems possible that such revegetation could also be beneficial in a nature conservation sense, there is little evidence that this would, in fact, be the case, and only basic guidelines are available as to how revegetation could best be planned and implemented to benefit conservation (Hobbs 1993a; Recher 1993; Saunders et al. 1993b). To be of conservation value, revegetation would need to be used to enhance the existing system of remnants. Revegetation could be used to provide buffer zones around remnant areas, corridors between remnants, or as additional habitat (Hobbs 1993a). Nevertheless, biodiversity conservation per se will rarely be the primary driver of revegetation efforts, and has therefore to be integrated with activities designed with the broader agricultural perspective in mind. An integrated approach to the planning and design of revegetation and its linkage with existing native ecosystems is therefore essential (Saunders 1994).

All of the above options are a distinct departure from traditional conservation management which has focused almost entirely within individual remnant/reserve areas, and has been conducted by conservation agencies largely in isolation from (or in opposition to) surrounding land managers. The suggested approach requires a broader perspective which encompasses the entire landscape, including the conservation and production components. It also requires increased cooperation between neighboring land owners and managers. Conservation agencies which have previously often ignored the human community surrounding nature reserves are now becoming dependent on the cooperation and goodwill of that community to ensure the persistence of the reserves and of the non-reserved remnants which form an integral part of the conservation network. This will be achieved only if the local community recognizes the value of retaining remnant vegetation and agrees with the objectives of conservation management. Since the local community will also carry out much of the planning and management, it is essential that it is well informed, well motivated, and has access to the relevant planning and management tools, and the expert assistance and guidance needed for it to develop programs which tackle problems in an integrated and efficient way.

There is thus a strong social component to the management of fragmented landscapes, the importance of which cannot be underestimated. In the Western Australian wheatbelt, under-resourcing of the main conservation agency means that it cannot effectively manage the areas under its jurisdiction (Wallace and Moore 1987): i.e. it cannot undertake the necessary fragment-focused management, far less the broader matrix-focused issues. The persistence of the native biota over much of the wheatbelt thus depends on the activities of local communities. The development of community Land Care Groups and catchment groups (Goss and Chatfield 1993) illustrates a growing awareness of the need for a landscape-scale approach to the problems facing land managers. This approach is not easy, since it involves cooperation, discussion, compromise and an additional layer of responsibility over and above the traditional responsibility for one's own land. The successful reversal of current trends of degradation of both conservation and agricultural values depends on the successful development of techniques for integrated landscape management. This in turn depends on the success of attempts to bring local communities and all concerned parties together to tackle the problems. The trends are encouraging, but there is still a long way to go, with many scientific, social and economic issues still to be resolved. These three sets of issues are inextricably linked, and have to be resolved together.

References

Abbott I, Christensen P (1994) Application of ecological and evolutionary principles to forest management in Western Australia. Aust For 57:109–122

Arnold GW, Weeldenburg JR (1991) The distributions and characteristics of remnant native vegetation in parts of the Kellerberrin-Tammin-Trayning and Wyalkatchem Shires of Western Australia. CSIRO, Western Australia

Attiwill PM (1994a) The disturbance of forest ecosystems: the ecological basis for conservative management. For Ecol Manag 63:247–300

Attiwill PM (1994b) Ecological disturbance and the conservative management of eucalypt forests in Australia. For Ecol Manage 63:301–346

Attiwill PM (1995) Managing Leadbeater's possum in the mountain ash forests of Victoria – reply. For Ecol Manage 74:233–237

Baker WL (1992) The landscape ecology of large disturbances in the design and management of nature reserves. Landscape Ecol 7:181–194

Bari MA, Smith NJ, Boyd DW, Ruprecht JK (1994) Generation of streamflow following clearfell logging and regeneration at March Road catchment. Rep WS119, Water Authority of Western Australia, Leederville

Beard JS, Sprenger BS (1984) Geographical data from the vegetation survey of Western Australia. Part 1 – Area calculations. Vegmap, Applecross, Western Australia

Borg H, King PD, Loh IC (1987) Stream and ground water response to logging and subsequent regeneration in the southern forest of Western Australia. Interim results from

paired catchment studies. Surface Water Branch, Water Authority of Western Australia Rep No WH 34, Water Authority of Western Australia, Perth
Bormann FH, Likens GE (1981) Pattern and process in a forested ecosystem. Springer, Berlin Heidelberg New York
Burbidge AA (1989) Banksia woodlands: summary and conclusions. J R Soc West Aust 71:117–118
Burbidge AA, McKenzie NL (1989) Patterns in the modern decline of Western Australia's vertebrate fauna: causes and conservation implications. Biol Conserv 50:143–198
Burrows ND, Ward B, Robinson AD (1995) Jarrah forest fire history from stem analysis and anthropological evidence. Aust For 58:7–16
Calver MC, Hobbs RJ, Horwitz P, Main AR (1996) Science, principles and forest management: a response to Abbott and Christensen. Aust For 59:1–6
Carlton JT, Geller JB (1993) Ecological roulette: the global transport of nonindigenous marine organisms. Science 261:78–82
Churchill DM (1959) Late quaternary eustatic changes in the Swan River District. J Roy Soc West Aust 42:53–55
D'Antonio CM, Vitousek PM (1992) Biological invasions by exotic grasses, the grass/fire cycle, and global change. Ann Rev Ecol Syst 23:63–87
Dargavel J (1995) Fashioning Australia's forests. Oxford University Press, Melbourne
Davis GW, Richardson DM (eds) (1995) Mediterranean-type ecosystems: the function of biodiversity. Springer, Berlin Heidelberg New York
Dell B, Hopkins AJM, Lamont BB (eds) (1986) Resilience in mediterranean-type ecosystems. Junk, The Hague
Dell B, Havel JJ, Malajczuk N (eds) (1989) The Jarrah forest: a complex mediterranean ecosystem. Kluwer, Dordrecht
Dell B, Malajczuk N (1989) Jarrah dieback – a disease caused by *Phytopthora cinammomi.* In: Dell B, Havel JJ, Malajczuk N (eds) The Jarrah forest: a complex mediterranean ecosystem. Kluwer, Dordrecht, pp 67–87
Dixon B, Keys K, Hopper S, Wycherley P (1995) A fifty year record of fire management in Kings Park bushland. In: Harris J (ed) Burning our bushland. Proc Conf Fire and Urban Bushland, Urban Bushland Council (WA), West Perth, pp 15–18
Dodd J, Moore JH (1993) Introduction and status of *Kochia scoparia* in Western Australia. Proc 10th Australian Weeds Conf and 14th Asian Pacific Weed Society Conf Vol 1. Weed Society of Queensland, Brisbane, pp 496–500
Environmental Protection Authority (1994) Natural resource zones of the South West Land Division. Environmental Protection Agency, Perth
Flannery T (1994) The future eaters. an ecological history of the Australasian lands and people. Reed Books, Port Melbourne
Friend JA (1990) The numbat *Myrmecobius fasciatus* (Myrmecobiidae): history of decline and potential for recovery. Proc Ecol Soc Aust 16:369–377
Froend RH, Heddle EM, Bell DT, McComb AJ (1987) Effects of salinity and waterlogging on the vegetaion of Lake Toolibin, Western Australia. Aust J Ecol 12:281–298
Froend RH, McComb AJ (1991) An account of the decline of Lake Towerrinning, a wheatbelt wetland. J Roy Soc West Aust 73:123–128
George RJ, McFarlane DJ, Speed RJ (1995) The consequences of a changing hydrologic environment for native vegetation in south Western Australia. In: Saunders DA, Craig J, Mattiske L (eds) Nature conservation 4: The role of networks. Surrey Beatty, Chipping Norton, New South Wales, pp 9–22
Goss KF, Chatfield J (1993) Landcare groups in Western Australia: the role of self help groups in restoring degraded farmland. In: Saunders DA, Hobbs RJ, Ehrlich PR (eds) Nature conservation. 3: Reconstruction of fragmented ecosystems, global and regional perspectives. Surrey Beatty, Chipping Norton, New South Wales, pp 281–293

Groombridge B (ed) (1992) Global biodiversity status of the earth's living resources. Chapman and Hall, London

Hallam SJ (1975) Fire and hearth: a study of aboriginal usage and European usurpation in south-western Australia. Institute of Aboriginal Studies, Canberra

Hamilton SD, Lawrie AC, Hopmans P, Leonard BV (1991) Effects of fuel-reduction burning on a *Eucalyptus obliqua* forest ecosystem in Victoria. Aust J Bot 39:203–217

Hedgpeth JW (1993) Foreign invaders. Science 261:34–35

Hester AJ, Hobbs RJ (1992) Influence of fire and soil nutrients on native and non-native annuals at remnant vegetation edges in the Western Australian wheatbelt. J Veg Sci 3:101–108

Hiller SM (1987) The Swan River and regional planning. In: John J (ed) The Swan River estuary: ecology and management. Curtin University of Technology, Bentley, Western Australia, pp 244–255

Hingston FJ, Galaitis V (1976) The geographic variation of salt precipitated over Western Australia. Aust J Soil Res 14:319–335

Hobbs RJ (1987) Disturbance regimes in remnants of natural vegetation. In: Saunders DA, Arnold GW, Burbridge AA, Hopkins AJM (eds) Nature conservation: the role of remnants of native vegetation. Surrey Beatty, Chipping Norton, New South Wales, pp 233–240

Hobbs RJ (1991) Disturbance as a precursor to weed invasion in native vegetation. Plant Prot Q 6:99–104

Hobbs RJ (1993a) Can revegetation assist in the conservation of biodiversity in agricultural areas? Pac Conserv Biol 1:29–38

Hobbs RJ (1993b) Dynamics of weed invasion: implications for control. Proc 10th Australian Weeds Conf and 14th Asian Pacific Weed Society Conf Vol 1. Weed Society of Queensland, Brisbane, pp 461–465

Hobbs, RJ (1993c) Effects of landscape fragmentation on ecosystem processes in the Western Australian wheatbelt. Biol Conserv 64:193–202

Hobbs RJ (1996) Integrating biodiversity conservation and forestry production: a challenge for professional forestry. In: Norton DA, Allen JC (eds) Professional forestry – evolving issues in forest management. Paper No 23. School of Forestry, University of Canterbury, Christchurch, New Zealand, pp 27–34

Hobbs RJ, Atkins L (1988a) Effect of disturbance and nutrient addition on native and introduced annuals in the Western Australian wheatbelt. Aust J Ecol 13:171–179

Hobbs RJ, Atkins L (1988b) Effects of disturbance and nutrient addition on native and introduced annuals in plant communities in the Western Australian wheatbelt. Aust J Ecol 13:171–179

Hobbs RJ, Atkins L (1991) Interactions between annuals and woody perennials in a Western Australian wheatbelt reserve. J Veg Sci 2:643–654

Hobbs RJ, Hopkins AJM (1990) From frontier to fragments: European impact on Australia's vegetation. Proc Ecol Soc Aust 16:93–114

Hobbs RJ, Mooney HA (1998) Broadening the extinction debate – population deletions and additions in California and Western Australia. Conserv Biol 12:271–283

Hobbs RJ, Saunders DA (1991) Reintegrating fragmented landscapes: a preliminary framework for the Western Australian wheatbelt. J Environ Manage 33:161–167

Hobbs RJ, Saunders DA (1993) Effects of landscape fragmentation in agricultural areas. In: Moritz C, Kikkawa J (eds) Conservation biology in Australia and Oceania. Surrey Beatty, Chipping Norton, New South Wales, pp 77–95

Hobbs RJ, Saunders DA, Arnold GW (1993a) Integrated landscape ecology: a Western Australian perspective. Biol Conserv 64:231–238

Hobbs RJ, Saunders DA, Lobry de Bruyn LA, Main AR (1993b) Changes in biota. In: Hobbs RJ, Saunders DA (eds) Reintegrating fragmented landscapes: towards sustainable production and nature conservation. Springer, Berlin, Heidelberg New York, pp 65–106

Hobbs RJ, Groves RH, Hopper SD, Lambeck RJ, Lamont BB, Lavorel S, Main AR, Majer JD, Saunders DA (1995) Function of biodiversity in the mediterranean-type ecosystems of southwestern Australia. In: Davis GW, Richardson DM (eds) Mediterranean-type ecosystems: the function of biodiversity. Springer, Berlin, Heidelberg New York, pp 233–284

Hopper SD (1979) Biogeographical aspects of speciation in the southwest Australian flora. Ann Rev Ecol Syst 10:399–422

Hopper SD (1989) Conservation status of Banksia woodlands on the Swan Coastal Plain. J R Soc West Aust 71:115–116

Hopper SD (1992) Patterns of plant diversity at the population and species level in south-west Australian mediterranean ecosystems. In: Hobbs RJ (ed) Biodiversity of mediterranean ecosystems in Australia. Surrey Beatty and Sons, Chipping Norton, NSW, pp 27–46

Hopper SD, Harvey MS, Chappill JA, Main AR, Main BY (1996) The Western Australian biota as Gondwanan heritage – a review. In: Hopper SD, Chappill JA, Harvey MS, George AS (eds) Gondwanan heritage: past, present and future of the Western Australian biota. Surrey Beatty, Chipping Norton, New South Wales, pp x–xx

Hopper SD, van Leeuwen S, Brown AP, Patrick SJ (1990) Western Australia's endangered flora and other plants under consideration for declaration. Department of Conservation and Land Management, Wanneroo, Western Australia

Humphries SE (1993) Environmental impact of weeds. Proc 10th Australian Weeds Conf and 14th Asian Pacific Weed Society Conf Vol 2. Weed Society of Queensland, Brisbane, Weed Society of Queensland, Brisbane, pp 1–11

Humphries SE, Groves RH, Mitchell DS (1991) Plant invasions of Australian ecosystems. A status review and management directions. In: Kowari 2. Plant invasions. The incidence of environmental weeds in Australia. Australian National Parks and Wildlife Service, Canberra, pp 1–127

Hutchison MJ, Armstrong PH (1993) The invasion of a south-western Australian river system by *Perca fluviatillis*: History and probable causes. Global Ecol Biogeogr Lett 3:77–89

Jarvis N (ed) (1986) Western Australia, an atlas of human endeavour. 2nd edn. Department of Lands and Surveys and Education Department of Western Australia, Perth

Keighery GJ (1991) Environmental weeds of Western Australia. In: Kowari 2. Plant invasions. The incidence of environmental weeds in Australia. Australian National Parks and Wildlife Service, Canberra, pp 180–188

Keighery GJ (1995) How many weeds? In: Burke G (ed) Invasive weeds and regenerating ecosystems in Western Australia. Institute for Science and Technology Policy, Murdoch University, Perth, pp 3–12

Kennedy M (ed) (1990) Australia's endangered species. Simon and Schuster, Brookville, New South Wales

Kinnear JE, Onus ML, Bromilow RN (1988) Fox control and rock wallaby population dynamics. Aust Wild Res 15:435–450

Kitchener DJ, Chapman A, Dell J, Muir BG, Palmer M (1980) Conservation value for mammals of reserves in the Western Australian wheatbelt. Biol Conserv 18:179–207

Kohen J (1995) Aboriginal environmental impacts. University of New South Wales Press, Sydney

Kruger FJ, Mitchell DT, Jarvis JUM (ed) (1983) Mediterranean-type ecosystems: the role of nutrients. Springer, Berlin Heidelbeg New York

Lamont BB, Downes S (1979) The longevity, flowering and fire history of the grasstrees *Xanthorrhoea preissii* and *Kingia australis*. J Appl Ecol 16:893–899

Lands and Forest Commission (1994) Forest management plan 1994–2003. Department of Conservation and Land Management, Perth

Lefroy EC, Hobbs RJ, Scheltema M (1993a) Reconciling agriculture and nature conservation: toward a restoration strategy for the Western Australian wheatbelt. In: Saunders DA, Hobbs RJ, Ehrlich PR (eds) Nature conservation 3: Reconstruction of fragmented ecosystems, global and regional perspectives. Surrey Beatty, Chipping Norton, New South Wales, pp 243–270

Lefroy EC, Salerian J, Hobbs RJ (1993b) Integrating economic and ecological considerations. a theoretical framework. In: Hobbs RJ, Saunders DA (eds) Reintegrating fragmented landscapes. towards sustainable production and nature conservation. Springer, Berlin Heidelberg New York, pp 209–244

Lindenmayer DB (1995) Forest disturbance, forest wildlife conservation and the conservative basis for forest management in the mountain ash forests of Victoria – comment. For Ecol Manage 74:223–231

Majer JD, Abbott I (1989) Invertebrates of the jarrah forest. In: Dell B, Havel JJ, Malajczuk N (eds) The Jarrah forest: a complex mediterranean ecosystem. Kluwer, Dordrecht, pp 111–122

Mawson PR, Long JL (1994) Size and age parameters of nest trees used by four species of parrot and one species of cockatoo in south-west Australia. Emu 94:149–155

McCaw WL (1993) Effects of fuel-reduction burning on a *Eucalyptus obliqua* forest ecosystem in Victoria. Comment and response. Aust J Bot 41:413–414

McCaw WL, Burrows ND (1989) Fire management. In: Dell B, Havel JJ, Malajczuk N (eds) The Jarrah forest: a complex mediterranean ecosystem. Kluwer, Dordrecht, pp 317–334

McFarlane DJ, Howell MR, Ryder AT, Orr GJ (1992) The effect of agricultural development on the physical and hydraulic properties of four Western Australian soils. Aust J Soil Res 30:517–532

McFarlane DJ, George RJ, Farrington P (1993) Changes in the hydrologic cycle. In: Hobbs RJ, Saunders DA (eds) Reintergrating fragmented landscapes: towards sustainable production and nature conservation. Springer, Berlin Heidelberg New York, pp 146–186

McGrath M (1985) Fire planning and management: an overview. In: Ford JW (eds) Fire ecology and management in Western Australian ecosystems. WAIT Environmental Studies Group, Bentley, Western Australia, pp 219–221

Noble JC (1993) Relict surface-soil features on semi-arid mulga (*Acacia aneura*) woodlands. Rangelands J 15:48–70

Norton DN, Hobbs RJ, Atkins L (1994) Fragmentation, disturbance and plant distribution: mistletoes in woodland remnants in the Western Australian wheatbelt. Conserv Biol 9:426–438

Nulsen RA (1993) Changes in soil properties. In: Hobbs RJ, Saunders DA (eds) Reintegrating fragmented landscapes: towards sustainable production and nature conservation. Springer, Berlin Heidelberg New York, pp 107–145

Nulsen RA, Bligh KJ, Baxter IN, Solin EJ, Imrie DH (1986) The fate of rainfall in a mallee and heath vegetated catchment in southen Western Australia. Aust J Ecol 11:361–371

Parsons WT, Cuthbertson EG (1992) Noxious weeds of Australia. Inkata, Melbourne

Pen LJ, Green JW (1983) Botanical exploration and vegetation changes on Rottnest Island. J Roy Soc West Aust 66:20–24

Pigott JP, Loneragan WA (1995) Fire and human disturbance regimes and impacts on plant communities at the Star Swamp Bushland Reserve. In: Harris J (ed) Burning our bushland. Proc Conference on Fire and Urban Bushland. Urban Bushland Council (WA), West Perth, pp 19–24

Pittock AB (1988) Actual and anticipated changes in Australia's climate. In: Pearman GI (eds) Greenhouse: planning for climatic change. CSIRO, Melbourne, pp 35–51

Rab MA (1994) Changes in physical properties of a soil associated with logging of *Eucalyptus regnans* forest in southeastern Australia. For Ecol Manage 70:215–229

Recher HF (1993) The loss of biodiversity and landscape restoration: conservation, management, survival – an Australian perspective. In: Saunders DA, Hobbs RJ, Ehrlich PR (eds) Nature conservation 3: Reconstruction of fragmented ecosystems. Surrey Beatty, Chipping Norton, New South Wales, pp 141–151

Resource Assessment Commission (1993) Ecological impacts of forest use: a survey of completed research. RAC Research Paper 9. Australian Government Printing Service, Canberra

Richardson DM, Williams PA, Hobbs RJ (1994) Pine invasions in the Southern Hemisphere: determinants of spread and invasibility. J Biogeogr 21:511–527

Rippey E, Rowland B (1995) Plants of the Perth coast and islands. University of Western Australia Press, Nedlands

Salama RB, Bartle GA, Farrington P (1994) Water use of plantation *Eucalyptus camaldulensis* estimated by groundwater hydrograph seperation techniques and heat pulse method. J Hydrol 156:163–180

Saunders DA (1989) Changes in the avifauna of a region, district and remnant as a result of fragmentation of native vegetation: the wheatbelt of Western Australia: a case study. Biol Conserv 50:99–135

Saunders DA (1990) Problems of survival in an extensively cultivated landscape: the case of Carnaby's Cockatoo *Calyptorhynchus funereus latirostris*. Biol Conserv 54:277–290

Saunders DA (1993) Community based observer scheme to assess avian response to habitat reduction and fragmentation in south western Australia. Biol Conserv 64:203–218

Saunders DA (1994) Can we integrate nature conservation with agricultural production? Land Urban Plann 28:63–71

Saunders DA, de Rebeira CP (1993) The birds of Rottnest Island. DAS and CPdeR, Guildford

Saunders DA, Hobbs RJ (1992) Impact on biodiversity of changes in land-use and climate. In: Hobbs RJ (ed) Biodiversity in mediterreanean ecosystems in Australia. Surrey Beatty, Chipping Norton, New South Wales, pp 61–75

Saunders DA, Ingram J (1995) Birds of southwestern Australia. An atlas of changes in the distribution and abundance of the wheatbelt fauna. Surrey Beatty, Chipping Norton, New South Wales

Saunders DA, Rowley I, Smith GT (1985) The effects of clearing for agriculture on the distibution of cockatoos in the southwest of Western Australia. In: Keast A, Recher HF, Ford H, Saunders D (eds) Birds of eucalypt forests and woodlands: ecology, conservation, management. Surrey Beatty, Chipping Norton, New South Wales, pp 309–321

Saunders DA, Hobbs RJ, Arnold GW (1993a) The Kellerberrin project on fragmented landscapes: A review of current information. Biol Conserv 64:185–192

Saunders DA, Hobbs RJ, Ehrlich PR (1993b) Reconstruction of fragmented ecosystems: problems and possibilities. In: Saunders DA, Hobbs RJ, Ehrlich PR (eds) Nature conservation 3: Reconstruction of fragmented ecosystems, global and regional perspectives. Surrey Beatty, Chipping Norton, New South Wales, pp 305–313

Scheltema M, Harris J (eds) (1995) Managing Perth's bushlands. Greening Western Australia, Perth

Schofield NJ, Stoneman GL, Loh IC (1989) Hydrology of the jarrah forest. In: Dell B, Havel JJ, Malajczuk N (eds) The Jarrah forest: a complex mediterranean ecosystem. Kluwer, Dordrecht, pp 179–201

Scougall SA, Majer JD, Hobbs RJ (1993) Edge effects in grazed and ungrazed Western Australian wheatbelt remnants in relation to ecosystem reconstruction. In: Saunders DA, Hobbs RJ, Ehrlich PR (eds) Nature conservation 3: Reconstruction of fragmented ecosystems, global and regional perspectives. Surrey Beatty, Chipping Norton, New South Wales, pp 163–178

Senate Standing Committee on Science and the Environment (1977) Woodchips and the environment. Australian Government Printing Service, Canberra

Smith RCG, Huang X, Lyons TJ, Hacker JH, Hick P (1992) Change in land surface albedo and temperature in southwestern Australia following the replacement of native perennial vegetation: satellite observations. Paper No. IAF-92-0117. 43rd Congr of the Int Astronautical Federation, Washington, DC

Stodart E, Parer I (1988) Colonisation of Australia by the rabbit, *Oryctolagus cuniculatus* (L.). CSIRO, Division of Wildlife and Ecology, Canberra

Strahan R (ed) (1983) The complete book of Australian mammals. Angus and Robertson, Australian Museum, Sydney

Tingay A (1985) Contemporary views of the voluntary conservation movement on the use of fuel reduction burns as a land management technique. In: Ford JW (eds) Fire ecology and management in Western Australian ecosystems. WAIT Environmental Studies Group, Bentley, Western Australia, pp 215–218

Twigg LE, King DR (1991) The impact of fluoroacetate-bearing vegetation on native Australian fauna: a review. Oikos 61:412–430

Underwood RJ, Sneeuwjagt RJ, Styles HG (1985) The contribution of prescribed burning to forest fire control in Western Australia. In: Ford JW (eds) Fire ecology and management in Western Australian ecosystems. WAIT Environmental Studies Group, Bentley, Western Australia, pp 153–170

Wallace KJ, Moore SA (1987) Management of remnant bushland for nature conservation in agricultural areas of Southwestern Australia – operational and planning perspectives. In: Saunders DA, Arnold GW, Burbidge AA, Hopkins AJM (eds) Nature conservation: the role of remnants of native vegetation. Surrey Beatty, Chipping Norton, New South Wales, pp 259–268

Williams JE, Gill AM (1995) The impact of fire regimes on native forests in Eastern New South Wales. New South Wales National Parks and Wildlife Service, Hurtsville, New South Wales

Wills RT (1993) The ecological impact of *Phytopthora cinnamomi* in the Stirling Range National Park. Aust J Ecol 18:145–159

Wills RT, Keighery GJ (1994) Ecological impact of plant disease on plant communities. J Roy Soc West Aust 77:127–131

Withers PC, Cowling WA, Wills RT (ed) (1994) Plant diseases in ecosystems: threats and impacts in south-western Australia. J R Soc West Aust 77(4):97–186

Wood WE (1924) Increase of salt in soil and streams following destruction of native vegetation. J R Soc West Aust 10:35–47

Yates CJ, Hobbs RJ, Bell RW (1994) Landscape-scale disturbances and regeneration of semi-arid woodlands of south-western Australia. Pac Conserv Biol 1:214–221

6 Land Use Conflicts in California

H.S. Walter

6.1 Introduction

California's astonishing richness in natural resources was mercilessly exploited as the Golden State became part of the American dream. Generations of writers have written on and lamented over the "destruction of California" (Dasmann 1965), "vanishing California" (Anonymous 1989), or "California – the endangered dream" (TIME 1991). Explosive population growth and land development have been accompanied by a plethora of environmental problems from air pollution to hazardous wastes and – most of all – large-scale landscape transformation and degradation (Palmer 1993) that have reduced some ecosystem types such as wetlands and riparian systems to less than 10% of their spatial expanse in less than 100 years.

This chapter examines the current land use conflicts affecting California's bio-landscape consisting of plant and animal taxa, natural communities, and habitats occupied by lower and higher plants as well as vertebrates and invertebrates. I emphasize the state's richness in biotic diversity because this landscape quality factor lies at the root of many land use conflicts. The analysis of current land use practices impacting the bio-landscape focuses on the ultimate and proximate driving forces behind a multitude of conflicts. Hope for resolving some of the latter exists today as awareness and understanding of the problem complex associated with land use issues have increased; scientists, politicians, and administrators have proposed strategies for the management of the remaining natural landscapes of California.

6.2 California: the Diversity State

California was a magnificent wildlife paradise before the taming of its rivers and the general land conversion of its plains and coastal wetlands

Ecological Studies, Vol. 136
Rundel et al. (eds.) Landscape Degradation and Biodiversity in Mediterranean-Type Ecosystems

transformed it into a state of economic promise and ecological decline (Dasmann 1965; Schoenherr 1992). Fittingly, the extirpated grizzly bear is the state mammal and – depicted on the flag of the California Republic – a constant powerful reminder of the unwise overexploitation of resources. What California lost forever in the last 150 years is the incredibly rich biotic fabric of its landscapes where salmon, waterfowl, elk, whales, seals, and eagles abounded in grandiose river valleys, marshes, redwood forests, and rocky coasts. This diversity can be grouped into twelve distinct bioregions (Fig. 6.1). What California offers today is only a fraction of that wild landscape; but it is still an enormous variety of natural habitat patches and corridors that extends over a remarkable gradient of topographic and climatic diversity from the cool Pacific shores to the hot inland deserts, and to the frosty Sierra peaks.

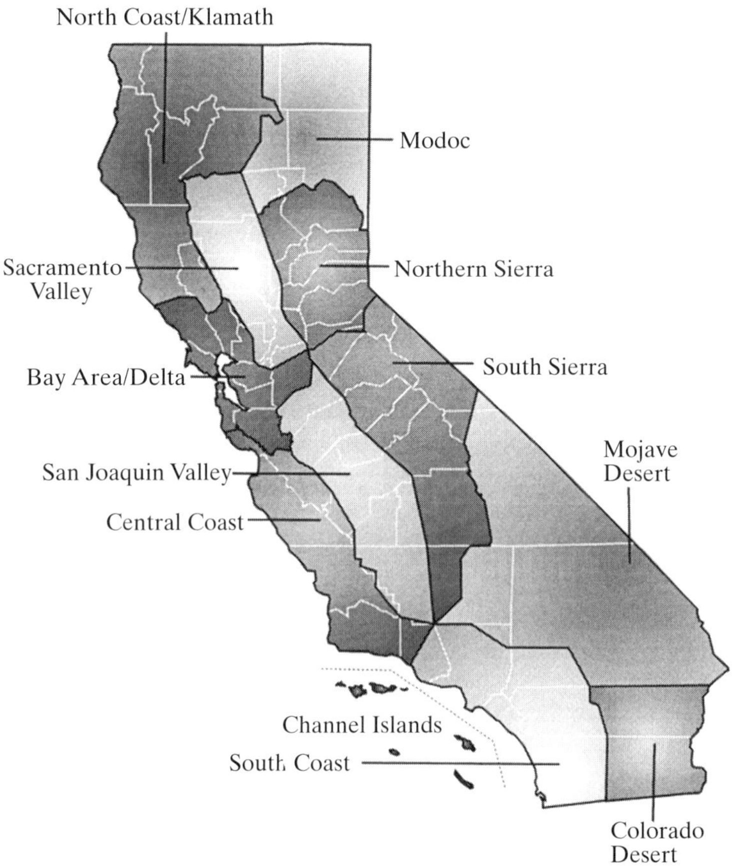

Fig. 6.1. The twelve bioregions of California. Bioregions of California (counties shown with white line)

Used to the television images of urban California, potential students from other regions often dismiss Los Angeles as a suitable place for ecological and conservation studies. Little do they know that within 15 minutes by car there are undisturbed and species-rich ecosystems of coastal sage scrub and chaparral in the nearby Santa Monica Mountains. In fact, Los Angeles County - infamous for smog, urban sprawl and freeways - appears to possess the greatest biotic diversity gradient in the entire country: in terms of habitat diversity and contrasting ecosystems this county has no rival: from the fog-shrouded oak woodlands of Catalina Island to the alpine heights of the San Gabriel Mountains and the dry alkali flats of the Mojave Desert more than 70 different natural plant communities have been described.

The high floristic richness - so characteristic of all Mediterranean-type regions - is accompanied by a high degree of endemism (Table 6.1). Many plants are naturally rare due to evolutionary climatic and edaphic adaptations; others are rare today as a result of land conversion and various forms of anthropogenic ecosystem degradation. The fauna of California is rich as well; many of its invertebrates still await discovery and scientific treatment. Relatively few species have gone extinct but hundreds of taxa are listed as threatened, endangered or of special concern by conservation authorities (Jensen et al. 1993, CDFG 1994, Thelander and Crabtree 1994). Were it not for the historic and current richness of the bio-landscape the land use conflicts in California would revolve around the customary problem of uncontrolled development and a few other forms of resource exploitation leading to local and regional ecosystem degradation. Instead, the unique mix of factors comprised of people, their cultural, social, and economic activities, and the historic introduction and turnover of technologies affecting the natural resources of California has created a complex scenario of factors and processes contributing to and aggravating many a land use conflict where the very survival of individual taxa, entire natural communities, and even entire ecosystems (involving unique physical and biotic components) is at stake.

6.3 Historic Landscape Degradation

The historic impacts on the land and waters of California are well known and documented (Dasmann 1965; Hart 1984; Howitt 1995; Jensen et al. 1993; Palmer 1993; Thelander and Crabtree 1994). An excellent introduction into the current status of the California landscape has been provided by the exquisite aerial photography of R. Cameron (1990a,b, 1994, 1995).

Table 6.1. California's plant and animal diversity

	Major group	Number of full species	Notes on species/taxa
Higher plants		5862	173 Families 1222 Genera 4839 Native spp. (of which 1416 are endemic) 1023 Naturalized alien spp. 26 Endemic spp. presumed extinct
Animals	Insects	>28000	Among them are: >7000 Beetles >5600 Flies >5000 Bees, wasps, ants >3000 Moths >600 True bugs >240 Butterflies >100 Dragonflies and damselflies 18 Taxa extinct in California
	Fishes	124	66 Native freshwater and anadromous spp. 13 Native euryhaline or marine spp. 45 Introduced, well-established spp. 8 Extinct spp. in California
	Amphibians	46	43 Native spp. 3 Introduced, well-established spp.
	Reptiles	83	73 Non-marine spp. 7 Marine spp. 3 Introduced, well-established spp.
	Birds	586	Recorded since 1900 325 Spp. breeding at one time or another 270 Regular breeding spp. 8 Introduced, well-established spp. 7 Extinct spp. in California
	Mammals	214	193 Native spp. 21 Introduced, well-established spp. 6 Extinct spp. in California

Table 6.2 provides some statistical data on land use. Public lands comprise 48.9% of the total surface area (410990 km^2) containing most of the forest, grassland, and sagebrush/chaparral formations; a large percentage of the public lands is used in some way or other (logging, snowmobile use, livestock grazing, mining). It is difficult therefore to find any truly wild landscape that has not been impacted by modern civilization. There probably is none left as air pollution has reached even the remotest places, and changes in soil acidity, soil mycorrhizae, insect pollinators, disease and parasite

Table 6.2. Status of land use in California 1990. (Data from Jensen et al. 1993)

		Area (km^2)	Relative area (%)
Total land and water area		410990	100
	Lands in public ownership	200851	48.9
Total forest		164807	40.1
	productive	73567	17.9
	unproductive	91240	22.2
Desert/arid scrub		98227	23.9
Grassland		49319	12.0
Agriculture		45209	11.0
Sagebrush/chaparral		25070	6.1
Urban/industrial		18084	4.4
Barren land		7389	1.8
Riparian/wetlands		2877	0.7

abundances, water table, and others have affected ecosystem components and processes.

Of principal concern from the point of view of the biogeographer and landscape ecologist are not simply the dry statistics on the number of extinct or threatened taxa in a given region or political space. We have admirable data sets on this qualitative decline of our biodiversity (CDFG 1994; Kecera and Barrett 1995). Much more significant is the quantitative picture: the state wide phenomenon of massive habitat loss and shrinkage experienced in the time frame of just a few human generations. Many California counties, particularly in the San Joaquin Valley, have simply lost their natural ecological heritage, and it is doubtful that they will be able to recover it. Several bio-landscape types (Table 6.3) have lost close to 90% of their distribution area; in addition, their remnant fragments have almost certainly been degraded by exotics, pollutants, etc. in some form or other. Thus, this catastrophic loss of several unique California bio-landscapes is reminiscent of the post-Columbus destruction of the forest lands along the Atlantic coast of Brazil. The near future will answer the question as to whether this history of qualitative and quantitative loss of California's biodiversity will continue, slow down, stop, or possibly reverse itself.

6.4 Driving Forces of Contemporary Land Use Conflicts

Concern and alarm over the declining biodiversity of California naturally leads to an inquiry into the nature of land use conflicts affecting the state's

Table 6.3. Estimated major habitat losses in California since 1850. (**A** Data from Kreissman 1991; Jensen et al. 1993. **B** Data from City of San Diego 1995)

Habitat	Losses (km^2)	Losses (%)
A. California		
Coastal wetland	817	80
Interior wetland	15169	96
Riparian wetland	3313	89
Valley grassland	88904	99
Vernal pools	11205	66
B. San Diego County		
Maritime succulent scrub		90
Southern maritime chaparral		82
Freshwater marsh		90
Coastal sage scrub		70
Native grassland		95
Vernal Pools		98

natural heritage. In particular, we should ask the question: what are the driving forces behind different types of land use conflicts? To find an answer, I have distinguished between a relatively few *ultimate driving forces* and a larger number of *proximate driving forces*. The latter are directly contributing or causing such conflicts either on a local level in parts of the state, at the regional or statewide level, or at the global level including California. The former can be understood as basic conditions or processes underlying or enabling the operation of the proximate forces.

6.4.1 Ultimate Driving Forces

In the end, most of the current land use conflicts are caused (Table 6.4) and energized by the rapid population growth of California and the highly clustered settlement patterns. The following list shows the ultimate driving forces of land use conflicts: (1) population growth and settlement patterns; (2) absence of sustainable ecosystem planning; (3) new technologies; (4) transformation of rural economies; (5) popularity of outdoor recreational activities; and, (6) point and diffuse sources of anthropogenic pollution. Figure 6.2 shows the relationship between human population growth, all California bank assets, and the degree of urbanization. The successful state economy and its promise of further growth put money in the banks which was used for further land conversion, "development", and another phase of population growth (Medvitz and Sokolow 1995). The large metropolitan superorganisms of Los Angeles, San Francisco-San Jose,

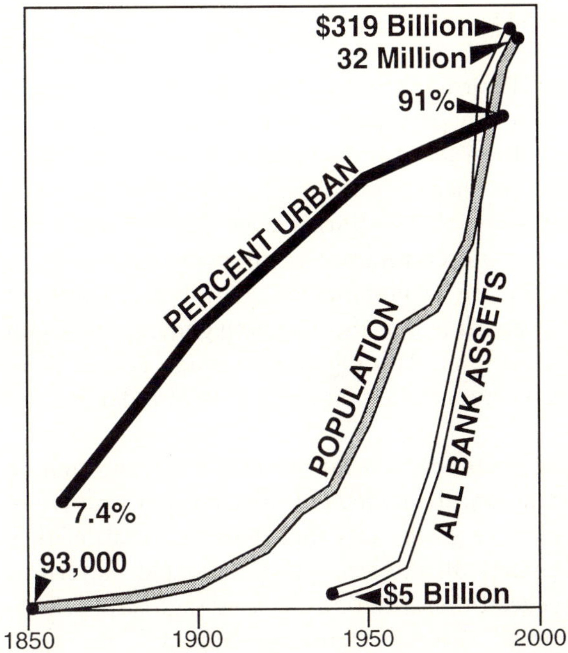

Fig. 6.2. California's growth in population, urbanization, and bank assets

and San Diego require massive modifications of landscapes that lie hundreds of kilometers away (lowering of groundwater tables, aqueducts, dams, transportation systems). The functioning of the Los Angeles ecosystem requires even more: energy, raw materials, and processed goods arrive here from all over the North American continent, and its harbor and airport process shipments from literally all countries on Earth (H.S. Walter, unpubl.).

The rapid spread and use of new technologies such as electronic information systems (Internet) has contributed to new settlement patterns of urbanized people: they have moved out of the congested cities into the rural backcountry, creating novel conflict situations at the wilderness edge. The outdoors have become ever more popular, and powerful industrial factors are dependent on the regular dispersion of millions of people into the open spaces of California (snowmobiles, off-road vehicles, wildlife photography, mountain biking, etc.). This love affair with "nature" has created a host of proximate driving forces. Finally, the *chemicalization* of the human environment is depositing ton after ton of man-made or man-generated molecules from industrial plants, automobiles, agricultural areas, and other pollution sources rather indiscriminately over the coast, the hills, the valleys, and even the much cherished mountain resorts of the Sierra Nevada (SNEP 1996).

6.4.2 Proximate Driving Forces

The increasing mobility of Californians and the global nature of imports and exports have resulted in the permanent invasion of California by hundreds of exotic plants (Jackson 1985) and animals (Mooney et al. 1986); parasites and disease organisms pose serious threats to wild and domestic taxa alike. Water pollution along the California coast has seriously degraded the intertidal and benthic communities; air pollution threatens mature coniferous forests in the mountains of central and southern California (SNEP 1996). It is possible that the alarming general decline of amphibians in the Yosemite National Park region (Drost and Fellers 1996) has been caused by cumulative impacts of pollution, exotic predators, and still unknown factors.

Habitat losses of valuable wetlands, coastal hillsides, etc. can often be attributed to financial speculation; the owners of the land may not live in California and may not be aware of the fact that their financial balance sheets have had serious impacts on the size and fragmented pattern of already severely shrunken vegetation formations (coastal sage scrub, riparian woodland, old growth forest). Other foreign-based impacts concern the pursuit of commercially valuable plant and animal products such as mushrooms or deer antlers. Ecotourism is not only an economic necessity for many communities, it has also become an ecological nightmare in tourist destinations such as Yosemite Valley and Lake Tahoe (Table 6.4).

The list of regional level forces contains many items familiar from other Mediterranean landscape systems. The transformation of the biolandscape has resulted in the proliferation of exotic and evergreen shrubs and trees in the urban setting as well as in agricultural ones (eucalypt groves and windbreaks, pines and palms everywhere in towns and cities). Peculiar to the California ecosystem is the recent range extension of the brown-headed cowbird (*Molothrus ater*), a brood parasite of endemic songbird taxa; this bird was able to utilize anthropogenic, often agricultural landscapes to spread from the eastern US into southern and central California; it causes massive brood losses in host populations that have not yet adapted to its presence. The high density of coyote (*Canis latrans*) and possibly also of mountain lion (*Felis concolor*) at the urban/wildland interface is a result of high prey abundance (cats and small dogs, deer) near human dwellings, and the lack of hunting in the urban zone.

In the city, the proliferation of pets and the California-style garden designs have had a negative impact on bird diversity and abundance; the latter because its preference for evergreen and hard-leaf shrubs and trees reduces the arthropod diversity enormously, the former because the lack of thickets and other places to hide makes it easy for cats to capture birds.

Table 6.4. Proximate driving forces of land use conflicts

Level of operation	Proximate driving forces
Global	Poverty and economic disorder (disregard for sustainable planning strategies)
	Global mobility and transportation (invasion of pests and disease organisms)
	Diffuse pollution (air and water degraded, serious biotic consequences)
	International financial speculation (landowners do not live on their land)
	Foreign markets for bioresource products (poaching on mushrooms, deer, bear)
	Ecotourism (disturbance of wildlife, trampling on resources)
Regional	Exotic landscapes (eucalyptus, oleander)
	Wetland drainage (tremendous loss of riparian, marsh, vernal pool acreage)
	Forest clearing (loss of ancient and mature forest ecosystems)
	Overgrazing and overbrowsing (soil compaction, lack of tree recruitment)
	Development of water edges (lake, coast, river; impact on aquatic biota)
	Proliferation of anthropogenic ecotones (spread of brood parasites)
	Automobile dependency (massive road construction, high number of roadkills)
	Fragmentation of habitats (loss of rare species; generally higher extinction risks)
	Acculturation of predators at urban-wildland interface (increase in carnivores)
	Coastal pollution (marine ecosystem simplification)
	Chemical pollution of rural landscape (loss of arthropods, birds)
	Coastal overcrowding (loss of beach and dune resources)
	Proliferation of pets (dogs, cats preying on songbirds)
	Immigrant insensitivity to California biodiversity (collecting protected species)
Local	Leveling mountaintops and landfills (loss of micro-habitats and -climates)
	Instant landscapes with exotic biota (do not provide food for insects or birds)
	Golf course development (notoriously poor in biodiversity)
	Overbuilding (lack of balance between open space and built-up area)
	Excessive gardening (removal of leaf litter eliminates native invertebrates)
	Lack of natural urban parks (cities lack native biotic components such as oak ecosystems)
	Freeway barriers, river dams (demise of migrating and linkage-requiring species)

Finally, the multiethnic and recent immigrant composition of large parts of the urban population creates additional pressure on precious or protected biodiversity elements because of unfamiliarity with state law and social customs or because of attitudes toward wildlife resources imported from abroad. For a brief period in the 1970s when some members of the Vietnamese refugee community in California began to trap birds and to hunt lizards in the nearby surroundings, these activities were quickly stopped after a media campaign explained California law and wildlife ethics.

Driving forces operating at the local level (Table 6.4) have been noted in southern California. The many planned communities such as Mission Viejo and Simi Valley often transform not just the bio-landscape, they also change the physical landscape. Huge earth-moving machines are used to level off mountain tops and to fill in ravines and valleys creating terraces and small mesas in the process which are then converted into instant residential or commercial landscapes with imported soil, grass, and the planting of large trees grown for this purpose in large boxes in special tree nurseries. Needless to say, the local extinction of all native soil biota and most other taxa has preceded such an artificial landscape.

Many land use conflicts in the southern part of the state result from uncontrolled urban sprawl creating overbuilt neighborhoods with no or little open and natural space; recreational facilities are mostly devoid of arboreal vegetation, designed for golf, baseball and football, with little thought given to passive recreation such as hiking or bird watching. The management of the precious turf grass requires fertilizer and pesticide applications which are harmful to native biota. The increasing insularization of natural habitat fragments contributes to local extinction events and the fenced and deadly freeways create total barriers to animal dispersal and movement. No wonder that proposals for further "development" touch a raw nerve in those who have witnessed the price that California's ecosystems have paid in the name of social and economic "progress".

6.5 Regional Case Studies

Each of the twelve bioregions (Fig. 6.1) is beset by numerous land use conflicts. Because of intense public participation and mandatory open hearings on all issues involving government policy and management in California a large body of data exists in unpublished environmental impact reports and conference proceedings. Published books relating policy to resource management are rare (but see Keeley 1993). In the following, five regional case studies will be singled out as examples of current land use conflicts involving threatened taxa, habitats, and even entire regional ecosystems.

6.5.1 Northern Spotted Owl and Marbled Murrelet

In the north coast-Klamath bioregion, issues of the survival of the world renowned old growth forests (Jimerson and Hoover 1992) and the persistence of water quality and quantity in highly modified watersheds (Higgins and Kier 1992) dominate. The increasingly valuable but shrinking groves of old-growth coast redwood and Douglas fir are also the homes to northern spotted owls (*Strix occidentalis caurina*) and marbled murrelets (*Brachyramphus marmoratus*). An intense conflict over the utilization and/or protection of these forests has been one of the hottest national environmental issues for almost a decade (Thomas et al. 1990; Dietrich 1992; Yaffee 1994). The price of lumber has increased markedly in recent years. Each individual tree from an old growth stand of Douglas fir (*Pseudotsuga menziesii*) represents a small fortune. Timber companies have been eager to cut their last remaining holdings but have run into stiff opposition from environmental organizations. Detailed and laborious research efforts supported by the government clearly showed the existence of a dependence of the northern spotted owl on old growth stands for a variety of reasons (microclimate, roost sites, nest sites, prey density, access to prey, etc.). In order to protect the "ancient" forests outside of National Parks comprehensive restrictions on logging and other impacts were imposed by the US Fish and Wildlife Service after the completion of the significant Thomas report (Thomas et al. 1990); the owl became the flagship species of this conservation effort. The even more specialized marbled murrelet is at high risk as well. It is a small auk that breeds in the southern zone of its range only in the upper branches of mature conifers, often some 20–40 km inland from its marine feeding zone; this unique biology makes the species utterly dependent on the persistence of ancient trees (Ralph et al. 1995). The "battle" for the last remaining ancient forests has increased industry sensitivity to wildlife issues; it has also focused the public's attention on the unique resources and high value of mature forest ecosystems (Verner et al. 1992); many urban citizens became aware of the consequences of unregulated clear cutting practices for the first time (Dietrich 1992).

6.5.2 Sacramento River and Delta

What has happened to the Sacramento River since 1850 is a mirror of modern California's recent history. It is almost surprising that there is still native life in this river system. Several races of the chinook salmon

(*Oncorhynchus tshawytscha*) are at a high extinction risk today because of stream diversions, dams, and thermally or chemically polluted water (Thelander and Crabtree 1994). The introduction of many exotic fish species and the lack of water during critical seasonal migration periods makes survival a risky gamble year after year (Marchetti and Moyle 1995). In 1995, the federal government decided to require private and public land owners and water users to take significant (and costly) measures to restore the water quality and flow in order to allow for the recovery of the endangered population of the winter run chinook.

There are also rich biota (including a small fish, the delta smelt *Hypomesus transpacificus*) dependent on the well-watered marsh lands of the Sacramento Delta ecosystem (Steinhart 1990; Thelander and Crabtree 1994) which are threatened by planned water diversions channeling additional water from the water-rich northern part to the water-starved southern part of the state. The former extensive riparian woodlands lining the flood plain of the Sacramento have almost entirely disappeared, victims of land conversion for agriculture and urban development (see Table 6.2).

6.5.3 Desert Tortoise in the Mojave Desert

The desert tortoise (*Gopherus agassizii*) is listed as an endangered or threatened taxon throughout large desert areas of the southwestern United States. As in the case of the northern spotted owl, the quality and quantity of the bio-landscape is tightly bound to the well-being of this large desert herbivore. Many tortoise populations have declined in the last 20 years at rates ranging between 3 and 59% per year, even in well protected reserves. The most serious problem facing the Mojave populations of this species is the cumulative load of human and disease-related mortality accompanied by habitat destruction, degradation, and fragmentation (Fish and Wildlife Service 1994). Unfortunately, former prime tortoise habitats to the northwest of Los Angeles in the western Mojave Desert have become increasingly urbanized. The city of Palmdale grew from 33 500 to 56 500 in population size from 1987 to 1990, a growth of 70% in 3 years (Fay 1991). A Desert Tortoise Recovery Team worked for almost four years to develop a recovery plan; if implemented, this will combine critical habitat areas into special tortoise reserves called recovery units. These have been selected using the latest data and simulation models available to conservation biologists. If successful, this plan will permit the delisting of the species by the year 2019 (Fish and Wildlife Service 1994). Total government expenditures for securing habitat and management purposes are expected to cost some 17 million US dollars between 1994 and 2006.

6.5.4 Coastal Sage Scrub

In the south coast bioregion, issues of urban sprawl (Keeley 1993) have reached the crisis point as more and more coastal slopes and canyons have been covered by homes, shopping centers, and golf courses. Most of the land is privately owned, and the real estate values are among the highest in the nation. The key bio-landscape affected is the southern coastal sage scrub which grows on dry coastal slopes and on interior plains to the west and south of the Transverse and Peninsular Mountains of southern California. Coastal sage scrub (CSS) is composed of a rich community of herbs, grasses and evergreen or summer-deciduous shrubs, often characterized by species of *Artemisia*, *Salvia*, and *Eriogonum*. While it may look unattractive at the end of the long and dry summer months, this community is remarkably rich in endemic plant and animal taxa. Much of the southern CSS was lost during the growth of Los Angeles after World War II; the recent surge of growth in urban suburbs in Orange and San Diego County has threatened the very viability of this community and its many associated species.

At the state level, a sophisticated Natural Communities Conservation Planning (NCCP) program was developed in order to curb the escalating land use conflicts between private developers and state and federal agencies administering endangered species regulations in southern California (CDFG 1993; Scott et al. 1995). The boiling point was reached when ornithologists determined that the coastal-sage dependent population of gnatcatchers in southern California was in fact a threatened endemic taxon (Mann and Plummer 1995). Any further grading and conversion of CSS had to be judged in light of the plight of the California gnatcatcher (*Polioptila californica californica*). The NCCP program has worked to enlist local governments and private land owners to mediate conflicts over "sensitive" habitats (containing rare or threatened taxa) and to set aside large patches of coastal sage scrub habitats linked by appropriate natural corridors; in return a small percentage of CSS can be developed. This program uses all currently available concepts and insights for maintaining the genetic variability, demographic stability, and spatial dynamics of biotic populations. It is still controversial at the political level, however, as the various interest groups are uncertain about their potential gains and losses compared to the well known but gridlocked status quo.

In San Diego County, diverse geological and climatic features have contributed to an even richer coastal bio-landscape. There is still some CSS left but there are also remnant grasslands, riparian corridors, and unusually diverse chaparral habitats between the sprawling suburbs (Fig. 6.3). Once again, however, land use conflicts have congested the judicial system and

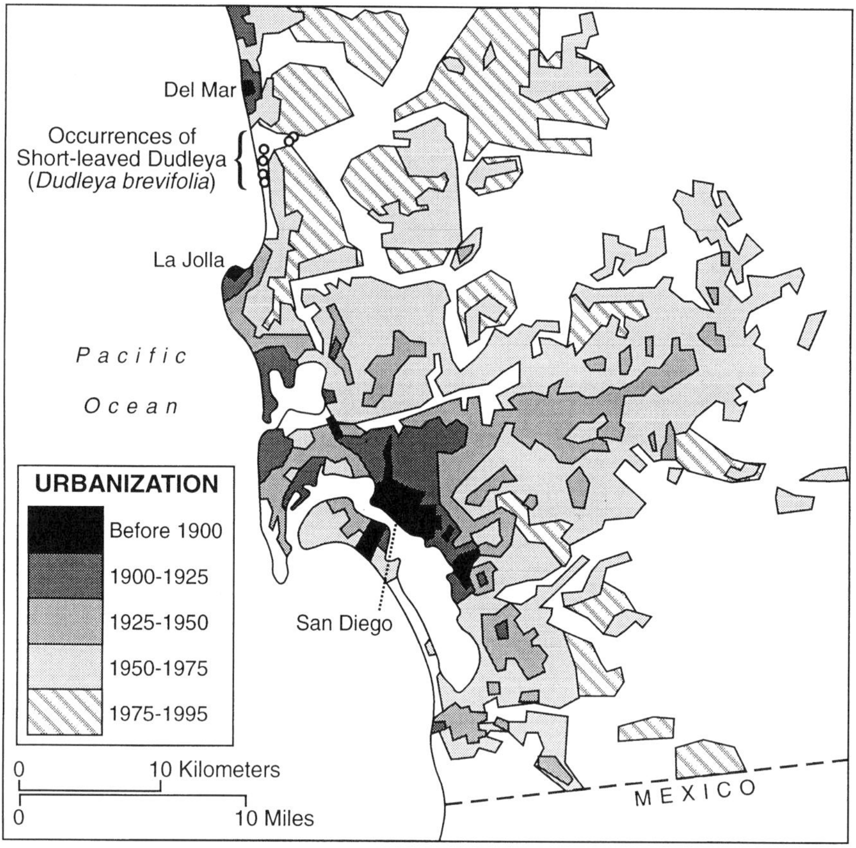

Fig. 6.3. Expansion of urbanized land area in San Diego (1875–1995)

threaten to stifle economic growth. In recognition of the need to preserve coastal lagoons and canyons for esthetic, recreational and conservation reasons, the City and County of San Diego have adopted a Multiple Species Conservation Program (MSCP); it is even more ambitious than similar habitat conservation plans in other states (Beatley 1994). Making good use of recent advances in computer-based natural diversity databases, scientists mapped out a comprehensive spatial system of habitat reserves and linkage zones to effectively protect all sensitive, threatened and endangered taxa as well as sensitive habitat complexes (see Holland 1986). Where field records were sparse, natural history, climatic, and vegetation data were used to assess individual species ranges, as in the case of the orange-throated whiptail lizard (*Cnemidophorus hyperythrus*); a database hypercube was developed resulting in a predictive distribution map (Hollander et al. 1994).

This program (MSCP) is designed to overcome the many individual conflict situations by endorsing a comprehensive plan; its application is dependent, however, on the availability of public funds needed to acquire private lands designated as open space or habitat reserves. In the meantime, some rare taxa have become more vulnerable to extinction. An example is a tiny succulent plant, the short-leaved dudleya (*Dudleya brevifolia*) which occurs only near the coast north of San Diego in Torrey Pines State Reserve and in a few nearby patches of hardpan soil rich in marble-like iron-titanium concretions. The entire range of this species, which shrivels to near invisibility after a short 6-week growing season, amounts probably to no more than one hectare. The last two remaining patches outside of the main patch near Del Mar lie on private land which has been proposed for residential development.

6.5.5 The Sierra Nevada Ecosystem

The Sierra Nevada region is what the Alps are to central Europeans: a unique and diverse mountain region with tremendous visitor appeal in winter and summer, and a diverse history and palette of bio-landscape problems. How to protect this resource from being overrun by tourists and further degraded by the activities of its resident human population (some 180 communities with a total of 650 000 people in 1990) was one of the questions asked in a pioneering multidisciplinary scientific program, the Sierra Nevada Ecosystem Project (SNEP). This three-year study by over 100 scientists made an assessment of the current status of the Sierra Nevada (core area 20 663 930 acres [83 588 km^2]). Grazing, timber harvest, and land type conversions were implicated by the study as the major factors causing changes in the plant and terrestrial vertebrate communities of the Sierra; of special concern was the loss of nearly 800 000 acres (3236 km^2) of oak woodlands in the last 40 years. The aquatic/riparian systems were found to be "the most altered and impaired habitats of the Sierra"; amphibians and aquatic invertebrates have severely declined, and anadromous fish (chinook salmon, steelhead *Oncorhynchus mykiss*) are now nearly extinct from Sierran rivers (SNEP 1996).

The SNEP objective was not to develop a plan for resolving the many land use and resource conflicts in the Sierra; SNEP was an educational and science-based project to make an assessment of the status of the Sierra and to offer suggestions for resolving conflicts or for further studies and debates. The technical aspects of the project were divided into three primary components: ecosystem assessments, analysis of policy strategies, and a geographic information system (GIS) database. The latter is expected to become part of a Sierra Nevada Information System which will be accessible to the public via the Internet.

SNEP conducted several case studies to investigate the efficacy of different institutional arrangements in managing ecosystems. The first concerns the Mammoth-June Ecosystem Management Project of the Inyo National Forest. The issue is the implementation of the new national landscape-analysis policy for ecosystem management of the Forest Service. Based on historic trends and dynamic models of existing system factors and processes, a desired condition, or a "word-picture" of the mostly forested landscape had to be developed. The team chose a desired condition that contradicted the historic condition for some resources because returning an ecosystem to past structure and composition may be highly inappropriate ecologically, given current and future "natural" environmental conditions (SNEP 1996).

The second case study focused on the Lake Tahoe basin, a highly altered environment with multiple anthropogenic impacts. Increased water pollution and import of sediment loads has decreased the clarity of Lake Tahoe waters significantly, and the well-being of wildlife and human resources depends on adaptive management and environmental restoration. This study proved the necessity for constant inter-agency consultation and for public-private collaborative efforts; there was broad recognition of the interdependency of economic and environmental concerns since the lake's economic base would collapse in a worst case scenario of deforested mountain slopes and further increases in air and water pollution.

A third case study investigated the tool of Formal Mediation employed by the federal government to resolve land issues concerning the long-term habitat needs of giant sequoia (*Sequoiadendron giganteum*). Giant sequoia occurs today in widely isolated groves, some remaining in private lands, and there is a strong national interest in their preservation. At debate were issues such as fuel hazard reduction, the logging of associated conifers, and the management of buffer zones. It was found that mediation provided a means of negotiating resolution of disagreement, not necessarily of sharing or implementing the best science, and despite these shortcomings, the parties involved have generally been able to work within the Mediated Settlement Agreement of 1990 (SNEP 1996).

This pioneering scientific effort will be of great benefit to many parties. Suspicious local government officials learned that university scientists are not agents of the Federal Government; scientists became more broad-minded about complex resource management problems after participating in public workshops and hearings, and forest and wildlife managers gained flexibility and interagency communication skills. SNEP was but a start in addressing the present and future of a unique regional ecosystem.

6.6 Conclusions

The land use conflicts discussed here provide promise and despair alike for the future. Despair because California is so large, complex, and regionally variable in its people and in its bio-landscapes. There will always be conflicts over land use somewhere even if we employ the best science and agree on comprehensive conservation plans. Policies and procedures are subject to change because of political and socio-cultural factors. In addition, government agencies are not infallible: there are books filled with case studies of failed and bungled conservation programs (Alvarez 1993). Scientists have, however, experienced a steep learning curve in recent years (Lawton and May 1995) and are much more confident about their predictions and recommendations today than just a decade earlier. Thus, science should be able to contribute much to local and regional planning efforts in the future.

Part of the promise lies in the fact that the large number of land use conflicts means that there is still much natural habitat to lose in California. That is a significant difference compared to Mediterranean countries in Europe where almost all of the endemic biotas are squeezed into small remnant habitat patches. The mediterranean landscapes of Chile have also been heavily impacted by human disturbance (Fuentes and Muñoz 1995). In California, by contrast, there are still thousands of square kilometers where no sheep, goat, or woodcutter has ever been active.

The recent state-wide rise in interest in environmental and regional planning by environmental activists and community groups bodes well for the future natural landscapes in California. Conflicts may persist but solutions are likely to incorporate the conservation of non-extractive resources. Urban sprawl itself has fallen into disfavor and may soon be mitigated if population pressures do not become too large. This is the assessment of an interdisciplinary policy group including the Bank of America; it issued a report on the "new California" declaring urban sprawl to be wasteful and unsupportable in the long run due to the excessive costs of infrastructure maintenance (Anonymous 1995a).

The other part of the promise lies in the fact that we have already witnessed significant conflict resolutions in many parts of California. Private nature reserves have sprung up in northern California, cities like Los Angeles and Palo Alto have developed urban wildlife reserves, and some highly degraded natural landscapes are rebounding (sometimes in unexpected ways) due to careful nurture and restoration work. A good example for the latter is Santa Cruz Island which had suffered from severe overgrazing and subsequent erosion due to sheep and cattle ranching. Its flora (Junak et al. 1995) was altered and impoverished as the ranges of endemic taxa began to shrink. Removal of the sheep in the early 1980s started the process of ecological recovery (Halvorson 1994). Several plant species believed to be locally extinct have reappeared on the island in the 1990s, an encouraging process for local restoration ecologists. Hopefully, these examples of ecological planning, conservation, and restoration will become state-wide trends and reduce the anthropogenic extinction risks to the priceless bio-landscapes of California.

References

Alvarez K (1993) Twilight of the panther: biology, bureaucracy and failure in an endangered species program. Myakka River, Sarasota, Florida

Anonymous (1989) Vanishing California. California 14(12):74–149

Anonymous (1995a) Beyond sprawl: new patterns of growth to fit the new California. EXEC-6776, Environmental Policies and Programs, Bank of America, San Francisco

Anonymous (1995b) The world almanac and book of facts 1996. World Almanac Books, Mahwah, New Jersey

Beatley T (1994) Habitat conservation planning: endangered species and urban growth. University of Texas Press, Austin, Texas

California Department of Fish and Game (1993) Southern California coastal sage scrub natural community conservation planning process guidelines. California Resources Agency, Sacramento

California Department of Fish and Game (1994) Special plants list. Natural diversity data base, Natural Heritage Division, Calif. Dept. Fish and Game, California Resources Agency, Sacramento

Cameron R (1990a) Above Los Angeles. Cameron, San Francisco

Cameron R (1990b) Above San Diego. Cameron, San Francisco

Cameron R (1994) Above Carmel, Monterey and Big Sur. Cameron, San Francisco

Cameron R (1995) Above Tahoe and Reno. Cameron, San Francisco

City of San Diego (1995) Multiple species conservation program: public review draft. MSCP Plan, City of San Diego

Dasmann, RF (1965) The destruction of California. Macmillan, New York

Dietrich W (1992) The final forest: the battle for the last great trees of the Pacific Northwest. Simon & Schuster, New York

Drost CA, Fellers GM (1996) Collapse of a regional frog fauna in the Yosemite area of the California Sierra Nevada, USA. Conserv Biol 10:414–425

Fay SJ (ed) (1991) California almanac, 5th edn. Pacific Data Resources, Santa Barbara

Fish and Wildlife Service (1994) Desert tortoise (Mojave population) recovery plan. U S Fish and Wildlife Service, Portland, Oregon

Fuentes ER, Muñoz MR (1995) The human role in changing landscapes in central Chile: implications for international comparisons. In: Zedler PD, Fox MD (eds) Ecology and biogeography of mediterranean ecosystems in Chile, California, and Australia. Springer, Berlin Heidelberg New York, pp 401–417

Halvorson WL (1994) Ecosystem restoration on the California Channel Islands. In: Halvorson WL, Maender GJ (eds) The fourth California Islands symposium: update on the status of resources. Santa Barbara Museum Natural History, Santa Barbara, California, pp 485–490

Hart J (1984) The new book of California Tomorrow: reflections and projections from the Golden State. W. Kaufmann, Los Altos, California

Hickman JC (ed) (1993) The Jepson manual: higher plants of California. University California Press, Berkeley, Los Angeles

Higgins P, Kier WM (1992) A plan for restoring biodiversity of anadromous fish resources of the Klamath River. In: Kerner HM (ed) Proceedings of the symposium on biodiversity of northwestern California. Rep 29, Wildland Resources Center, University of California, Berkeley, pp 217–225

Holland RF (1986) Preliminary descriptions of the terrestrial natural communities of California. Dept Fish and Game, State of California

Hollander AD, Davis FW, Stoms DM (1994) Hierarchical representations of species distributions using maps, images and sighting data. In: Miller RI (ed) Mapping the diversity of nature. Chapman & Hall, London, pp 71–88

Howitt RE (1995) How economic incentives for growers can benefit biological diversity. California Agriculture 49(6):28–33

Jackson LE (1985) Ecological origins of California's Mediterranean grasses. J Biogeogr 12:349–361

Jensen DB, Torn MS, Harte J (1993) In our own hands: a strategy for conserving California's biological diversity. University California Press, Berkeley, Los Angeles

Jimerson TM, Hoover LD (1992) Old-growth forest fragmentation: changes in amount, patch size and edge as a result of logging. In: Harris RR, Erman DC (Tech Coord) Kerner HM (ed) Proc Sympos biodiversity of northwestern California. Rep 29 Wildland Resources Center, University of California, Berkeley, pp 168–174

Junak S, Ayers T, Scott R, Wilken D, Young D (1995) A flora of Santa Cruz Island. Santa Barbara Botanic Garden, Santa Barbara

Keeley JE (ed) (1993) Interface between ecology and land development in California. Southern California Academy Sciences, Los Angeles

Kreissman B (1991) California: an environmental atlas and guide. Bear Klaw, Davis, California

Kucera TE, Barrett RH (1995) California wildlife faces uncertain future. California Agriculture 49(6):23–27

Lawton JH, May RM (eds) (1995) Extinction rates. Oxford University Press, Oxford New York

Mann CS, Plummer ML (1995) California vs. gnatcatcher. Audubon 97(1):38–104

Marchetti MP, Moyle PB (1995) Conflicting values complicate stream protection. California Agriculture 49(6):73–78

Medvitz AG, Sokolow AD (1995) Population growth threatens agricultural open space. California Agriculture 49(6):11–17

Mooney HA, Hamburg SP, Drake JA (1986) The invasions of plants and animals into California. In: Mooney HA, Drake JA (eds) Ecology of biological invasions of North America and Hawaii. Springer, Berlin Heidelberg New York, pp 250–272

Palmer T (ed) (1993) California's threatened environment: restoring the dream. Island Press, Washington, DC, Covelo, California

Pavlik BM, Muick PC, Johnson S, Popper M (1991) Oaks of California. Cachuma, Los Olivos, California

Ralph CJ, Hunt GL, Raphael MG, Piatt JF (tech eds) (1995) Ecology and conservation of the marbled murrelet. Gen Tech Rep. PSW-GTR-152. Pacific SW Res Stn, Forest Service, Albany, California

Schoenherr AA (1992) A natural history of California. University California Press, Berkeley, Los Angeles

Scott T, Standiford R, Pratini N (1995) Private landowners critical to saving California biodiversity. California Agriculture 49(6):50–57

Sierra Nevada Ecosystem Project (SNEP) (1996) Sierra Nevada ecosystem project: final report to Congress.Wildland Resources Center Rep 36. University California, Davis, California

Steinhart P (1990) California's wild heritage: threatened and endangered animals in the Golden State. Sierra Club Books, San Francisco

Thelander CG, Crabtree M (eds) (1994) Life on the edge: a guide to California's endangered natural resources: wildlife. BioSystems Books, Santa Cruz, California

Thomas JW, Forsman ED, Lint JB, Meslow EC, Noon BR, Verner J (1990) A conservation strategy for the spotted owl: report of the interagency scientific committee to address the conservation of the northern spotted owl. U. S. Forest Service, Portland, Oregon

Time (1991) California: the endangered dream. TIME 138(20):31–110

Verner J, McKelvey KS, Noon BR, Gutierrez RJ, Gould GI Jr, Beck TW (1992) The California spotted owl: a technical assessment of its current status. Gen Tech Rep PSW-GTR-133, Pac SW Res Stn, Forest Service, Albany, California

Yaffee SL (1994) The wisdom of the spotted owl: policy lessons for a new century. Island Press, Washington, DC, Covelo, California

7 Abandoned Lands and Land Use Conflicts in Southern France

M. Etienne, J. Aronson and E. Le Floc'h

7.1 Introduction

Across southern France, and in adjacent countries, large areas of formerly populous, productive agricultural lands are being abandoned (González Bernáldez 1989; Baudry & Bunce 1991; Farina 1993). In many of these areas, native mediterranean woodlands are making a dramatic comeback, albeit in modified forms due to faunistic, floristic and ecological changes induced during the multi-millennial history of human occupation, exploitation and resource management. At present, some 1% of the overall area of Languedoc-Roussillon, for example, is "reverting" to woodland each year.

For some observers, long-term fallows are more than welcome. They are seen as a sign of and vehicle for the return of wildlife and wildflowers to the countryside. For others, however, it signals a failure to maintain a rural way of life rich in knowhow, traditions and local lore. Comprehensive data are lacking on whether biodiversity is going up or down as a result, but at the landscape level it is certain that a former mosaic or patchwork quality is being lost. No doubt this will favor the return or spread of some (Mediterranean) organisms while limiting others. Finally, the long-term economic potential of the numerous resources previously exploited is being neglected and degraded, if not definitively lost. Similarly, options for modified land use in the face of unpredictable changes in regional, national or global realities are being lost.

In this paper, we propose some terminology and a general approach for planning, evaluating and, hopefully, reorienting rural and regional planning in southern France. For this task, we combine ecological, historical and socio-economic considerations, and we argue that all three are necessary. Specifically, we argue that it is desirable, not to say necessary, to "pilot" Mediterranean ecosystems and landscapes, even in the absence of a short-term profit motive. Like many ecologists, we reject current market economics to the extent that monetary values are not customarily attached

Ecological Studies, Vol. 136
Rundel et al. (eds.) Landscape Degradation and Biodiversity in Mediterranean-Type Ecosystems

to nonmarketed, but tangible and important, functions of ecosystems, i.e. "free nature services" (Westman 1977).

After defining our terms and providing a brief historical overview of ecosystem trajectories in the study area, over the past two millennia, we present four case studies where conflicting interest groups contribute to the difficulty of integrated land use planning. Each current human activity is analysed in terms of its ecological impact, the political force of the social group involved, and its relative spatial and temporal importance. Various scenarios of revised land use are then presented and compared for one study area, prior to offering some recommendations for those among decision-makers seeking a truly integrated land use plan. We conclude with a general discussion of the future of mediterranean ecosystems and landscapes in the wider theatre of southern France.

7.2 General Processes Affecting Landscape Transformation

Chorematic models (Brunet 1986) of changes in rural landscape structure are given for four types of land use dynamics typical of current trends in the French Mediterranean region (Fig. 7.1). Two processes correspond to fallow land development: (1) abandonment of marginal crop fields or of agricultural activities on high labour-consuming soils (e.g. terraces in Corsica, the Cévennes or the Pyrénées; wheat fields in the Préalpes); (2) end of forest management, especially in the holm oak coppices, but also in the chestnut groves of Corsica and the Cévennes, and in the maritime pine forests of the Maures and Esterel mountains.

Two processes deal with use diversification: (1) the substitution of traditional forest management by new activities linked with the demand for recreation in the coastal forests of Provence and the eastern Pyrénées; (2) the recuperation of rangelands for developing new agricultural or industrial activities, especially in the plains of La Crau or eastern Corsica.

These processes are illustrated below by four case studies wherein novel land planning objectives are proposed, taking into account ecological criteria and the functional role of biodiversity.

7.3 Three Responses to Ecosystem Degradation or Abandonment

We have previously presented (Aronson et al. 1993a) a general model describing the degradation of ecosystems in arid and semi-arid lands,

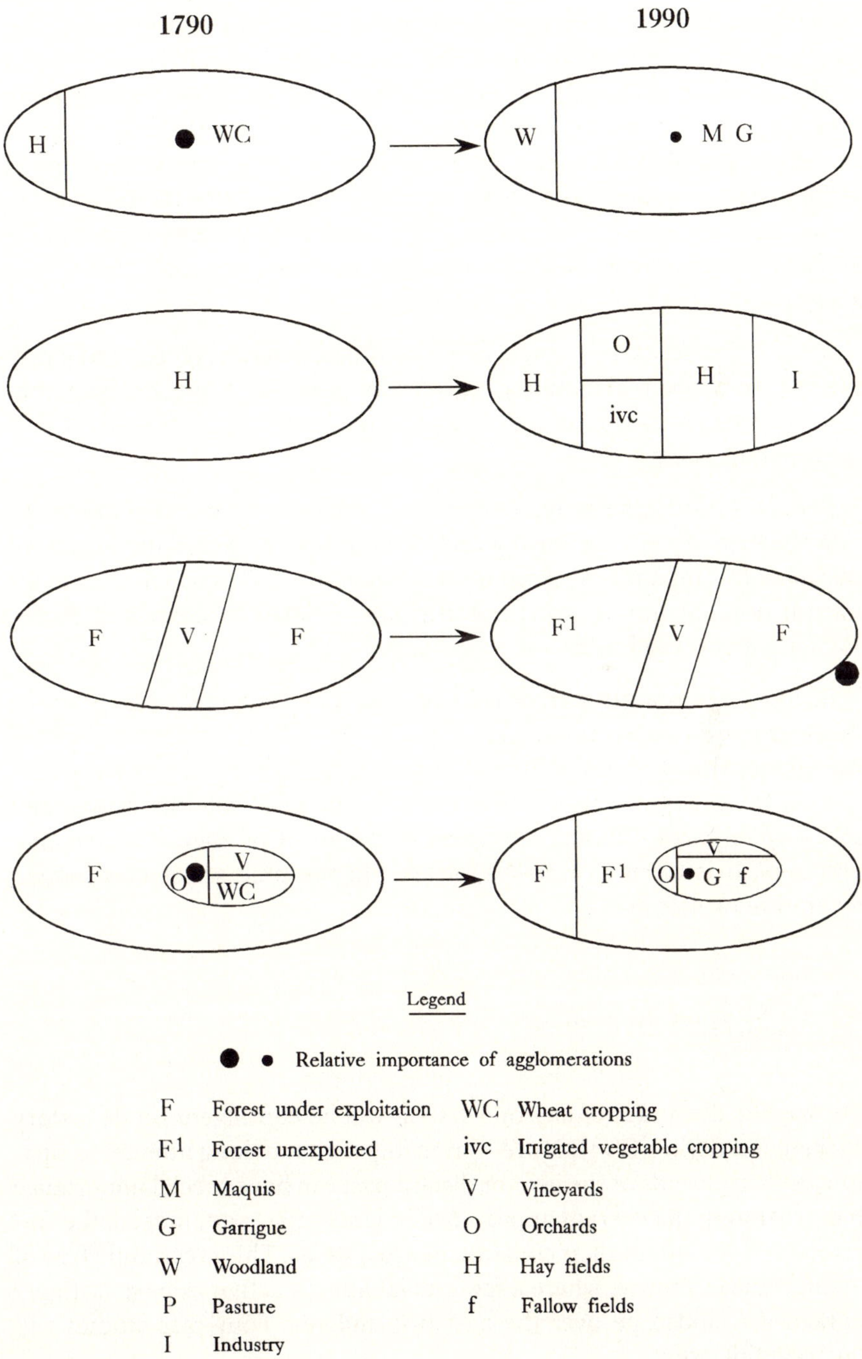

Fig. 7.1. Four typical land use dynamics in the French Mediterranean region during the last century

under the effects of prolonged, anarchic exploitation of resources and, more recently, rapidly growing human and, often, livestock density. In that model, we also distinguished between the three principal responses to degradation: restoration, rehabilitation and reallocation. In the special context of the Mediterranean Basin, we here propose that abandonment of long-standing agro-ecosystems is tatamount to "degradation". In so doing, we also broaden our enquiry to consider landscapes as an equally important hierarchical level as ecosystems. Now, we shall define our terms.

Restoration seeks a complete or near-complete return of a site to a pre-existing state, by reassembling, insofar as possible, a species inventory corresponding to the fauna and flora presumed to have constituted the pre-existing ecosystem.

Rehabilitation seeks to repair damaged or blocked ecosystem *functions*, with the primary goal of raising ecosystem *productivity* for the benefit of local people. Emphasis is placed on sustainability of all production systems and on functional and structural attributes related to flows and fluxes, nutrient cycles, food webs.

Reallocation is a general term describing what happens when part of a landscape, in any state, is assigned a new use that does not necessarily bear any relationship with the structure or functioning of the pre-existing ecosystem. It usually assumes a permanent managerial role for people and normally requires ongoing subsidies in the form of energy, water, and fertilizers. It is often *irreversible*, at least in the time scales of a human generation or lifetime.

7.4 Historical Overview

The specific dynamics of any one system will be contingent on its history (Pickett and Parker 1994). Moreover, in any region of long human occupation, certain periods of the near or distant past can have special importance in determining the current appearance of landscapes and, indeed, the trajectory and content of their constituent ecosystems. This is certainly true of mediterranean France, where a few crucial human activities have lastingly marked the landscape over the past two millenia. Four case studies will illustrate this point.

7.5 Case Studies

In analysing the case studies that follow, we employ the terms and concepts laid out above, along with additional multiple criteria of evaluation. Each case study will be briefly introduced with a description of the bio-physical environment, an historical review of the exploitations and "development" undertaken, and of the current conflicts among those involved locally. Current human activities will then be analysed by means of a grid designed to allow a comparison of the ecological impact and spatio-temporal and sociological dimensions of each.

The principal ecological criteria employed are as follows: 1) longevity (*long*) of the human activity; 2) degree of artificialisation (*artif*) involved, a useful indicator of the transformations imposed upon the indigenous ecosystem (and landscapes) by successive reallocations; 3) the degree of modification of basic structures (*mod*) at the ecosystem or landscape levels; and 4) reversability (*rev*), or relative feasability of returning to a pre-existing ecosystem trajectory.

Each activity is also situated in time, space and social context with the aid of the following three criteria: 1) spatial occupation (*space*), as the percentage of the landscape concerned by the activity; 2) temporal occupation (*time*), as the duration of the activity on the scale of one calendar year; 3) social force (*soc*), which indicates the size and political clout of the social group concerned.

7.5.1 The Terraces of Interior Corsica

Bearing testimony to the colossal efforts of a rural society that until recently had to provide all its own food in a steep, mountainous region, the terraces of central Corsica are visable today only after wildfires remove the invading, all-covering maquis. Prato and Piedigriggio provide good examples of picturesque, perched villages floating in a sea of *Cistus* (rockrose), with here and there a few stands of *Quercus ilex* (holm oak) on the more inaccessible slopes. A few rare flocks of sheep or cows make a meagre living from these pasture lands continually ravaged by fire. By contrast, 200 years ago a totally different image prevailed, as revealed by the Land Use Map ("Plan terrier") drawn up by Napoleon's geographers between 1770 and 1795 (Albitreccia 1942). At that time, 70% of the land surface in these two townships was devoted to cereal production, while the rest was either reserved for grazing lands or unused maquis (Etienne 1977). In the ensuing century, nearly all agricultural activities disappeared, except for a bit of viticulture on the flattest, most fertile lands.

Animal husbandry influenced the landscape only by means of the periodic fires set by the shepherds to renew shoot growth in the low shrub and herbaceous strata.

Table 7.1 shows the principal activities which still remain, hunting and extensive sheep rearing being the two major land uses and shrubland maquis the major landscape feature. The only potential conflict is related to the shepherd pratice of setting fire to open the maquis vegetation.

7.5.2 The Township of Puéchabon

The township of Puéchabon, 30 km NW of Montpellier, occupies 2270 ha, mostly in holm oak garrigues, and a grand total of 274 residents. For at least 5000 years, fire has been an important land management "tool" in this region. By the end of the 10th century, however, important settlements began to arise, under the aegis of the abbeys of Aniane and St. Guilhem-le-Désert, both of which were created around the beginning of the 9th century (Bougette 1950).

As a result, clearing and planting, as well as forestry activities, accelerated greatly during this period. Moreover, throughout the late Middle Ages, glassworks and charcoal manufactures consumed huge quantities of wood over vast areas (Saint-Quirin 1904). By the middle of the 17th century, woodcutting had become so important that governmental ordinances requiring reforestation were established in 1669 by Colbert, minister of Louis XIV.

In the 1790s, there were some 650 inhabitants at Puéchabon (Lalanne 1989). From that time, and up to the first third of the 20th century, there was a highly varied land use occupation including cereals, vineyards, olive

Table 7.1. Evaluation[a] of human activities in interior Corsica according to multiple criteria

date[b]	Activity	long	artif	mod	rev	space	time	soc
ca.1600	Hunting	+	–	–	+	+	–	+
ca.1600	Wood-cutting	+	0	–	+	–	–	–
ca.1800	Livestock rearing	+	–	0	–	+	+	0
1970	Viticulture	0	+	+	–	–	+	0

[a] +, 0 or – indicates an intensity gradient for the ecological criteria (longevity, artificialization, structure modification, reversability) as well as the ratio of space and time concerned or the social pressure of human activities, i.e. for time + = 12 months, 0 = 6 months and – = 1–2 months per year.

[b] Dates correspond to the first appearance of each activity in the region but the evaluation is based on the current situation.

Table 7.2. Evaluation[a] of human activities in the township of Puéchabon according to multiple criteria

Date[a]	Activity	long	artif	mod	rev	space	time	soc
50000 B.P.	Hunting	+	0	−	+	−	−	+
5000 B.P.	Sheep rearing	+	−	+	+	0	0	0
ca. 1100	Cerealiculture	−	+	+	+	−	−	−
ca. 1100	Woodcutting	+	−	+	−	+	−	+
ca. 1800	Viticulture	+	+	+	+	−	−	−
ca. 1980	Ecotourism	−	0	0	+	−	−	−

[a] *see* comments at the bottom of Table 7.1.

groves, grazing lands and, above all, intense forestry activities. Several light industries, including distilleries, tanneries, cotton and silk manufacturers, co-existed with and depended on agriculture, animal husbandry and forest activities in a more or less integrated fashion.

Table 7.2 shows the principal activities which have transformed the township, with approximate dates of their initiation. Certain activities, e.g. animal husbandry, have been completely abandoned. Others, such as hunting and woodcutting, continue, but in an entirely new socio-economic context. Hunting for small game is a very popular weekend activity in winter, while collective hunts for wild boars are organized to reduce the number of these unruly predators on vineyards. The woodcutting responds to an increasing demand for chimney firewood in the bedroom communities of Montpellier. Fully one third of the township's revenues derive from the annual sale of woodlots.

7.5.3 The La Crau Plain

Near the Camargue, the Plain of La Crau is a site of great biological and ecological interest. A former delta of the Durance River, this is the last surviving large example in France of a steppe formation (Devaux et al. 1983) and associated avifauna (Blondel 1970). This steppe, locally known as "coussoul", is the result of a dry, windy climate and low relief of rounded silicaceous rocks overlying an alluvial water table lying at a depth between 2 and 5 m.

A zone of intensive sheep rearing from very ancient times (Darluc 1783), the La Crau plain has been host to a range of cereal and animal feed cropping systems since the construction of the Craponne canal in the late 16th century. The installation of a dense network of irrigation canals and the concomitant influx of silt have led to a system of high quality feed-production with itinerant animal husbandry based on seasonally variable forage resources. This combination was further enhanced by the possibility of early-season feed production and the exceptional phytosanitary conditions. Starting in the 1950s, direct pumping from the water table permitted the development of early-season melon production, later replaced by fruit orchards. Huge amounts of capital in land clearing and levelling are now being invested in efforts to expand fruit production to all areas with shallow water tables. However, wildlife conservation and ecotourism have started to develop. In particular, *Tetrax tetrax*, the "canepetière", a protected bird species, provides a rallying point for environmentalists.

The history of land use and resource exploitation in La Crau is not limited to agriculture. This large, flat and largely unoccupied space near the sea attracts the attention of industrial and military concerns, including the iron and steel industry of Fos-sur-Mer and the air base at Istres. Finally, the innumerable round stones, lying in places in layers several dozen meters thick, have kept a number of quarry companies busy since the end of the second World War.

This rapid overview and the qualitative analysis of the different activities presented in Table 7.3 indicate a rather sharp rupture around 1945–1950, between an earlier phase of diverse agricultural activities leading to a diversified landscape organised in bands or zones and a second phase of proliferation and intensification of activities. This latter phase has brought about a near-total restructuring of the plain, which now resembles a sort of archipelago (Fig. 7.2). Severe fragmentation of a formerly homogenous region is taking place, with little or no consideration being given to the interactions among the different "islands" of the archipelago.

Based on the data presented in Table 7.3 and Fig. 7.2, three more or less realistic scenarios can be constructed, as shown in Fig. 7.3. These indicate the varying trends of land use occupation that could evolve if restoration, rehabilitation, or continued reallocation were most favored in the coming years. Broadscale restoration of the original steppic vegetation and ecosys-

Table 7.3. Evaluation[a] of human activities in the Crau plain according to multiple criteria

Date[a]	Activity	long	artif	mod	rev	space	time	soc
ca. 1200	Sheep rearing	+	–	–	+	+	0	–
ca. 1800	Hay making	+	+	0	–	0	+	+
1950	Melons	–	0	–	+	–	+	–
1950	Quarries	+	+	+	–	–	+	+
1970	Steel industry	0	+	+	–	+	+	+
1980	Ecotourism	–	–	–	+	–	–	+
1990	Orchards	0	0	+	0	–	+	0

[a] *see* comments at the bottom of Table 7.1.

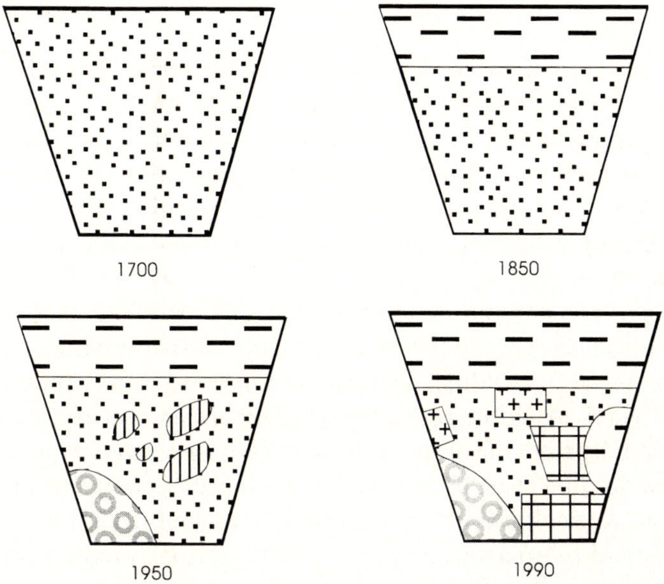

Fig. 7.2. Schema of evolving land use occupation in the plain of La Crau in the 20th century

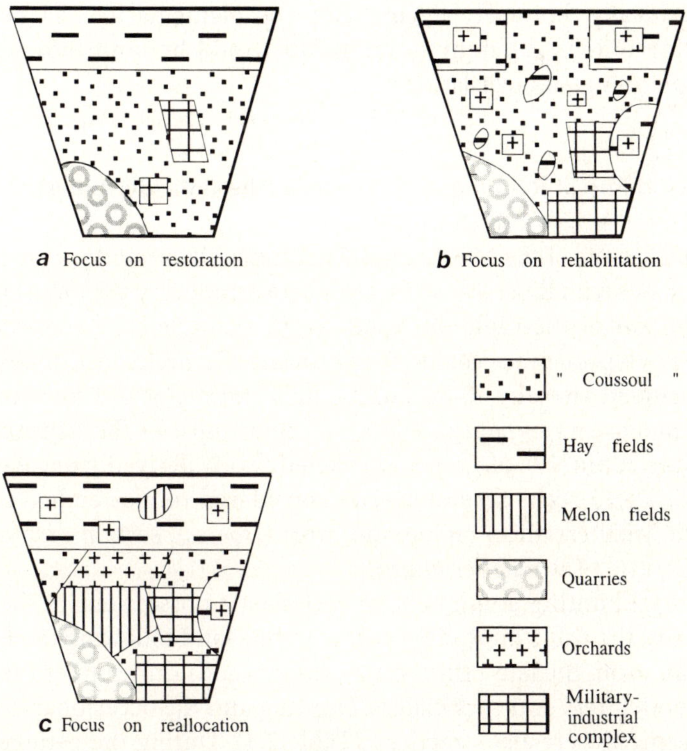

Fig. 7.3a–c. Three scenarios for La Crau based on the notions of restoration, rehabilitation and reallocation

tem (Fig. 7.3a) is perhaps not a serious possibility today, given the heavy socio-economic pressures in La Crau. Nevertheless, it is presented here for purposes of contrast and in view of the fact that social priorities might conceivably change in the coming decades. Representing an extreme conservationist position, this scenario would logically call for the elimination of many existing activities considered incompatible with the goals of restoration.

In contrast, a rehabilitation program (Fig. 7.3b) would seek a compromise between conservation and development; certain areas of coussoul would be reserved for wildlife, particularly migratory birds. Industrial or military activities would be restricted to areas already so degraded as to appear to have crossed one or more thresholds of irreversibility and, therefore, to have become more or less incompatible with agriculture or nature conservation. Agricultural activities would be encouraged, but would be reorganised, and relegislated. New extractive activities would be discouraged or prohibited altogether.

Finally, a reallocation scenario (Fig. 7.3c) would encourage diversification of agricultural production and all other short-term economic activities. Ideally, however, a search for complementarity and peaceful cohabitation among different social groups could be built into regional land use planning and legislation.

7.5.4 Coastal Forest Remnants in Provence (the Colle du Rouet)

The coastal forests of the Maures and Esteral massifs provide another case of a rural space with diverse historical and contemporary uses. We take the forest remnant of the Colle du Rouet as an example. The ecosystem of reference is a typical pine-oak forest of Provence. In prehistoric times, such forests occupied virtually all the coastal hills and plains of Provence-Côte d'Azur. This forest remnant (1600 ha) occupies most of the Argens River Plain, about 20 km NW of Fréjus, on skeletic soils derived from Permian sandstone. There occur several species considered rare or endangered in the French Mediterranean region: quillwort (*Isoetes*), several orchids and Herman's turtle (*Testudo hermanni*).

From royal hunting dominions, to ecclesiastic possessions in the High Middle Ages, the Colle du Rouet became public land following the French Revolution. From the late 18th century, these woods came under the pressure of woodcutters and cork-takers. Hunting and the occasional installation of beehives were also practised (Table 7.4). During the 19th century however vineyards came to occupy most of the area except 1600 hectares remaining for the time being in the public domain.

Table 7.4. Evaluation[a] of human activities in coastal forests according to multiple criteria

Date[a]	Activity	long	artif	mod	rev	space	time	soc
ca. 1600	Hunting	+	–	–	+	+	–	+
1800	Woodcutting	+	0	–	+	–	+	+
1800	Bee-keeping	+	–	–	+	0	0	–
1800	Plant collecting	+	–	–	+	–	0	+
1950	Walking	+	–	–	+	–	–	+
1970	Shrub mowing	–	0	0	0	–/0	+	+
1980	Plant harvesting	0	–	–	0	–	–	–
1980	Sheep rearing	0	0	0	+	0	0	–
1990	Hiking	–	–	–	0	–	–	–

[a] *see* comments at the bottom of Table 7.1.

In the 1960s, immense stands of maritime pine, *Pinus pinaster*, were decimated by the cochineal *Matsuccocus feytaudi.* At the same time, the international market for natural cork experienced a price collapse. As a result, forestry activities declined sharply. Independently, the Côte d'Azur became rapidly overdeveloped creating the demand for green belts near the beaches; where tourists could engage in mushroom or dried flowers harvesting, and hiking.

These new forest "users" and the growing problem of human-caused summer fires, during the 1970s and 1980s, led forest managers to set up a network of firebreaks that now occupies some 20% of the surface area of the forest (Etienne et al. 1989). Initially, these firebreaks were created and maintained by foresters with the aid of heavy machinery. Since 1981, the maintenance task has been partially conferred to a shepherd, the Forest Service imposing some management rules but also providing subsidies according to the area grazed by the flock. Additional operations like oversowings or perennial fodder crops have been simultaneously developed to promote an emerging sylvo-pastoral system.

The evolution of coastal forests is exactly the reverse of the far more extensive areas of holm oak-dominated garrigues. Here, the top public priority has become the prevention of forest fires, but it is still possible to propose contrasting scenarios.

7.6 Discussion

In the simulated example, the restoration and rehabilitation scenarios proposed permit maintenance of a traditional Mediterranean "moving mosaic" landscape (Blondel and Aronson 1995), wherein mosaic units correspond to, and are affected primarily by, only one or two principal economic activities. In two case studies, increasing redistribution of land for disparate economic and recreational uses underlies the need for achieving widely-based social consensus on fundamental issues relating to ecosystem "health" and landscape integrity. In practice, the problem will be to reconcile nature conservation with sustainable development and to assure ecological and socio-economic compatibility of unrelated activities in a shared "commons" (sensu Hardin 1968). It also raises crucial issues related to the social and economic means available, and the scientific and managerial tools needed to achieve skillful piloting of ecosystems. More challenging still will be the reintegration (or even redesigning) of landscapes to meet new needs and realities in an increasingly crowded region.

In the two other case studies, by contrast, land abandonment and the narrowing of economic activities over a relatively short period have favored relatively homogeneous areas and lead to a dangerous situation of high fire risk, among other things. It is, however, possible to imagine different strategies for reintegrating these Mediterranean landscapes by a co-ordinated series of restoration, rehabilitation and reallocation operations. Different portions of each landscape would be preferentially reserved for one of these, based on preliminary studies in situ, and an analysis of the probable modes and pathways of connectivity between and among ecosystems.

As mentioned above, we envision such a reintegration as being a collective enterprise. In general terms, restoration should be attempted for those sites of particular importance for conservation of threatened species, e.g., marshes and other wetlands, as well as for maintaining high beta and gamma diversity. Moreover, some ecotones, movement corridors, and other natural areas should be candidates for restoration, as might sites of particular cultural significance or historical importance, such as springs, pre-historic villages and burial sites, medieval abbeys, etc.

Secondly, rehabilitation should be given priority for those degraded areas where agricultural, pastoral or sylvo-pastoral productivity can conceivably be profitable, in selected portions of the area, long-, mid-, or short-term. Additionally, in some areas of no interest to farmers or animal breeders, rehabilitation could be favored as a means to assuring long-term productivity, and flexibility in the face of an unpredictable future. This assumes that maintaining high biodiversity and "healthy", resilient ecosystems is the best guarantee of long-term productivity.

We do not pretend to have a recipe for managing all the varied land, water and bio-resources in the area. Our goal is merely to suggest the need for applying multiple criteria (ecological, economical and sociological) in rural planning, and to favor ecological ones when planning activities of widescale, long-lasting consequences for ecosystems and landscapes. We maintain that it is essential to collectively direct the trajectories of ecosystems and landscapes so as to maintain high biodiversity, ecosystem resilience and flexibility, and high quality of life. When using the terms "restoration of ecosystems" or "reintegration of landscapes", moreover, we do not seek a blind, or emotional adherence to past patterns and uses. This would be foolhardy and doomed in advance, except in the dubious context of "living museums".

We seek an approach to land use planning and resource development that is more audacious and effective than anything currently practised under existing French law and custom. We see it also as being more conservative of biodiversity, or ecodiversity, i.e. biological, ecological and cultural diversity (Naveh, Chap. 2 this vol.), while at the same time profoundly subversive, to the extent that current economic "rules" are not respected. In so saying, we follow Garret Hardin (1985) in calling ecology itself a "conservative, subversive science". Many, if not most, of the ecological problems of modern society simply cannot be solved by technical solutions alone.

Acknowledgements. We thank the drawing studio of the CEFE-CNRS of Montpellier for the figures.

References

Albitreccia A (1942) Le plan terrier de la Corse au XVIII° siècle. Paris, Presses Universitaires de France

Aronson J, Le Floc'h E, Floret C, Ovalle C, Pontanier R (1993) Restoration and rehabilitation of degraded ecosystems in arid and semiarid regions. I. A view from the south. Restoration Ecol 1:8–17

Baudry J, Bunce RGH (eds) (1991) Land abandonment and its role in conservation. Options Méditerranéenes. 15. CIHEAM, Zaragoza

Blondel J (1970) Biogéographie des oiseaux nicheurs en Provence occidentale, du Mont Ventoux à la Mer Méditerranée. L'oiseau 40:1–47

Blondel J, Aronson J (1995) Biodiversity and ecosystem function in the Mediterranean Basin. In: Davis G, Richardson DM (eds) Biodiversity and ecosystem function in mediterranean-type ecosystems. Springer, Berlin Heidleburg New York, pp 43–119

Bougette E (1950) Histoire de Puéchabon. Imprimerie des Beaux Arts, Lodève

Brunet R (1986) La carte-modèle et les chorèmes, Mappemonde 4:2–6

Darluc M (1783) Histoire naturelle de la Provence, vol 2. Avignon

Devaux JP, Archiloque A, Borel L, Bourrelly M, Pallluel JL (1983) Notice de la carte phytoécologique de la Crau. Biologie-Ecologie Méditerranéenne 10(1–2):5–54

Etienne M (1977) Bases phytoécologiques du développement des ressources pastorales en Corse. Thèse Ecologie, USTL Montpellier, 175 p

Etienne M, Napoleone M, Jullian P, Lachaux M (1989) Elevage ovin et protection de la forêt méditerranéenne contre les incendies. Etud Rech Syst Agraires Dév 15:1–45
Farina A (1993) L'ecologia dei sistemi ambientali. CLEUP, Padova, Italy
González Bernáldez FG (1989) Des communautés rurales aux complexes métropolitains: exemple de la déstabilisation des systémes pastoraux traditionnels et évolution des paysages pâturés méditerranéens dans le centre de l'Espagne.In: Clayson WJ (ed) Landscape ecology: study of mediterranean grazed ecosystems. University of California, Davis, pp 62–77
Hardin G (1968) The tragedy of the commons. Science 162:1243–1248
Hardin G (1985) Human ecology: the subversive, conservative science. Am Zool 25:469–476
Lalanne JF (1989) Un canton des garrigues de 1789 à 1799 (Aniane, Argelliers, La Boissière, Puéchabon, St.-Guilhem-le-Désert). Imprimerie de la Charité, Montpellier
Pickett STA, Parker VT (1994) Avoiding the old pitfalls: opportunities in a new discipline? Restoration Ecol 2:75–79
Saint Quirin (1904) Les verriers du Languedoc 1290–1790. Réédité par l'Association "La Réveillée", Montpellier, (1985)
Westman WE (1977) How much are nature's services worth? Science 197:960–964

8 Land Use Changes and Conflicts in the Mediterranean-Type Ecosystems of Western Crete

V.P. Papanastasis and A. Kazaklis

8.1 Introduction

Mediterranean-type ecosystems (MTEs) of the Mediterranean region have undergone significant land use changes over the last few decades. After having been extensively used for centuries for traditional human activities, such as agro-pastoralism, tree and woodcutting and hunting, which have led to the establishment of a man-maintained dynamic equilibrium, they now are facing clashing demands by modern developers and classical conservationists. Depopulation and abandonment are conflicting with intensification of agro-pastoral utilization, expansion of urban-industrial sprawl as well as mass recreation and tourism resulting in functional, structural and visual landscape degradation (Naveh and Lieberman 1994).

Crete has gone through sweeping changes in the last 20–30 years following the pattern already established in other Mediterranean regions. She has been transformed from a traditional, closed and self-sufficient economy to an open market with millions of tourists flushing in every year. As a result, large upland areas have been abandoned by people who have moved to urban centers and coastal areas in search of a better life, mainly by accommodating the ever-increasing massive tourism. This development has dramatically changed the people-land relationship and balance in the lowlands and resulted in severe land use conflicts. Similar conflicts have been created in upland Mediterranean ecosystems, namely phrygana, maquis, garrigues and coniferous forests, where traditional agro-pastoral activities have been replaced by contemporary intensive agro-industrial practices.

In this chapter, the nature, extent and pattern of land use changes and conflicts in the MTEs of western Crete are analysed, the principles and processes involved are discussed, and management strategies are formulated and proposed.

Ecological Studies, Vol. 136
Rundel et al. (eds.) Landscape Degradation and Biodiversity in Mediterranean-Type Ecosystems

8.2 A Representative Area of Western Crete

A strip of 424 km^2, stretching across the island a few kilometers to the west of Chania, was chosen to represent western Crete (Fig. 8.1). From the built up north coast it extends up to the Keritis river through the Alikianou basin, then broadens on the White Mountains to embrace the Omalos plateau and ends up on the south coast after including the Sougia river basin and the Samaria National Park. It contains all the different types of Mediterranean ecosystems, touristically developed and underdeveloped areas on the north and south coasts respectively, intensively cultivated lands, a large part of the White Mountains where altitudes exceed 1800 m, three major and minor gorges and rugged coastal areas in the Libyan Sea. In addition to several land uses, the study area also includes several geological substrates, such as hard limestones, phyllites and schists, marls and alluvia (Fig. 8.1).

This study area has been the pilot zone of a 2-year project (1991–93) entitled "A threatened Mediterranean landscape: West Crete", which was financed by the European Union (EV4C-CT90-0112) and carried out by the Universities of Cambridge, UK and Thessaloniki, Greece (Grove et al. 1993).

8.3 Mapping and Identifying Land Uses

Land use/cover types were mapped from panchromatic aerial photographs taken in 1989 by using a Bausch and Lomb Stereo Interpretation Systems SIS-95 instrument. The classification scheme was based on the US Geological Survey system modified according to the Greek experience (Kazaklis and Karteris 1993). From the air photos with a scale of 1:30 000, the interpretation data were transferred to conventional topographic maps at a scale of 1:15 000 by means of a Bausch and Lomb Transfer Scope (ZTS). This and the previous instrument were subsequently used to compare the 1989 map with the 1945 photographs and to construct tables of land use changes between the two years. Only the land use changes related to MTEs are discussed in this chapter.

Based on geology and topography, the study area was divided into landscape systems and then into subsystems or landscape types by identifying characteristic land use and land use change patterns. The minimum mapping unit was set at 1 km^2 and a manual method was chosen for their identification. Ridge lines, streams, and riverbeds were used whenever possible as boundaries (Grove et al. 1993). Among the several identified

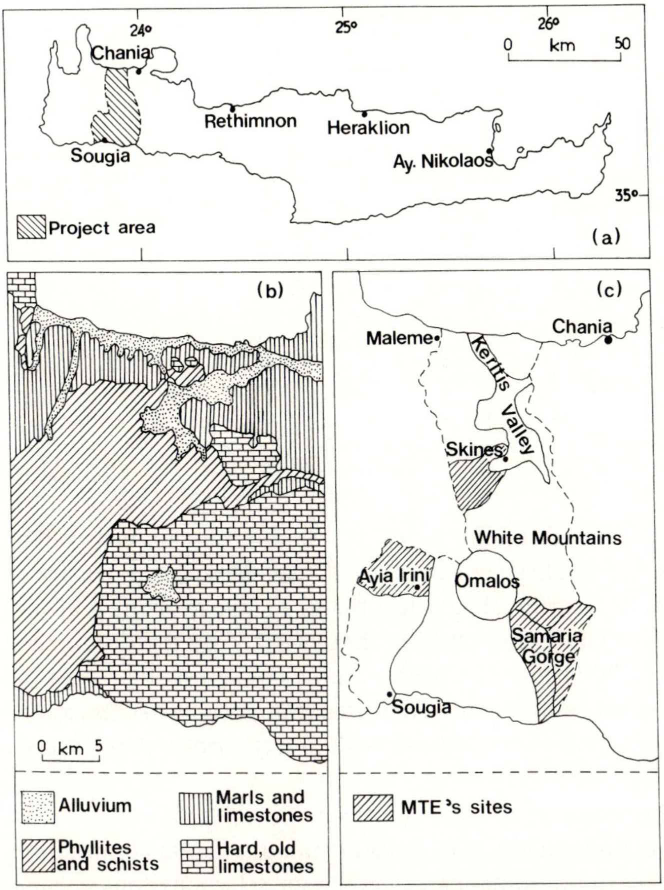

Fig. 8.1. The study area in western Crete: *a* location, *b* geology, and *c* places of selected sites with typical Mediterranean-type ecosystems (MTEs). (Adapted from Grove et al. 1993)

landscape types, three of them dominated by MTEs were selected (Fig. 8.1) and their land use changes and conflicts were studied in detail.

8.4 Land Use Changes

Out of the 42369 ha of the mapped area, MTEs were found to represent 27676 ha or 65% of the total in 1989, indicating their dominance in the study area. Other land uses included agricultural lands (22%) and miscellaneous uses (13%) such as barren and rocky lands, residential areas, wetlands and minor vegetation types (e.g. riverine vegetation).

Table 8.1. Changes of the MTEs in the whole study area of western Crete between 1945 and 1989

Type of ecosystem	Area (ha)		
	1945	1989	Change (%)
Phrygana	7934	6544	−17.5
Garrigue	6020	4832	−19.7
Maquis	4222	3668	−13.1
Coniferous forests	10504	12632	+20.3
Very open (<30%)	2184	1920	−12.1
Open (30–70%)	3672	2576	−29.9
Dense (>70%)	4648	8136	+75.0
Total	28680	27676	−3.5

Table 8.1 shows the different types of Mediterranean ecosystems mapped. We classified as phrygana all of the dense (80% crown density or more) scrub communities with shrubs less than 1 m high growing typically on several soil types. This community is characterized by the dominance of seasonally dimorphic dwarf shrubs, spiny or aromatic, such as *Sarcopoterium spinosum* and *Thymus capitatus* (Margaris 1980). Its expansion is favored by wildfires combined with overgrazing (Papanastasis 1977).

We classified the dense scrub communities with tall shrubs (more than 1 m high), grown mainly on siliceous soils as maquis. These communites are characterized by the dominance of evergreen sclerophyllous shrubs, such as *Arbutus unedo* and *Erica arborea* (Margaris 1980; Liacos 1982). We classified open scrub communities grown on both calcareous and siliceous soils with phryganic or evergreen screlophyllous shrubs less than 1 m high as garrigues. These are transitional communities between phrygana and maquis.

Finally, we classified *Pinus brutia* and *Cupressus sempervirens* forests with different crown densities (<30%, 30–70% and >70%) as coniferous forest. The conifers invade and replace phrygana and most areas of garrigue once wildfires and overgrazing are discontinued.

It is clear from Table 8.1 that the total area of the native communities (MTEs) did not significantly change between 1945 and 1989. Phrygana, garrigue and maquis, however, were impressively decreased while forests substantially increased. Looking at the different crown densities of forests, it becomes evident that the open stands decreased in area while the dense stands increased, indicating that the coniferous forests of the region became denser in 1989 than they were in 1945.

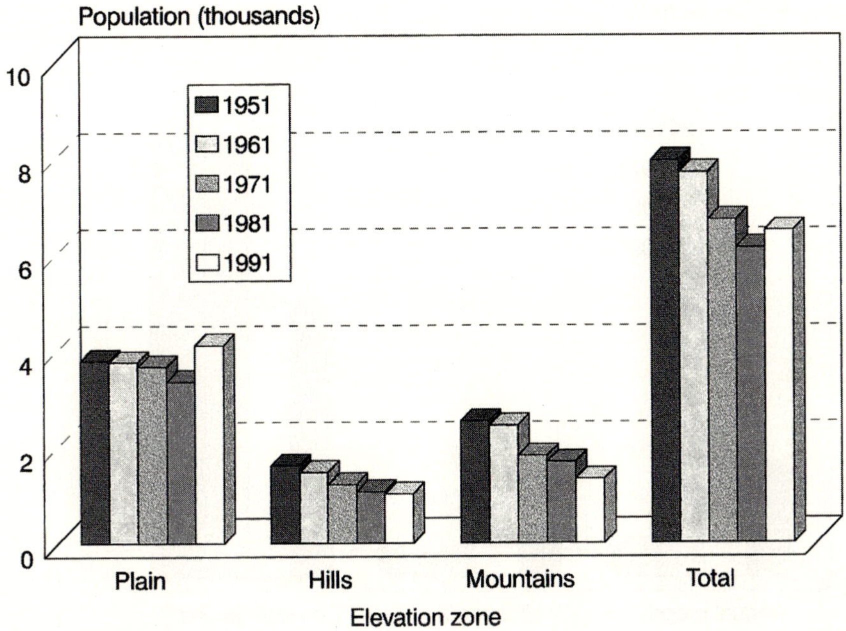

Fig. 8.2. Population distribution and changes by elevation zone between 1951 and 1991. (After Ispikoudis et al. 1993)

All these changes were caused by the reduction of the traditional human activities in the hilly and mountainous zones, where MTEs are mainly distributed, as a result of the migration of the rural people to the plains and urban centers of the study area (e.g. the city of Chania). Human population decreased by 36% in the hills and by 47% in the mountainous village communities from 1951 to 1991 while it changed very little in the plain areas (Fig. 8.2).

Traditional human activities in the study area chiefly included extensive farming of cereals and vineyards on terraces built in the limestone zone (mainly in phryganic areas) and extensive grazing with sheep and goats in all MTEs. This grazing was often combined with rotational burning of the Mediterranean vegetation by the shepherds themselves to suppress the woody plants unpalatable to animals (pastoral wildfires). Secondary activities were firewood and charcoal collection, timber harvesting and honey production. Since 1961, when reliable census data became available, cereal and vineyard cultivations in the hilly and mountainous areas were sharply declined to be partly replaced by more intensively cultivated olive and citrus groves (Fig. 8.3) thus leading to the abandonment of terraces and favouring the expansion of forests. This latter expansion was facilitated by

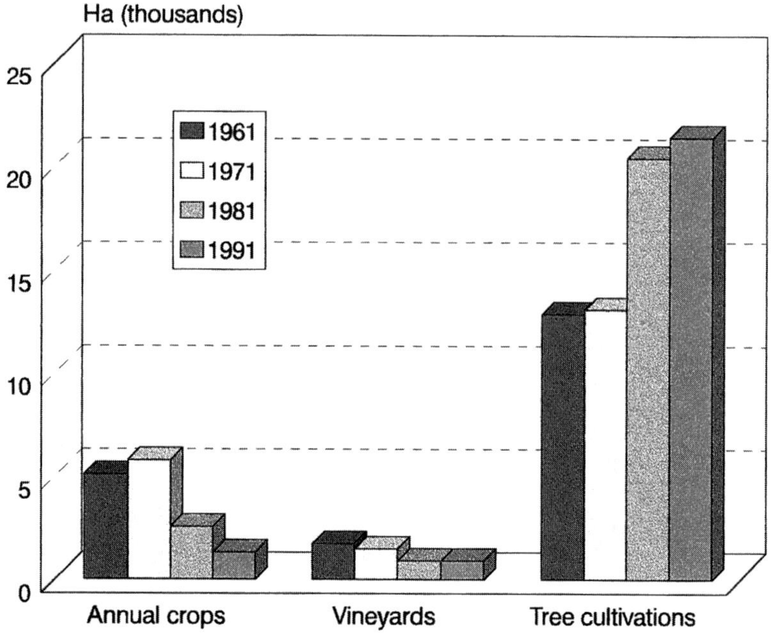

Fig. 8.3. Changes of annual crops (e.g. cereals), vineyards and tree cultivations (e.g. olive groves) in hilly and mountainous areas between 1961 and 1991. (After Ispikoudis et al. 1993)

the reduction of grazing pressure and other activities such as firewood, charcoal and timber harvesting.

Livestock husbandry has been restricted to a few sites and became more intensive over the same period by using feed supplements coming from the market (outside the study area), semi-improved breeds of sheep and goats, more pastoral wildfires to control the bushes, and improved infrastructure such as access roads and stables. Since 1961, the number of sheep and goats has been reduced in the plain and hilly areas but has started to increase in the mountainous areas since the mid 1980s (Fig. 8.4). The latter increase should be attributed to the subsidies provided by the European Union in order to support the income of the mountain people. Such developments in livestock husbandry have resulted in undergrazing of some areas of MTEs or overgrazing of others, with variable impact on land use changes.

Further analysis of the land use changes in the four MTEs of western Crete within the 44-year period indicates that they were multi-directional and therefore dynamic. Processes both increased and decreased the area covered by each land use and operated concurrently or sequentially on the same or on different pieces of land. Table 8.2 shows the changes in the area

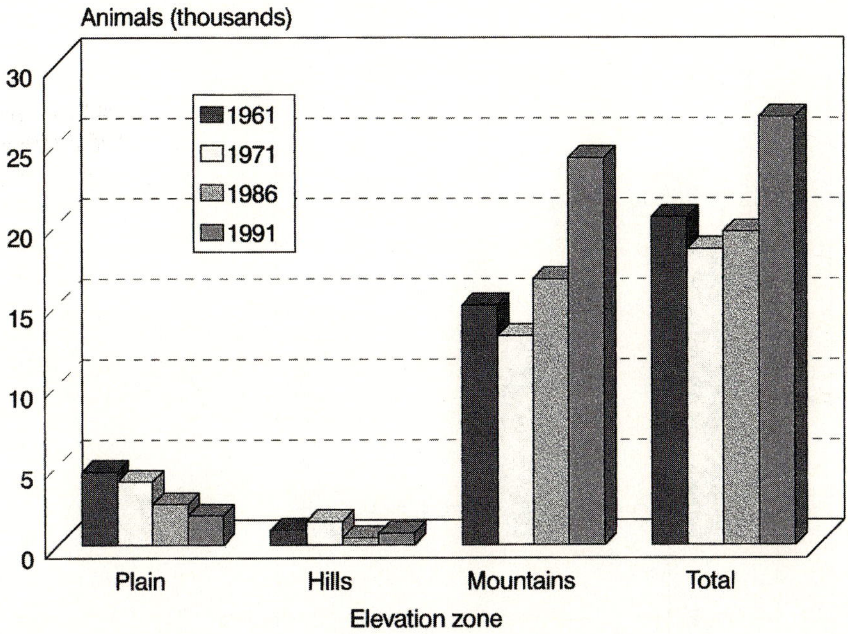

Fig. 8.4. Total livestock (sheep and goats) distribution and changes by elevation zone between 1961 and 1991. (After Ispikoudis et al. 1993)

covered by the four MTEs in relation to the different land uses. From this table the following information may be extracted:

1. The reduction of phrygana from 1945 to 1989 was mainly caused by the expansion of forests and olive groves, while a minor part of this community was converted to garrigue. Phygana expansion in the abandoned cereal fields as well as in burned maquis was not enough to counterbalance reductions in cover;
2. The reduction of garrigue in the same period occurred because of their conversion to maquis or coniferous forests through succession, while a significant area of garrigue was also converted to olive groves;
3. The reduction in area of maquis was mainly caused by its conversion to olive groves (more than 1000 ha), while only a small part retrogressed to phrygana through the wildfires;
4. The increase in area of forests, finally, was mainly caused by their expansion in areas which formerly supported phrygana, and to a lesser degree in the garrigues.

It may be concluded that land use changes in MTEs of western Crete since World War II were closely linked to the socio-economic changes occurred during the same period.

Table 8.2. Analysis of land use changes (ha) within the four MTEs of the study area in western Crete between 1945 and 1989

Land use/cover type	Mediterranean-type ecosystems							
	Phrygana		Garrigues		Maquis		Coniferous forests	
	Increase	Decrease	Increase	Decrease	Increase	Decrease	Increase	Decrease
Olive groves	8	360	12	204	32	1092	12	16
Orange groves	0	0	0	4	0	24	0	0
Cereals and vineyards	400	24	144	32	140	4	60	0
Phrygana	–	–	144	148	154	316	1368	40
Garrigue	148	188	–	–	820	240	676	144
Maquis	316	154	240	820	–	–	68	20
Coniferous forests	36	1364	150	676	20	72	–	–
Other forests[a]	0	40	4	0	0	0	0	0
Residential	0	4	0	0	0	12	0	0
Riverine vegetation[b]	4	4	0	0	8	0	0	0
Barren and rocky land	0	72	0	0	0	4	68	16
Clouds[c]	–	72	–	16	–	–	–	–
Total	912	2282	694	1900	1174	1764	2252	236

[a] Chestnuts, ash, etc.
[b] Plain trees, poplars, etc.
[c] Area covered by clouds when the airphotos were taken.

8.5 Land Use Conflicts

Land use conflicts can be better detected and explained if they are associated with the land use policy applied in a particular area. Such policy is not uniform in western Crete because the local governmental agencies have different, very often conflicting, objectives in management of the Mediterranean landscapes (Grove et al. 1993). For this reason three specific sites representative of the MTEs dominating in western Crete were selected to illustrate land use conflicts in relation to local policies of resource management.

8.5.1 Conservation vs Agricultural Development

Maquis are the dominant ecosystems covering the hills surrounding the southern and southwestern part of the Keritis valley (Fig. 8.1). Their traditional uses, especially extensive grazing and woodcutting, have been almost abandoned and the current policy applied by the Forest Service, to which their management belongs, is conservation (Papanastasis 1993), namely protection from clearing for cultivation and urbanization, so that their multiple environmental role to society is ensured. On the other hand, Keritis valley is intensively cultivated with citrus and olive groves which provide a high income to the farmers. As a result, the neighbouring maquis hills are receiving a strong pressure from local people to be converted to arable lands because this would result in a higher income than from traditional uses. The practice is to intentionally burn the maquis so that the vegetation is removed and then to establish monocultures of olive trees on terraces constructed by bulldozers. Burning is uncontrolled and it very often becomes a big wildfire.

In such an area located near to the township of Skines (Fig. 8.1), the land use changes between 1945 and 1989 were studied (Table 8.3). It was found that: (1) phrygana were increased at the expense of garrique and maquis after apparently intentional burning; (2) garrigue were decreased because they were converted either to maquis through abandonment of grazing or to olive groves; and, (3) maquis were decreased because they were converted to olive groves or to garrigues by wildfires. Altogether, more that 500 ha of garrigue and maquis were converted to olive groves at this particular site (i.e. 25% of the whole area). This suggests that agricultural development has been the main force behind these land use changes.

It should be pointed out that agricultural development of the maquis ecosystems in western Crete has been subsidized in the past by the government (Ministry of Agriculture), when olive oil had a good market. Cur-

Table 8.3. Changes of the MTEs in selected areas of western Crete between 1945 and 1989

Type of ecosystem	Skines			Ayia Irini			Samaria Gorge		
	Area (ha)		% change	Area (ha)		% change	Area (ha)		% change
	1945	1989		1945	1989		1945	1989	
Phrygana	76	252	+231.6	512	536	+4.7	1104	752	−31.9
Garrigue	556	108	−80.6	796	720	−9.6	244	188	−22.9
Maquis	1316	1084	−17.6	244	252	+3.3	0	0	0.0
Coniferous forests	0	0	0.0	96	172	+79.2	3420	3896	+13.9
Very open (<30%)	0	0	0.0	0	0	0.0	688	340	−50.6
Open (30–70%)	0	0	0.0	36	60	+66.7	588	492	−16.3
Dense (>70%)	0	0	0.0	60	112	+86.7	2144	3064	+42.9
Total	1948	1444	−25.8	1648	1852	+12.4	4768	4834	+1.4

rently, prices are reduced and subsidies suspended but conversion of maquis to tree cultivation is tolerated by the Agricultural Service because people still earn more than by extensive grazing. This development is in conflict with the conservation policy applied by the Forest Service (another service of the Ministry of Agriculture) on forest lands of Crete where maquis are also included according to the forest law (Papanastasis 1993) because farming on hillsides has resulted in serious soil erosion and degradation and is a potential threat to the stability of maquis ecosystems.

In order to solve the conflict between agricultural development and conservation in the maquis ecosystems, a landscape ecology approach to management should be adopted. This approach favours opening up of maquis to establish mixed tree crops (no monocultures) so that landscape diversity and stability is increased. Such agricultural development, however, should be done after careful planning and in co-operation between the different governmental agencies and the local people. In addition, it should avoid wildfires and use small instead of big machines (small tractors) so that the intervention is moderate thus preventing irreversible disturbances and accelerated soil erosion.

8.5.2 Extensive vs Intensive Livestock Husbandry

Livestock grazing, a dominant traditional use in MTEs of western Crete, has been reduced and restricted to certain, mainly mountainous, areas. One of the few sites where it is still practised is Ayia Irini located on the northern part of the Sougia Basin (Fig. 8.1). In the past, livestock husbandry was practised extensively. Animals were communally grazed but their numbers were kept in equilibrium with the grazing capacity of MTEs which were rotationally burned every 5–10 years to control the density and spread of the unpalatable woody plants. In recent years, mainly since Greece joined the European Union (EU) in 1981, the situation has changed. Although the communal system of grazing is still practised, the number of sheep and goats has been increased by 50% and new infrastructure such as access roads and stables has been built because of the subsidies provided by the EU (Ouled Belgacem 1993). The subsidies helped shepherds not only to increase the number of animals but also to buy hay and concentrates (e.g. grain) from the market and feed their stock during critical periods of the year, thus increasing animal production and, therefore, making the production process more intensive.

The site is covered by all four types of MTEs, especially garrigues and phrygana. Compared with 1945, 1989 shows relatively small changes amounting to a general increase in the area of garrigues and phrygana of about 12% (Table 8.3). Phrygana have slightly increased in area largely because of the abandonment of cereal fields and the retrogression of garrigues through wildfires; garrigues have been reduced in area because they were either converted to phrygana through wildfires or to forests through reduction of grazing; maquis remained almost stable; and forests have almost doubled in area chiefly because of the colonization of new lands.

Changes in Ayia Irini indicate a dynamic system with rotational burning being a vehicle in controlling shrubs and favouring the establishment of vegetation palatable to animals, especially sheep. On the other hand, the rotational burning seems to get concentrated on increasingly smaller areas, although the number of animals raised in the site is increasing. This suggests an intensification of livestock husbandry leading to more burning, heavier grazing and, therefore, accelerated degradation (Ouled Belgacem 1993). Such an intensification is in conflict with the traditional, extensively practiced, livestock grazing which was in dynamic equilibrium with conservation of MTEs.

The conflict between extensive and intensive livestock husbandry can be resolved by applying the principles of range management and respecting the tradition and culture of the local people. The landscape of Ayia Irini is adapted to the combination of grazing and rotational burning, but this combination should be kept in equilibrium with the natural resources, which means that any intensification of the system should be avoided.

8.5.3 Conservation vs Massive Tourism

Landscapes of western Crete are of outstanding beauty with the gorge of Samaria (Fig. 8.1) being world famous. In 1962, this gorge was declared as a National Park after expropriating and removing the remaining inhabitants as well as banning traditional human activities such as livestock grazing, woodcutting and bee keeping. The objective was to manage it for conservation including the protection of the agrimi, the wild goat of Crete, and for recreation. For the last twenty years, Samaria has been the most famous mountain attraction in Crete being visited annually by more than 300 000 people (Ispikoudis et al. 1993).

In 1989, Samaria gorge was almost exclusively covered by MTEs, especially coniferous forests of *Pinus brutia* and *Cupressus sempervirens*; their area was slightly increased as compared with 1945 (Table 8.3). The only type of Mediterranean community missing is maquis, apparently because the geological substrate is calcareous (Fig. 8.1). From 1945 to 1989, there has been a drastic decrease in the area of both phrygana and garrigue with a succession to pine forests, while most of the open and very open pine stands became dense (Table 8.3). This evolution indicates that the Park became more forested in 1989 and with denser stands than it was 44 years ago.

The conservation policy that is applied by the Forest Service in Samaria is virtually identical with a wilderness policy since no human interventions are allowed. This is the reason that the gorge changed from a heterogeneous and diversified landscape in 1945 to a highly homogeneous one in 1989. The expansion of pine forests has also greatly increased the fire risk. Furthermore, the large number of visitors going through the gorge in the summer has not only seriously damaged the walking path but, most importantly, has become a major source of fire ignitions. This means that conservation as a land use of Samaria gorge is in conflict with recreation in the form of massive tourism.

This conflict can be resolved by reconsidering the management policy of the National Park of Samaria. The wilderness policy applied so far is disastrous; it should be changed to allow management interventions so that the landscape becomes more heterogeneous. On the other hand, the number of visitors should be kept to a maximum by diverting excess people to other gorges in western Crete, very attractive but not yet well advertised.

8.6 Conclusions

1. Land use changes in MTEs of western Crete have been very dynamic since the Second World War when the traditional agro-pastoral equilibrium became disrupted through socio-economic changes which continue today.

2. Land use changes within MTEs of western Crete have been multi-directional with both negative and positive impacts on ecological processes operating concurrently or sequentially in space and in time. Due to the cessation of the traditional activities by a reduced rural population, however, dwarf scrub communities such as phrygana have been replaced by taller shrub communities and by pine forests through succession, thus leading to a homogeneous and less diversified landscape which is very vulnerable to wildfires.
3. Conservation in the form of overprotection and abandonment in MTEs of western Crete is in conflict with "development" applied either as intensive agriculture in the maquis ecosystems by clearing and establishing terraced olive groves, or as intensive livestock husbandry or as recreation in the form of massive tourism. All these conflicts can be resolved by applying a landscape ecology approach to management.

8.7 Summary

Mediterranean-type ecosystems of Crete have experienced dramatic changes in their traditional uses over the past few decades. These changes have been studied in a representative area of western Crete, where the land use/cover types were mapped from aerial photographs taken in 1989 and compared with aerial photographs taken in 1945. It was found that phrygana, garrigue, and maquis ecosystems substantially decreased in area over this period. These communities were replaced by coniferous forests, especially of *Pinus brutia*, which expanded and became denser. This change in vegetation cover was caused by the reduction of the traditional agro-pastoral activities as a result of the migration of rural people to urban centers. In the meantime, contemporary land use conflicts developed in several areas, mainly between conservation and development in the form of intensive terraced olive culture, intensive livestock husbandry and massive tourism. These conflicts have resulted in significant landscape instability and degradation.

References

Grove AT, Ispikoudis J, Kazaklis A, Moudy JA, Papanastasis VP, Rackham O (1993) Threatened Mediterranean landscapes: West Crete. Mimeo report submitted to European Commission, Brussells

Ispikoudis J, Lyrintzis G, Kyriakakis S (1993) Impact of human activities on Mediterranean landscapes in western Crete. Landscape Urb Plann 24:259–271

Kazaklis A, Karteris M (1993) Land use types: patterns and relationships with abiotic landscape factors in western Crete. Landscape Urb Plann 24:249–257

Liacos L (1982) Grazing management of evergreen brushlands in Greece. In: Conrad E, Oechel WC (eds) USDA General Technical Report PSW-58. Pacific SW Forest and Range Research Station, Berkeley, California, pp 270–75

Margaris NS (1980) Structure and dynamics of Mediterranean-type vegetation. Port Acta Biol 16:45–48

Naveh Z, Lieberman AS (1994) Landscape ecology: theory and application. Second Edition. Springer, Berlin Heidelberg NewYork

Ouled Belgacem A (1993) Grazing practices in western Crete and their impact on rangeland vegetation. Ms Thesis, Mediterranean Agronomic Institute of Chania, Crete

Papanastasis VP (1977) Fire ecology and management of phrygana communities in Greece. In: Mooney HA, Conrad CE (tech coord) USDA, General Technical Report WO-3 Washington, DC, pp 476–482

Papanastasis VP (1993) Legal status of land tenure and use and its implication for open landscapes of western Crete. Landscape Urb Plann 24:273–277

9 Land Use Changes and Conflicts in Central Chile

J. Aronson, A. Del Pozo, C. Ovalle, J. Avendaño, A. Lavin
and M. Etienne

9.1 Introduction

In mediterranean-climate regions (MCRs), historical and geographical as well as ecological approaches are needed to elucidate processes and patterns occurring at that rarely defined level of complexity called "landscape". In the case of Chile, a glance at the past five centuries of history is particularly crucial to the understanding of the various impacts of landscape degradation. Furthermore, to aid in our attempt to combine both ecological and human geographical considerations, we will borrow the "three waves" paradigm of sociologist Alvin Toffler (1980). Toffler was, of course, dealing with all of human history, and at the full planetary scale. Here we will be zooming in on the so-called *secano interior*, or "interior drylands", of the subhumid region of the Chilean MCR.

9.1.1 "Landscape"?

According to the definition of Forman and Godron (1986) which most ecologists seem to accept, landscapes comprise "a recognizable cluster of ecosystems" interacting somehow with one another in an undefined space predicted to be about several km^2. This definition is attractive because it fits so well with prevailing notions of "ecosystem" (Likens 1992). Yet, a purely materialistic definition of "landscape" hinders collaboration with most geographers and other social scientists, since it fails to address past and present human activities, customs, sentiments, etc. In what follows, we will address human as well as non-human determinants of change, degradation and diversity.

Ecological Studies, Vol. 136
Rundel et al. (eds.) Landscape Degradation and Biodiversity
in Mediterranean-Type Ecosystems

9.2 The Secano Interior of Central Chile

9.2.1 Overview

Within the *secano interior* (Fig. 9.1), we will deal only with the subhumid zone, the area lying some 250 to 400 km south of Santiago, equal to some 40% of the total *secano interior* area. In this subhumid zone (ca. 35°–37° S; 600–1000 mm mean annual rainfall), corresponding roughly to the central valley of the 7th and 8th Administrative Regions of Chile, over 144 000

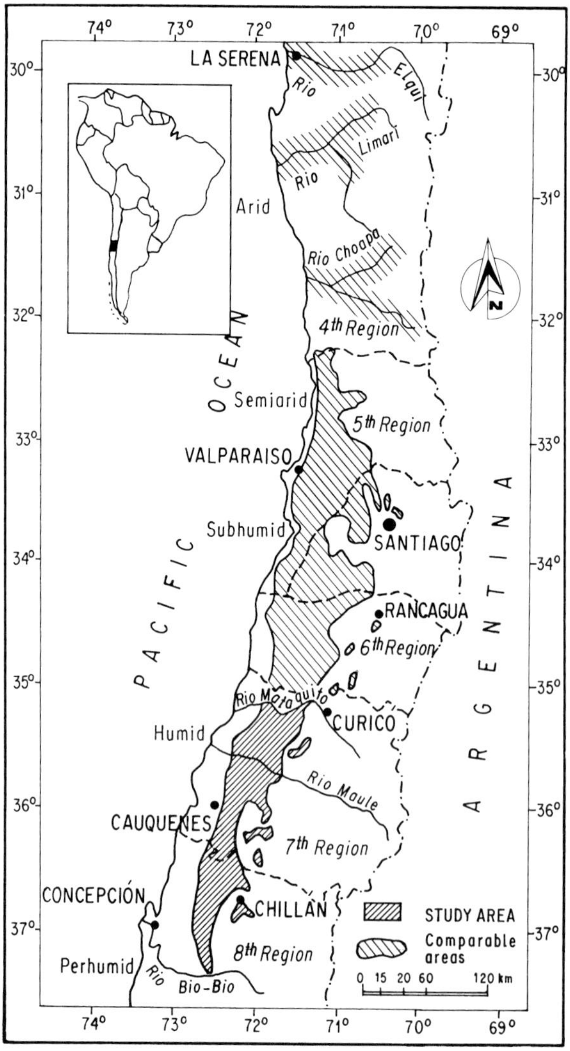

Fig. 9.1. The secano interior of central Chile

people live and farm some 800 000 ha of non-irrigated land (Ovalle et al. 1990).

About 61% of the human population in the subhumid *secano interior* live on farms, and actively engage in farming, while only 39% live in villages or small cities (INE 1990). Cauquenes is the largest agglomeration in the area with 38 141 inhabitants. Total population in these secano communes has increased by over 36% in the past 25 years, and some 18.5% of the population live below Chile's official poverty line (INE 1994).

The major pre-conquest vegetation type of central Chile (30°–37° S) was apparently a dense, biologically diverse woodland that resembled those of other mediterranean-climate regions in their physiognomy and the preponderance of sclerophyllous trees and shrubs (di Castri et al. 1981; Balduzzi et al. 1981). According to local conditions, these woodlands were dominated by *Maytenus boaria, Quillaja saponaria, Cryptocarya alba* and *Peumus boldus*. Today, however, as the result of 450 years of European colonization, the unirrigated portions of the secano are almost entirely dominated by a monotonous, synanthropic plant formation called *espinal* that could be called an anthropogenic savanna (Ovalle et al. 1990, 1996).

The *secano* includes both *llanos* (plains) and *lomas* (hills) in the central Depression, and the inland portions of the coastal foothills on the one hand, and the Andean pre-Cordillera on the other (Fig. 9.2). Subhumid secano ecosystems have been sufficiently transformed as to bear little or no relationship to their prehistoric predecessors. Exotic fauna and flora abound and, overall, landscapes have been dramatically homogenized and "banalized" over the past 4.6 centuries. As suggested above, woodland has been to a very large extent replaced by *Acacia caven*-dominated *espinales*, and now appears to be incapable of autogenic restoration (Aronson et al. 1993a,b,c; Ovalle et al. 1996). A few dozen fragments of woodland degraded to matorral (open coppice stand of sclerophyllous species more or less invaded by tall shrubs) do occur in the subhumid *secano interior*, especially on large land-holdings (C. Lusk, unpubl. data), but they are all very small, i.e., under 100 ha, and in imminent danger of final destruction.

In addition to soil erosion, one of the major consequences of past ill-constrained land use on ecosystems and, particularly, landscapes appears to have been the diminution of biological zonation along toposequences and the virtual disappearance of ecotones. The nearly ubiquitous presence of several species of lichens, and even mosses, occurring on the branches of *Acacia caven* trees in the seasonally inundated *llanos* bears vivid testimony to the microclimatic differences between *llanos* and *lomas*. Yet, apart from epiphytes, absolutely no vegetational differences can be observed between these two habitats. Clearly, it is in the area of land use history that the origins of this monotony must be sought. Further evidence of degradation

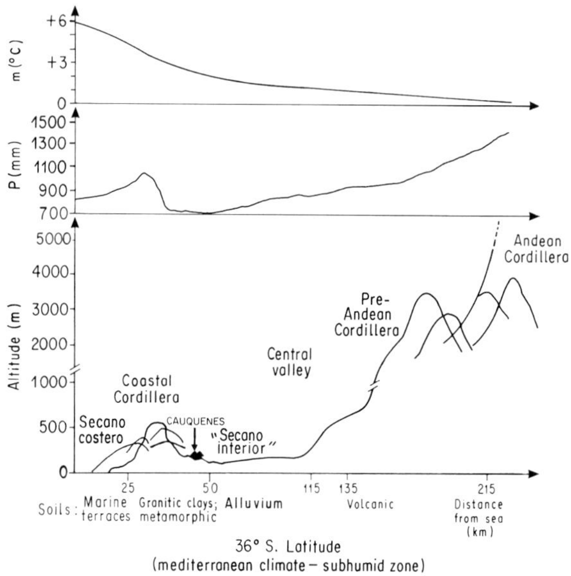

Fig. 9.2. A typical longitudinal cross section of central Chile near Cauquenes in the 7th Region. *m* average lowest temperature of the coldest month, July; *P* mean annual precipitation

is provided by the fact that it is mainly on the lomas that traditional *espinal* wheat-pasture-wheat rotation is practised today, whereas in the past, both *llanos* and *lomas* were used for cropping.

9.2.2 The Green Tide

Traditional agroforestry systems and matorral fragments in the *secano interior* are both threatened as a new land use is spreading rapidly throughout the region. We refer to the monospecific plantations of Monterey pine (*Pinus radiata*) or *Eucalyptus globulus* that some observers refer to as "the green tide" or the "green desert". This trend – which will be discussed in more detail below – represents the first major sociological and ecological "conflict" in recent times for the *secano interior*. Of primary relevance here is that if current economic trends continue, biodiversity and human presence may diminish even further. However, it should be mentioned that no sociological study has yet been undertaken to determine what local people think of contemporary land use, and landscape, changes.

9.3 The Three Waves

Before proceeding further, we shall now introduce the "Third Wave" paradigm. Toffler (1980) called the invention of agriculture the "First Wave", and the advent of modern manufacturing the second (ca. 1860 to 1920). Toffler then went on to suggest as a comparably important event the post-World War II phenomena related to high-tech, high-speed information processing and transfer that is sometimes called the "information age", or the "third industrial revolution" (Rifkin 1995); Toffler called it simply the "Third Wave".

Writing in 1980, however, Toffler probably was not aware of the emerging view shared by a growing number of biologists and physicists that life itself is essentially a process of genetic information flow, driven by what Richard Dawkins (1995) has called the "replication bomb" of digitally-coded DNA. Toffler failed to consider information from a biological or evolutionary point of view. At the end of this chapter, we will therefore explore the notion of the "third wave", as it applies to central Chile, with a post-Darwinian sense of "information" thrown in.

9.3.1 The First Wave

In Fig. 8.3, we have traced an abbreviated landscape history for the relevant part of the *secano interior* over the past ten millennia, based on the available data and speculations of historians, archeologists, et al. Pre-Columbian peoples in central Chile, nowadays called Araucanos, were organised into different tribes according to the type of environment in which they lived. The current 7th and 8th Regions were occupied by Ranquelche ("reed man") in the coastal swamps, Mapuche ("earth man") in the central valley and Pehuenche ("Chile pine man") in the coastal foothills and the Andean Cordillera, wherever *Araucaria auraucana* grew. These peoples kept llamas, set fires to clear areas, and cultivated some irrigated crops in the transversal river valleys. In short, they began the gradual transformation and elimination of natural vegetation. However, Gay (1865) and Bahre (1979) concluded that these cultures had comparatively little impact on ecosystems and landscapes, at least as compared to what happened much later, following the arrival of a small but powerful group of southern Europeans. In our study area this is certainly the case since the extension to central Chile of the mighty Incan empire was stopped around 1470 by Mapuche warriors north of the Rio Maule.

The beginning of the First Wave in central Chile coincided with the arrival of Europeans in the mid-16th century (1536–1541). The first stage of this colonial period, however, had only few and very scattered effects on secano interior landscapes until the end of the 17th century. Indeed, during the Arauco War which lasted throughout the 17th century, Spanish settlements were limited to small fortified villages with some cultivated lands. It was only somewhat later, when outside economic forces came into play, that a truly revolutionary force unfurled on central Chilean ecosystems and landscapes. There were two such periods of accelerated landscape transformation corresponding to intense immigration processes – in the late 17th and the late 19th centuries. We will call these First Wave, Parts 1A and 1B (Fig. 9.3). Such episodes of the major landscape transformation of this part of Chile are important to identify since they may aid in comparative landscape studies and preliminary diagnoses related to any proposal of restoration or rehabilitation.

Starting in the late 1680s (Gay 1865; Bauer 1970), the Spanish Viceroy had his headquarters in Lima – a hyperarid region – and it made good colonial sense to encourage cereal production in Chile, and to bring in, by ship, all the wheat that Chile could produce. Judging from the available records, the consequences for central Chile were a short-term, economic bonanza and a long-term, ecological calamity.

The second wheat "boom" took place two centuries later, and half a century after Chile won its independence in 1818. The nearly simultaneous gold rushes in California and Australia created huge markets for wheat in those two countries. From 1848 to the early 1880s, Chile was not only the world's leading copper producer, but also one of the two or three leading exporters of wheat (Bahre 1979; Domic 1979). As primary productivity was higher in the subhumid region than in drier portions of the mediterannean climate zone, it is here that intensive cereal and vineyard cropping was

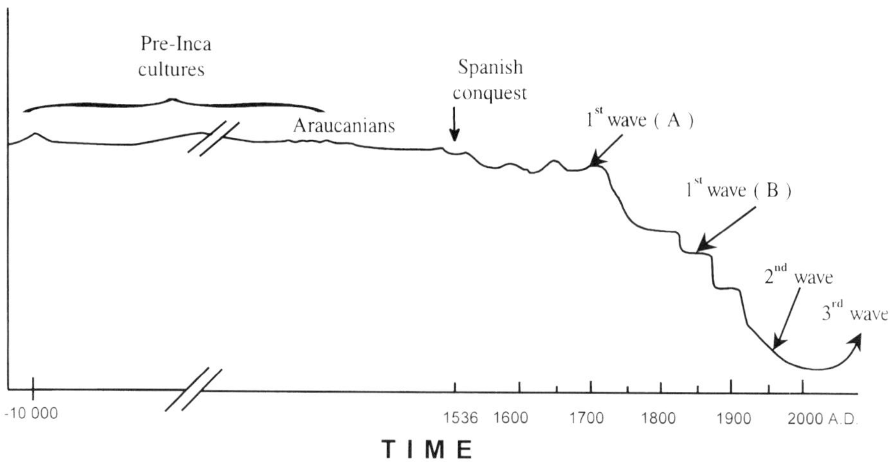

Fig. 9.3. Schematic representation of the impact of human activities and demographic fluctuations on matorral woodlands and other ecosystems in the 7th and 8th Regions of central Chile over the last 10 millenia. The term "landscape integrity" is intended to refer to overall stability and other indices of ecosystem and landscape "health"

concentrated. Accordingly, the long-term impact of the First Wave (Parts A and B) was more important here than further north, especially since the area south of the Rio Bio-Bio river remained largely inaccessible to Europeans until this century, due to the presence of belligerent Mapuches.

9.3.2 The Second Wave

9.3.2.1 Changing Land Use

The industrial revolution had – as elsewhere – far-reaching consequences in Chile, e.g., the centralization, reorganization and mechanization of the means of production in all parts of the country. In the case of the *secano interior*, this has entailed massive plantations of fast-growing trees such as *Pinus radiata* and *Eucalyptus globulus*, the so-called "green tide" referred to above. Among other uses, the most important outlet is as pulpwood sold to the paper mills along the Pacific Coast.

A government subsidy for tree planting was initiated in 1974 to encourage the planting of trees (and shrubs, in the semiarid region). But, as this subsidy has no surface threshold, large forestry companies have been the major beneficiaries, as they buy up former farmlands in the *secano interior* and adjacent regions for large scale tree planting. This strategy has been favoured by the opening of Chile to the world market with immediate consequences on prevailing wheat and lamb prices.

It might be thought that with higher rainfall, forest trees should grow faster in the temperate regions further south. Yet, of the annual average of 66 500 ha of pine or eucalyptus planted over the whole country each year throughout the 1980s, and nearly double that in each year from 1990–1995, more than half each year were situated in the 7th and 8th Regions (ODEPA 1996; CONAF unpubl. data). The explanation lies in the fact that south of Chillán (37° S) winter temperatures drop drastically and only cold-tolerant temperate zone tree species survive and perform well. In comparison, the mediterranean-climate region north of Chillán offers adequate rainfall for tree growth along with relatively lower risks of killing frosts. Moreover, in the secano interior of the 7th and 8th Regions, there are, for the time being, abundant croplands and rangelands for sale at modest prices that have already been cleared of tree cover. Thus, it is not surprising that relatively large-scale forestry plantations now occupy over 20% of land area in the MCR since 1974, the year that the Chilean government took decisive steps to facilitate private enterprise. The mild and equable coastal zone and coastal foothills are preferred for tree planting, but a growing proportion of plantations occur in the *secano interior* itself (Pizarro 1993; CONAF unpubl. data).

That the "green tide" is gaining speed is well illustrated in Table 9.1, which compares district and regional tree planting since 1974. Notably, the annual area of tree plantations each year since 1993 surpasses the totals for the preceding decade (ODEPA 1996).

9.3.2.2 Land Tenure

In 1979, 1.4% of farm properties in the *secano interior* exceeded 500 ha and these occupied a disproportionate 33.5% of land area (Sarah 1979). Since 1979, however, there has been a steady sale of portions of these large properties to Chilean and foreign forestry companies. Less visible than the dismemberment of large land holdings is the impact which forestry operations have had on small-scale farmers and their land management practices. A recent study in the 7th Region (Velasco 1993) indicates a large gap between the relative viability of smaller (21–150 ha) and medium-sized farms (151–>350 ha) in the *secano interior*. Since over 85% of *secano* farms are under 35 ha, and of these over 57% are under 10 ha (Sarah 1979), and since the cost of living is increasing rapidly, it is not surprising that *secano* farmers are tempted to sell off portions of their lands. In some cases, of

Table 9.1. Area of annual tree planting over a 20-year period in representative districts of the 7th Region lying in the *secano interior*, as compared to the entire 7th and 8th Regions, and all of central Chile (Pizarro 1993; CONAF, unpubl. data)

Region	Total area (ha)	Tree plantation area (ha × 1000)		
		1974–1982	1983–1992	Total[b]
7th Region				
Licantén	45	4.9	4.7	8.6
Vichuquen	49	5.0	9.5	14.5
Hualañe	54	4.0	6.7	10.7
Pencahue	91	3.2	6.2	9.4
Curepto	108	9.4	11.9	21.3
Empedrado	112	10.7	11.1	21.8
Cauquenes	225	24.0	18.8	42.8
Total	3480	106	151	257
8th Region (total)	>3800	324	289	613
Central Chile (4th–10th Regions)	24485	688	761	1449

[a] 85–90% *Pinus radiata*, 5.5–7.8% *Eucalyptus globulus* and *E. nitens*.

[b] Prior to 1974, there were already over 100000 ha of tree plantations in both the 7th and 8th Regions, but annual growth in forestry activity was much slower. Since 1992, area of average annual planting of trees in the 7th and 8th Regions has been 15000 and 48000 ha, respectively.

course, *secano* land owners have reallocated portions of their land to monospecific forestry as part of their own land use activities.

It should be recalled at this point that traditional land use and farming systems in central Chile leave much to be desired in the modern day context. Typically, the Spanish settlers who followed in the wake of Diego de Almagro (1536), and Pedro de Valdivia – founder of Santiago, in 1541 – failed to bring with them the agro-sylvo-pastoral systems of southern Spain and Portugal (cf. Bahre 1979; Fuentes 1988, 1990; Ovalle et al. 1990). As Dansereau (1973) and others have pointed out, wherever Europeans settled in the New World, Asia or Africa, they tended to leave their folklore and rural *savoir-faire* behind them. Consequently, in virtually all their colonies, European settlers tended to farm and steward resources less intelligently than had their ancestors back in Europe.

Partly as a result of poor management practices, and also as a result of ongoing landscape and ecosystem degradation, the traditional farming systems of the *secano interior* are increasingly in crisis. Consequently, monospecific tree plantations yield an annual per hectare cash return higher than most traditional farmers can achieve, and the relatively short 25-year rotation required for these trees gives the farmer the opportunity to benefit himself from at least one cash windfall.

9.3.2.3 Demographic Trends

Demographic trends over the past 30 years (INE 1991) reveal a nearly steady rural exodus of about 2% per annum in the 7th and 8th Regions. From 1960 to 1990, when the annual growth rate of Chile as a whole averaged 1.7%, most *secano* districts of the 7th and 8th Regions showed a negative growth rate (INE 1970, 1980, 1990). Over the past decades, only three districts in the 7th and 8th Regions showed positive population growth, and in all three cases this resulted from small industrial developments rather than agricultural advances. Thus, the *secano* contrasts sharply with many other parts of the country where rural economic growth based on irrigated agriculture, and aquaculture, has been quite spectacular, allowing large influxes of workers and their families.

9.3.3 The Third Wave

9.3.3.1 Geographical Aspects

It may legitimately be argued that the pine and eucalyptus plantations cited above as part of the Second Wave are better understood as part of the Third

Wave, especially since some of the lumber companies active in Chile are multinationals and most of the pulpwood produced is exported abroad. The "green tide" can perhaps best be understood as a bridge between the Second and Third Waves. This will depend in part on what local people make of the trend.

If the Second Wave in Chile, as elsewhere, induced a marked trend towards simplification and nationalization of production, and hence an economic and political scale change, from local to national, the Third Wave carries within it yet another scale change, this time towards globalization. From an ecological as well as humanitarian point of view, this can be a positive or a negative trend, depending on how it is managed. Along with other nations, Chileans (and Chilean companies) are increasingly reorienting themselves to participate in international networks of various kinds. They are also gaining awareness that the sustainability of the planet is very much in question as a result of past degradation and mismanagement. What remains to be seen is whether resource and landscape conservation, not to mention the patrimony of "information" stored in existing biodiversity, will receive national attention along with that old standby, economic growth.

9.3.3.2 Ecological Impact

As mentioned above, there are a number of matorral fragments surviving in the Chilean secano, despite the overwhelming preponderance of *espinales*. There are also a small number of *espinales* especially well-managed (see Ovalle et al. 1990). It is interesting to compare these two situations with the much more common ones of degraded or badly degraded espinales. In Table 9.2, we have done this for seven indicators or vital attributes characterizing relative ecosystem health. The point to determine is under which conditions the highest values obtained for the first six attributes can balance the low level of productivity measured in matorral stands. An evaluation of the use efficiency of soil water and fertility combined with the social usefulness of the outputs produced by each ecosystem will help to answer this basic question for managers seeking sustainable development for their farm or region.

9.4 Discussion

Under current management systems, the *espinales* appear to be ecologically and economically "blocked" (Ovalle et al. 1990). Relatively few gen-

Table 9.2. Vital attributes, in late spring, of a Chilean matorral and the successive *espinal* stages in the subhumid zone near Cauquenes (Aronson et al. 1993b, modified; reprinted with permission of Blackwell Scientific Publications)

Vital attribute	Stages of Degradation[a]			
	Matorral	Esp1	Esp2	Esp3
No. annual species	15	46	27	17
No. perennial species	30	16	8	4
Total plant cover (%)	170	95	70	10–30
Soil organic matter (%)	3–5	2.7	1.0	0.2
Available soil water reserves (mm)[b]	100–120	70	50	30
Length of water availability (months)	8	7	6.5	5.5
Aboveground phytomass (tons DM $ha^{-1} yr^{-1}$)[c]	0.1–0.5	3.5–4.5	2.0–3.0	0.7–1.0

[a] Somewhat disturbed matorral fragments near Cauquenes (7th Region) (average or range of data for 4 sites); Esp1, mixed *espinal* with *Acacia caven*, *Maitenus boaria* and other tree species; Esp2, degraded *espinal* (50–75% tree cover); Esp3, badly degraded *espinal* (10–25% tree cover).

[b] Upper 30 cm of soil profile.

[c] Herbaceous plants only.

erations of burning and clearing followed by a century of mining, plough agriculture combined with overgrazing, and charcoal-making appear to have led to one or more "human-mediated vegetation switches" (Wilson and King 1995) that are probably reversible only through direct, and massive, interventions. The alternative is to let the entire area "go" to the forestry companies and let them manage resources as they see fit. Landscape diversity would be written off as a thing of the past and soil fertility would decline at even greater depths than under rotating cereal culture.

An alternative path would be an integrated management plan combining mixed farming infused with new techniques, and new germplasm, "new forestry" and, insofar as possible, conservation of the best surviving examples of semi-natural or cultural landscapes. This alternative requires that local populations become aware of the interest of environment preservation either in relation to sustainable productivity, for improving their way of life, or both. It has a chance to develop only if market rules take into account product quality and origin, and if some lumber companies prove receptive to new approaches to tree farming. These constraints and opportunities should all be taken into account when tackling the ecological and economic rehabilitation of the *secano interior* (Ovalle et al. 1990).

For example, increasing productivity in some landscape units of a given farm or watershed should have beneficial effects on nearby areas set aside for conservation or designated for long-term rehabilitation efforts. At the same time, improving water and nutrient use efficiency and stopping surface erosion should receive high priority via the reorganization, and reintegration, of farming and tree cropping systems at the landscape level. In particular, a broader array of farm products and production strategies is needed to take advantage of innate landscape diversity in terms of hydrological and edaphic conditions particularly. Exotic woody species such as *Pinus radiata* or *Chamaecytisus proliferus* ssp. *palmensis* may have a key role to play in this new strategy because of their high growth potential and their ability to protect or even rehabilitate degraded soils. But their silviculture must be managed so as to avoid monoculture over wide patches and to promote the combination of timber and firewood production with other shorter-term crops such as forage, game, aromatic plants, fresh fruits and so on. In addition, if more irrigation water is supplied to *secano* farms (N.B. the Chilean government is now making available subsidies to private individuals to cover some 50% of the costs of smallscale, one-farm irrigation projects), cash crops can certainly be envisioned in the area. Such a development would permit small and medium-sized farms to remain viable, provided that land owners look for new investment and management options, new potential sources of revenue, and a more diversified land use and marketing strategy in a fiercely competitive international market. Otherwise, the region will undergo more landscape "banalization" with only two main land uses – pine forestry and irrigated crops.

Yet the fundamental problem is – as usual – social and political, rather than strictly ecological or technical. In other words, a socio-economic revival is required if we hope to see any long-term improvement in ecological terms. Additionally, a certain change in mentality is probably required, such that biodiversity, landscapes, and "the rural way-of-life" – or all of the above – are given new importance at local and national levels.

We have argued that *secano interior* landscapes have been radically and irreversibly transformed, not to mention biologically impoverished by the impact of First and Second Wave activities altogether unaccompanied by ecological prudence. If this region is to avoid becoming further desertified – in the social and biological senses – traditional farming systems must be reinforced and renewed with an array of crops, and know-how made available precisely as a result of the Third "information" Wave. What is particularly needed as well is a change in attitude about the Third Wave to include the biological aspects of information we have alluded to. Instead of passively submitting to a so-called "third industrial revolution", local people need to decide what kind of "information" is really important to them, and for their descendants.

Acknowledgements. We wish to express our warm thanks to Francesco di Castri, Richard Groves, Richard Hobbs, Edouard Le Floc'h and Chris Lusk for discussions and criticisms of this manuscript. We also thank the CNRS/CEPE drawing studio for preparation of the figures.

References

Aronson J, Floret C, Le Floc'h E, Ovalle C, Pontanier R (1993a) Restoration and rehabilitation of degraded ecosystems. I. A view from the South. Restoration Ecol 1:8–17

Aronson J, Floret C, Le Floc'h CE, Ovalle C, Pontanier R (1993b) Restoration and rehabilitation of degraded ecosystems. II. Case studies in Chile, Tunisia and Cameroon. Restoration Ecol 1:168–187

Aronson J, Ovalle C, Avendaño J (1993c) Ecological and economic rehabilitation of degraded "Espinales" in the subhumid mediterranean-climate region of central Chile. Landscape Urb Plann 24:15–21

Bahre C (1979) Destruction of the vegetation in northern central Chile. Univ Calif Publ Geog 23:1–117

Balduzzi A, Tomaselli R, Serey I, Villaseñor R (1982) Degradation of the mediterranean type of vegetation in central Chile. Ecol Méditerr 7:223–240

Bauer AJ (1970) Expansión económica en una sociedad tradicional: Chile central en el siglo XIX. Historia (Instituto de Historia, Universidad de Chile) 9:137–235

Dansereau P (1973) Inscape and landscape. CBC Learning Systems, Toronto

Dawkins R (1995) River out of Eden: a Darwinian view of life. Basicbooks, New York

di Castri F, Goodall DW, Specht RL (eds) (1981) Mediterranean-type shrublands. UNESCO, Elsevier, Amsterdam

Domic L (1979) Geodemografía. Perspectivas de desarrollo de los recursos de la VII Region. Intendencia de la Region del Maule/IREN-CORFU, Santiago

Forman RTT, Godron M (1986) Landscape ecology. John Wiley & Sons, New York

Fuentes E (1988) Landscape development in mountainous habitats with similar climates and cultural backgrounds: central Chile and Andalusia. Mt Res Develop 8:75–77

Fuentes ER (1990) Landscape change in Mediterranean-type habitats of Chile: Patterns and processes. In: Zonneveld IS Forman RTT (eds) Changing landscapes: an ecological perspective. Springer, Berlin Heidelberg New York, pp 165–190

Gay C (1865) Flora Chilena. Historia fisica y politica de Chile. Toma I, Botanica (reissued 1973 by ICIRA, Santiago)

INE, 1960, 1970, 1980, 1990, 1991. Instituto Nacional de Estadisticas. Compendios Estadisticos. Santiago, Chile

Likens G (1992) An ecosystem approach: its use and abuse. Excellence in ecology, book 3. Ecology Institute, Oldendorf/Luhr, Germany

ODEPA (1996) Estadiaticos temporada agricola 1996. Min Agricultura, Santiago

Ovalle C, Aronson J, Del Pozo A, Avendaño J (1990) The espinal: agroforestry systems of the mediterranean-type climate region of Chile. Agrofor Syst 10:213–239

Ovalle C, Avendaño J, Del Pozo A, Aronson J (1996) Land occupation patterns and vegetation structure of the anthropogenic savannas (espinales) of central Chile. For Ecol Manage 86:129–139

Pizarro R (1993) El sector forestal chileno y sus relaciones comerciales con la comunidad europea. Chile Forestal, Doc Técnico 69, CONAF, Santiago

Rifkin J (1995) The end of work. Tarcher/Putnam, New York

Sarah C (1979) Gravitación y perspectivas del minifundo. Perspectivas de desarrollo de los recursos de la VII Región. Intendencia de la Región del Maule/IREN-CORFU, Santiago
Toffler A (1980) The third wave. William Morrow, New York
Velasco R (1993) Factibilidad economico del mejoramiento de la agricultura en el secano interior. INIA, Chillán (Chile)
Wilson JB, King WM (1995) Human-mediated vegetation switches as processes in landscape ecology. Landscape Ecol 10:191–196

Part III

Landscape Disturbance and Plant Diversity

10 Local Endemism and Plant Conservation in the Cape Floristic Region

R.M. Cowling and D.J. McDonald

10.1 Introduction

Relative to their area, the five mediterranean-climate regions of the world harbour a disproportionate number of plant species, including exceptionally high numbers of rare and locally endemic species (Greuter 1994; Cowling et al. 1996a). Foremost amongst these is the Cape Floristic Region (CFR) where 8550 plant species, of which 68% are endemic, are crammed into 90 000 km^2 (Bond and Goldblatt 1984). Landscape degradation, especially in the Cape lowlands, has transformed vast areas of natural habitat, resulting in the highest concentration of threatened plant species and subspecific taxa (1435) in the world (Rebelo 1992a; Cowling and Hilton-Taylor 1994). Indeed, in Myers' (1990) parlance, the CFR is the "hottest" of the world's hot-spots of plant biodiversity and endemism.

Recent reviews of global patterns of extinction suggest that losses will be highest in endemic-rich areas, and that a knowledge of patterns of local endemism will be essential for understanding current and future extinctions (Nott et al. 1995; Pimm et al. 1995). Small range size is used as an attribute in almost all systems to characterise threatened taxa (Mace 1994), and local endemics with small populations are regarded as the group most vulnerable to extinction (Lawton 1993; Johnson et al. 1995). However, the geography and biology of range-restricted rarity are poorly studied across all groups of organisms (Gaston 1994; Cowling and Samways 1995). In particular, very few studies have systematically compared the ecological characteristics of locally endemic (or otherwise rare) species with widespread species, either within or across lineages (but see Kunin and Gaston 1993). Indeed, most research has been on common species and we have no basis for extrapolating this knowledge to rare species, since we know little of the differences between the two groups (Gaston 1994).

Over the past few years some progress has been made with the study of local endemism in the CFR (Cowling and Holmes 1992a; Cowling et al. 1992; McDonald and Cowling 1995; McDonald et al. 1995; Trinder-Smith et

Ecological Studies, Vol. 136
Rundel et al. (eds.) Landscape Degradation and Biodiversity in Mediterranean-Type Ecosystems

al. 1996a; Willis et al. 1996a). These studies have provided profiles of locally endemic species in terms of habitat requirements, taxonomic affinities, and biological traits, by comparing them with more widespread species. This has resulted in insights on the geography, biology and management of local endemics, as well as processes associated with their origin. Data on patterns of endemism have also been used in the design of efficient and representative reserve systems in the CFR (Trinder-Smith et al. 1996b; Willis et al. 1996b; Lombard et al. 1997).

In this chapter we review research on the geography, taxonomy and biology of local plant endemism in the CFR. In keeping with the theme of this volume, which stresses the pervasive and rapidly escalating landscape degradation in mediterranean-climate regions, our discussion focuses on the conservation implications of this research.

10.2 Study Sites and Approach

There are detailed data on the patterns and correlates of local endemism from four sites in the southwestern CFR, namely the Cape Peninsula, Agulhas Plain, Langeberg Mountains and the Southern Cape limestones (Fig. 10.1; Table 10.1). At all sites the dominant vegetation is fynbos, a fire-

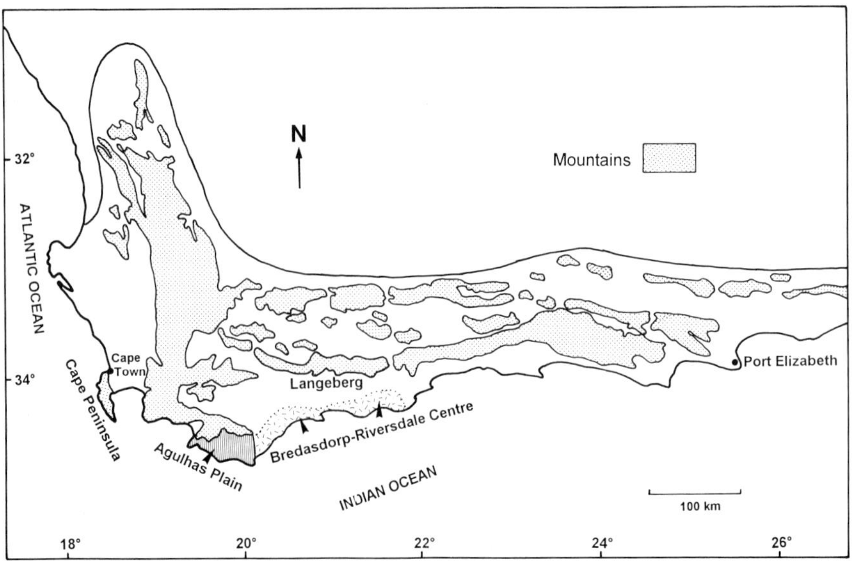

Fig. 10.1. Location of study sites. The boundary of the Cape Floristic Region is demarcated by the bold line. S. Cape limestone occurs as various-sized fragments within the Bredasdorp-Riversdale centre of endemism whose boundary is shown on the map

Table 10.1. Characteristics of study sites

Site	Area (km^2)	No. species (% endemic)	Heterogeneity			Vegetation	Reference
			Topographical	Climatic	Edaphic		
Cape Peninsula	471	2285 (3.9)	Very high	Very high	High	Fynbos Renosterveld	Cowling et al. (1996b) Trinder-Smith et al. (1996a)
Agulhas Plain	1609	1751 (5.7)	Moderate	Low	Very high	Fynbos Renosterveld	Cowling et al. (1988) Cowling and Holmes (1992a)
Langeberg Mts	1748	1228 (8.1)	High	High	Moderate	Fynbos Afromontane forest	McDonald and Cowling (1995)
S. Cape limestone	ca. 2000	ca. 1900[a] (5.8)	Moderate	Moderate	Low	Fynbos Subtropical thicket	Cowling et al. (1988) Rebelo et al. (1991) Willis et al. (1996a)

[a]Estimated from species-area regressions in Cowling et al. (1992).

prone shrubland associated with nutrient-poor soils and which includes most of the species (and endemics) in the CFR (Cowling et al. 1992). Renosterveld, a fire-prone shrubland on moderately fertile, shale-derived soils, and fire-free Afromontane forest and subtropical thicket, are the subdominant vegetation types at most of the sites. The Cape Peninsula is an area of very varied topography and climate, and has an extremely rich flora for its size. Owing to their location on the coastal forelands, the Agulhas Plain and Southern Cape limestone area have relatively low topographic and climatic heterogeneity. Edaphic heterogeneity on the Agulhas Plain is, however, very high. The Langeberg Mountain site is typical of the east-west trending Cape Fold Mountains of the southern CFR: a topographically varied landscape of moderate elevation and steep rainfall gradients. Details on the physiography, vegetation and flora of each site are given in the corresponding references shown in Table 10.1.

A similar research approach was adopted for each of the study sites. The fundamental question was as follows: when compared with more widespread species, are narrow endemics a random subset of the biota with regard to biotic and abiotic factors? The question was addressed by using chi-square analysis to test the null hypothesis that the frequency of local endemics within a particular category would not be significantly different from the frequency in the remaining flora. In each case the endemics (i.e. species whose extent of occurrence (Gaston 1994) was nested within a particular site) were a subset of the total flora of that site. The Southern Cape limestone was an exception: here the limestone-endemic flora was compared with the Agulhas Plain flora, after removal of limestone endemics (Willis et al. 1996a).

10.3 Geography of Endemism

10.3.1 Endemism and Area

Endemism and area are inextricably linked (Major 1988; Anderson 1994). Simply put, a taxon is regarded as endemic to an area if it occurs only in that area. Thus, the concept of endemism, as presently conceived, is a relative one, and the endemic status can have varying biological significance depending on the size and geographical area under consideration (Cowling and Samways 1995).

Range sizes for defining local endemism are often arbitrarily set. For example, Terborgh and Winter (1982), Gentry (1986) and ICBC (1992) define local endemics (birds and plants) to be those with geographical ranges (extent of occurrence) of $<50\,000\,km^2$. In the CFR, locally endemic plants are defined as those with extent of occurrence $<2000\,km^2$ (Cowling and Holmes 1992a). Indeed, a few of the endemics reported on in this chapter have natural range sizes $<5\,km^2$, thereby qualifying as point endemics. This categorisation of local endemism seems appropriate for the CFR, a small region with a high concentration of point endemics (Cowling and Samways 1995).

However, this simple categorisation (local endemic vs non-endemic) does have its problems. Range size is more accurately measured as a continuous variable defined as a taxon's area of occupancy (Gaston 1994). It seems reasonable to expect that species occupying very few $1 \times 1\,km$ quadrats within a region will differ in some non-trivial way from species which occupy several hundreds of such quadrats within the same region. In the scheme used here, species with vastly differing areas of occupancy are lumped together and endemism is analysed and modelled as a categorical variable. In defence of this folly, more accurate distribution data for the floras considered here would be extremely expensive to procure; and, as we shall show below, reasonably clear relationships emerge with this level of accuracy.

10.3.2 Habitat aspects

Local plant endemics are often associated with particular habitats, especially nutritionally unusual and/or regionally rare substrata (Raven and Axelrod 1978; Kruckeberg and Rabinowitz 1985; Major 1988). This is certainly true of the CFR where local edaphic endemism is particularly well developed on the coastal lowlands. For example, on the Agulhas Plain, only 15 of the 100 local endemics are associated with more than one substratum (Cowling and Holmes 1992a). A massive 37% of Agulhas Plain local endemics were associated exclusively with limestone (a chemically unusual substratum in an essentially acidic landscape), despite this rock type occupying only about 2.4% of the area.

Most Cape Peninsula endemics occur in the most widespread habitats, namely low altitude plateaux and mountain slopes (Trinder-Smith et al. 1996a). However, in terms of their aerial extent, endemics are significantly under-represented in these habitats. Endemics are over-represented in high altitude and wet sites (including wetlands) as well as in low altitude (and relatively dry) Pleistocene sand sites. The latter comprises a relatively unusual substratum on the Cape Peninsula. The patterns for the Langeberg Mts are similar to the Cape Peninsula: here, most endemics are associated with the most widespread habitat, namely mesic, pole-facing slopes

(McDonald and Cowling 1995). However, endemics are significantly over-represented in high-altitude wet habitats where almost twice the number occur than expected on the basis of aerial extent. Interestingly, relative to their area, endemics were significantly under-represented in the most xeric habitats. On both the Cape Peninsula and at the Langeberg, very few endemics occur in more than one habitat.

10.4 Taxonomy and Phylogeny of Endemism

Are local endemics in the CFR a heterogeneous group taxonomically or do certain endemics have a higher than expected probability of being an endemic? Evidence from our sites (Table 10.2) as well as others elsewhere in southern Africa (Cowling and Hilton-Taylor 1994) and the world (e.g. Major 1988; Cowling and Samways 1995 and references therein) indicate that local endemics are significantly over-represented in certain lineages (but see Auerbach and Shmida 1985). The existence of discernible phylogenetic correlates of endemism implies that range size may be an evolutionary stable character of a lineage (Ricklefs and Latham 1992; Lawton 1993). For example, Ericaceae are over-represented among endemics in both fynbos and summer rainfall temperate grassland habitats; and locally endemic Mesembryanthemaceae are over-represented in semi-arid succulent karoo as well as certain fynbos floras (Table 10.2; Cowling and Hilton-Taylor 1994).

Although some variation among sites is evident, there are consistent taxonomic correlates of local endemism in the CFR. Ericaceae, Proteaceae, Polygalaceae and Rutaceae are invariably associated with higher than expected levels of endemism, whereas Asteraceae, Poaceae and Cyperaceae are often under-represented among endemics (Table 10.2). Interestingly, families in both groups consistently appear among the ten largest families in regional floras from the CFR (see Cowling and Holmes 1992b).

10.5 Biology of Endemism

In this section we review the associations between local endemism and biological traits and attributes of species in order to determine whether endemics are a random assemblage with respect to population size, growth form, dispersal mode and means of fire survival.

Table 10.2. Relative frequency of local endemics in four regional floras from the Cape Floristic Region. Chi-square analysis tests the null hypothesis that the frequency of endemics within a family would not be different from the frequency within the total flora, excluding that family (independent test). Only larger families were used in order to avoid excessively low predicted cell frequencies. * = $P < 0.05$, ** = $P < 0.01$, *** = $P < 0.001$

Cape Peninsula[a]	Agulhas Plain[b]	Langeberg Mts.[c]	S. Cape limestone[d]
	Families significantly over-represented among endemics		
Ericaceae***	Ericaceae***	Ericaceae***	Rutaceae***
Proteaceae***	Proteaceae***	Restionaceae*	Ericaceae**
	Mesembryanthemac.***		Polygalaceae**
	Polygalaceae***		Fabaceae*
	Rutaceae***		Sterculiaceae*
	Families significantly under-represented among endemics		
Asteraceae**	Asteraceae**	Cyperaceae**	Restionaceae*
	Cyperaceae*	Asteraceae*	
	Poaceae*	Poaceae*	
	Families with proportional representation among endemics		
Brassicaceae	Campanulaceae	Iridaceae	Asteraceae
Bruniaceae	Fabaceae	Fabaceae	Campanulaceae
Cyperaceae	Iridaceae	Orchidaceae	Cyperaceae
Fabaceae	Liliaceae	Proteaceae	Iridaceae
Iridaceae	Orchidaceae		Liliaceae
Mesembryanthemac.	Restionaceae		Poaceae
Poaceae	Scrophulariaceae		Proteaceae
Restionaceae	Thymeleaceae		Rhamnaceae
Rosaceae			
Thymeleaceae			

[a] Data from Trinder-Smith et al. (1996a).
[b] Data from Cowling and Holmes (1992a).
[c] Data from McDonald and Cowling (1995).
[d] Data from Willis et al. (1996a).

10.5.1 Population Size

There is a broad but not universal positive correlation between range size and local population size (Willis 1922; Gaston and Lawton 1990; Lawton 1993). This aspect of local endemism in the CFR has not been studied in any detail. Many local endemics occur in very small populations (<100 individuals) (e.g. Boucher 1981; Moll and Gubb 1981; Rebelo 1992b), whereas others are extremely abundant within their limited ranges (S. Privett and R.M. Cowling, unpubl. data; see also Fiedler 1986; Rabinowitz et al. 1986; Gaston 1994). The fact that most Red Data Book plant species in the CFR are local endemics (Rebelo 1992a) provides some indirect support for a general relationship between range restriction and low population size in the region. However, many widely distributed species also have small local populations (S. Privett and R.M. Cowling, unpubl. data). Clearly, more research is required to clarify relationships between range size and population size for CFR plants.

10.5.2 Growth Form

There are few data on the relationships between plant growth form and endemism. In the mediterranean-climate regions of California and the Mediterranean Basin, herbs, and to a lesser extent shrubs (especially those that are killed by fire; see below), are over-represented among endemics (Raven and Axelrod 1978; Auerbach and Shmida 1985; Cowling et al. 1996a). Figure 10.2a, showing significant over-representation of low shrubs among Langeberg endemics, is typical of the other CFR sites (Cowling and Holmes 1992a; Trinder-Smith et al. 1996; Willis et al. 1996a). A broadly similar pattern exists for mediterranean southwestern Australia (Cowling et al. 1995).

10.5.3 Dispersal Mode

A generalisation for both plants and animals is that small range size is associated with short dispersal distances or low vagility (Kruckeberg and Rabinowitz 1985; Kunin and Gaston 1993; Cowling and Samways 1995). Patterns in the CFR are consistent with this generalisation: plants with ant-dispersed seeds, with dispersal distances of ca 10 m (Slingsby and Bond 1985), are consistently over-represented among endemics whereas species with wind- or vertebrate-dispersed propagules are under-represented (Cowling and Holmes 1992a; McDonald and Cowling 1995; Trinder-Smith et al. 1996a; Willis et al. 1996a; Fig. 10.2c).

10.5.4 Fire Survival

The relationship between mode of fire survival and range size has not been studied outside of mediterranean-climate regions. In a pioneering paper, Wells (1969) noted that non-sprouting species of the Californian chaparral shrub genera, *Arctostaphylos* and *Ceanothus*, had smaller range sizes and more specialised habitat requirements than post-fire resprouters. More

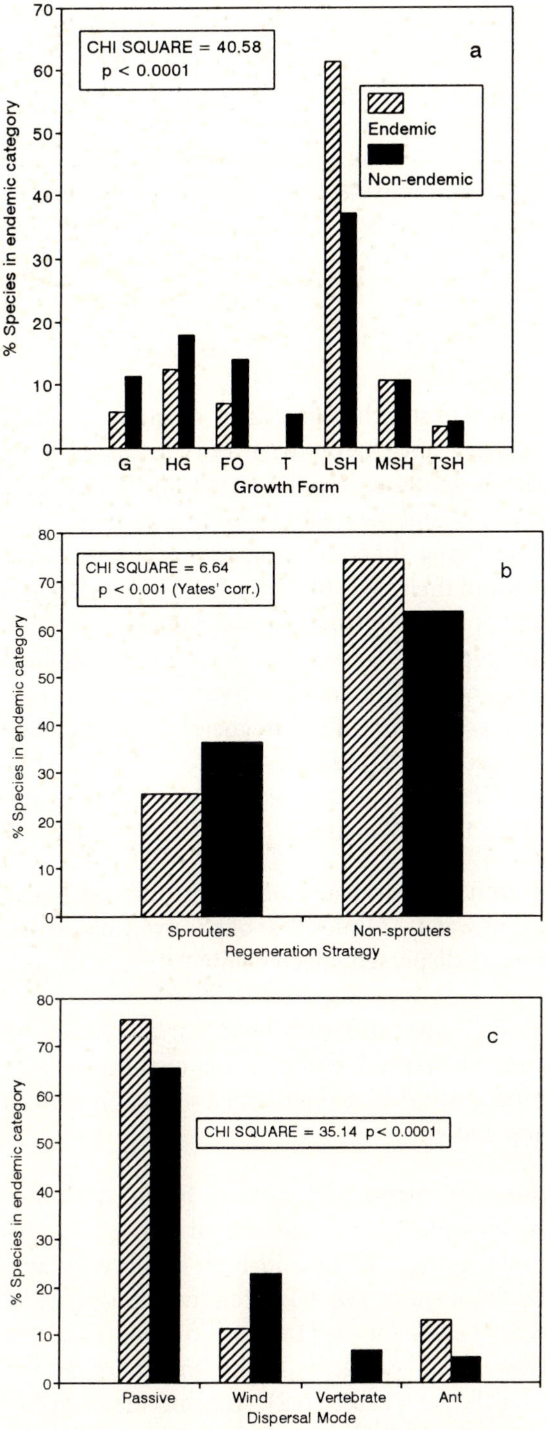

Fig. 10.2a–c. Percentage of endemic and non-endemic species in the Langeberg Mountain flora in: *a* seven growth form classes (*G* geophyte, *HG* graminoid, *FO* forb, *T* tree, *LSH* low shrub, *MSH* mid high shrub, *TSH* tall shrub); *b* two regeneration strategy classes; and, *c* four dispersal mode classes. Chi-square analyses were performed on untransformed data. (Reprinted with permission from McDonald and Cowling (1995)

recently, Cowling et al. (1992) have shown that similar patterns hold for many fynbos genera. In south-western Australia, resprouting *Banksia* species have greater range sizes and broader habitat requirements than non-sprouters (Lamont and Markey 1995).

In the CFR, both McDonald and Cowling (1995) (Langeberg flora) and Trinder-Smith et al. (1996a) (Cape Peninsula flora) have shown that non-sprouters are significantly over-represented among local endemics (Fig. 10.2b).

10.5.5 Trait Interactions

The correlates of local endemism (and other measures of rarity) are numerous and complex (Kruckeberg and Rabinowitz 1985; Gaston 1994), resulting in many co-linearities and interactions between traits (Kunin and Gaston 1993). Simple correlations and two-way contingency tables will seldom prove adequate for clarifying these relationships since the joint effect of a number of attributes and their possible interaction on endemism cannot be detected (McDonald et al. 1995; Cowling and Samways 1995).

McDonald et al. (1995) used logistic regression to model the biological aspects (growth form, dispersal mode, fire survival mode) of local endemism in the Langeberg flora (see Fig. 10.2 for categories and frequency data). Thus, they were able to assess the simultaneous effects and interactions between these traits and the occurrence of endemism. Some of the important inferences made from the model are:

1. For all growth forms and both fire survival modes, the odds on endemism in ant-dispersed species was 1.72 times that of passively dispersed species, and the odds for wind-dispersed species only 0.43 that of passively dispersed species;
2. Ant-dispersed, non-sprouting, low and mid-high shrubs have the largest probability (35 and 25%, respectively) of being endemic;
3. For all growth forms, sprouting, wind-dispersed species had almost zero (0–6%) probability of being endemic.

This study confirms an earlier (Cowling and Holmes 1992a) notion of the most commonly occurring biological profile associated with local endemism in the CFR, i.e. a non-sprouting, low to mid-high shrub with ant-dispersed seeds. Species with these traits also have relatively large seeds, low seed production, and weakly persistent seed banks (Cowling et al. 1994 and references therein). They are also likely to be members of the Proteaceae, Rutaceae and Polygalaceae (Table 10.2). Another major group of endemics are non-sprouting, low shrubs with passively dispersed seeds

(23% probability), of which most belong to the Ericaceae (McDonald et al. 1995). Both of these groups are predominantly insect-pollinated (Johnson 1992). The possibility exists that gene flow between populations of these species is limited as a result of restricted seed and pollen dispersal (Linder 1985), although evidence for the latter is lacking.

The odds are high that species with larger ranges will be sprouting and non-sprouting shrubs with wind-dispersed seeds (largely Asteraceae), and wind-dispersed graminoids (Poaceae), forbs and geophytes (McDonald et al. 1995). Many of these species produce large quantities of small seeds which form persistent seed banks. Another group of species with wide ranges in the CFR, and which was not included in McDonald model (McDoald et al. 1995), comprises very long-lived, sprouting, tall shrubs and trees with vertebrate-dispersed propagules (Cowling and Holmes 1992a; Cowling et al. 1992). These species are largely associated with fire-free forest and thicket vegetation and belong to tropical families including Celastraceae, Ebenaceae and Oleaceae. They are analogous to Keeley's (1986) obligate resprouters in Californian chaparral, and Herrera's (1992) "Cluster 1" (*Olea*, *Osyris*, *Pistacia* etc.) in Spanish mattoral. These lineages dominated pre-mediterranean-climate assemblages in the CFR and have persisted in suitable habitats to this day (Linder et al. 1992; see also Herrera 1992 for the Mediterranean Basin).

10.6 Speciation

What are the relationships, if any, between the biological correlates of local endemism discussed above, and the evolution of endemics in the CFR? For a wide range of lineages, life history characteristics that promote diversification may also promote extinction, since both processes are associated with reduced gene flow and short generation times (Vrba 1980; Endler 1989; Cowling and Holmes 1992a; Johnson et al. 1995). Thus, a discussion on the origination of local endemics in the CFR may have some significance for assessing contemporary and future rates of extinction (see below).

Most local endemics in the CFR are neoendemics i.e. they are low ranking members of clusters of closely related species and sub-species that have evolved relatively recently (Cowling and Holmes 1992a). This categorisation, based on perceived relationships, and geomorphological and palynological data, has been upheld by phylogenetic analysis (Linder and Vlok 1991; Linder 1995). Lineages with biological profiles typical of local endemics would have been unusually prone to diversification since:

1. Fire-induced plant mortality (non-sprouting) increases generation turnover, thus providing potential for more rapid evolution than sprouters (Wells 1969);
2. Small and weakly persistent seed banks result in non-overlapping generations, thereby increasing the probability of the manifestation of genetic novelties associated with each generation, as well as increasing the probability of population fragmentation via fire-induced local extinction (Lewis 1962; Cowling and Holmes 1992a);
3. Restricted gene flow promotes isolation and hence diversification of populations in new habitats (Linder 1985).

The result is rapid speciation due to catastrophic or disruptive selection acting on populations either at the margin of a species' distribution or in unusual (edaphic) habitats (Lewis 1962; Raven 1964). Over time, the massive differential diversification of these lineages, coupled with persistence in small populations (see below), resulted in the very fine-scale discrimination of habitats between closely related congeners, and the unusually high beta diversity and species-rich landscapes of the CFR (Cowling et al. 1992). Other lineages, such as many asteraceous genera with wind-dispersed seeds, underwent less diversification whereas long-lived and vertebrate-dispersed trees of tropical origin have probably persisted in their present form since the fire-free days of the late Tertiary (cf. Herrera 1992).

10.7 Conservation Implications

In this section we discuss the implications of the geographical and biological correlates of local plant endemism in the CFR for the conservation of the region's biodiversity. First, we discuss endemics in the context of present and future extinction rates, and then we evaluate the importance of endemics for reserve design.

10.7.1 Extinctions

It is widely accepted that gross habitat destruction is the major force driving the world's current extinction spasm (e.g. Soule 1987; Pimm et al. 1995). Reduced populations that remain in isolated fragments of natural habitat are also at risk of going extinct because of: demographic and environmental stochasticity, habitat deterioration, and loss of genetic diversity due to drift and inbreeding depression (Soule 1987; Boyce 1992; Ellstrand

and Elam 1993; Lande 1993). Species most vulnerable to extinction are those with low and variable population sizes, limited geographical range, low dispersal capabilities, and high habitat specificity (Lawton 1993; Mace 1994; Johnson et al. 1995). The majority of local endemics in the CFR have all or most of these characteristics.

There are an estimated 1435 Red Data Book plant species and subspecific taxa in the CFR (Cowling and Hilton-Taylor 1994), including 36 species that have gone extinct in the last 100 years, and some 618 species that are critically endangered (C. Hilton-Taylor, pers. comm.). Thus, the current and future extinction rates in the CFR are one extinction per 2.36×10^4 species years and 1.37×10^3 species-years, respectively (Nott et al. 1995). These rates are in the order of 10^2–10^4 above the estimated background extinction rate of one extinction per 10^6–10^7 species-years (Pimm et al. 1995). Most of the CFR Red Data Book species are local endemics and the major threats facing them are alien plant invasions and, to a lesser extent, agriculture (Rebelo 1992a). According to Richardson et al. (1996), unless steps are taken to contain and control alien plant invasions on the Cape Peninsula, almost 80% of the area's 161 Red Data Book species (including 82 of the area's 90 endemic species) will have their entire ranges impacted by alien plants within the next 100 years. The relatively high numbers of local endemics in mesic habitats within the CFR is of great concern since these habitats are most vulnerable to alien plant invasions (McDonald and Cowling 1995).

Clearly, wholesale habitat transformation represents a major threat to the CFR's locally endemic flora. But what of the more insidious impacts of habitat (and population) fragmentation, a pervasive problem on the more densely populated lowlands? What problems do these plants face when their populations are reduced to small isolates? Despite a plethora of theory, there is no consensus on minimum viable populations for almost all groups of organisms (Nunney and Campbell 1993); these are ambiguous results for low genetic diversity of range-restricted and rare plants (Cowling and Samways 1995) and very little evidence that inbreeding depression and reduced heterozygosity affects fitness (Caughley 1994). Many CFR local endemics apparently persist in natural populations of less than 10 individuals (several hundreds probably have populations of less than 100 individuals) (Rebelo 1992b), and highly localised limestone endemics on the Agulhas Plain persist on natural habitat fragments of <4 ha (Cowling and Bond 1991). It is highly possible that local endemics – species that are phylogenetically predisposed to low population densities and small geographic range sizes – will be easier to maintain in small and isolated reserves than formerly abundant and widespread species (McIntyre 1992; Rebelo 1992b; Lawton 1993).

10.7.2 Reserve Design

Despite the ongoing polemic regarding which taxonomic units are the most appropriate candidates for conservation (e.g. Vane-Wright et al. 1991; Williams and Humphries 1994; Linder 1995), endemics have, and continue to be used to identify areas for protection (Cowling and Samways 1995 and references therein). Usually, the maximum number of species will be represented in a reserve system that is based on habitat representation (Margules et al. 1988; Pressey and Logan 1994). However, in landscapes such as those of the CFR, where species have subdivided habitats at an extremely fine scale (beta diversity), and where turnover within habitats along geographical gradients is very high (gamma diversity) (Cowling et al. 1992), endemic plant localities can identify sites that would otherwise be overlooked (Willis et al. 1996b). Thus the approach used in the CFR has been to apply reserve selection algorithms that select sites to represent a specified area of habitats (vegetation types) as well as target point localities of rare and locally endemic plant species. This approach has been used to identify representative and efficient reserve systems for the Cape Peninsula (Trinder-Smith et al. 1996b) and the Agulhas Plain (Lombard et al. 1997). Both reserve systems are currently being implemented.

10.8 Conclusions

A number of conclusions salient to the conservation of plant biodiversity in the increasingly degraded landscapes of the CFR can be extracted from this review. Some of these are pertinent to other endemic-rich mediterranean-climate regions, especially southwestern Australia (Hopper 1992).

1. As most endemics are habitat specialists, emphasis should be on the identification and conservation of endemic-rich habitats. Distribution data for local endemics can be rapidly assembled from data in monographs and herbaria (Cowling and Holmes 1992a), phytosociological survey data (McDonald and Cowling 1995), as well as strategically directed fieldwork (Willis et al. 1996a).
2. Since local endemics have a reasonably predictable array of biological traits, it should be possible to generalise management protocols for population maintenance (McDonald and Cowling 1995).
3. Although detailed data on minimum viable populations and habitat size requirements of local endemics are lacking, preliminary accounts suggest that these species can persist in small populations in isolated

and tiny fragments of habitat for a very long time. Provided appropriate management protocols are applied (Bond et al. 1988), small fragments of natural habitat can effectively conserve endangered endemics (Rebelo 1992b).

Finally, an effective conservation strategy must include consideration of currently abundant and widespread species. This group includes many species in our "non-endemic" category. As landscapes continue to be degraded by habitat transformation, the populations of these species will become increasingly fragmented (Richardson et al. 1996). Under these circumstances, historically widespread species may be at greater risk of extinction than naturally rare and localised ones (McIntyre 1992; Lawton 1993). Furthermore, some of these wide-ranging habitat generalists may be ancestral to local endemics and have the evolutionary potential for speciating in the new environments created by environmental change (Linder 1995).

References

Anderson S (1994) Area and endemism. Quart Rev Biol 69:451–471

Auerbach M, Shmida A (1985) Harmony among endemic littoral plants and adjacent floras in Israel. J Biogeogr 12:175–187

Bond P, Goldblatt P (1984) Plants of the Cape Flora: a descriptive catalogue. J S Afr Bot Suppl 13:1–455

Bond WJ, Midgley JJ, Vlok J (1988) When is an island not an island? Insular effects and their causes in fynbos shrublands. Oecologia 77:512–521

Boucher C (1981) Autecological and population studies of *Orothamnus zeyheri* in the Cape of South Africa. In: Synge H (ed) Biological aspects of rare plant conservation. Wiley, New York, pp 343–353

Boyce MS (1992) Population viability analysis. Annu Rev Ecol Syst 23:481–506

Caughley G (1994) Directions in conservation biology. J Anim Ecol 63:215–244

Cowling RM, Bond WJ (1991) How small can reserves be? An empirical approach in Cape fynbos. Biol Conserv 55:243–256

Cowling RM, Hilton-Taylor C (1994) Patterns of plant diversity and endemism in southern Africa: an overview. In: Huntley BJ (ed) Botanical diversity in southern Africa. National Botanical Institute, Kirstenbosch, pp 31–52

Cowling RM, Holmes PM (1992a) Endemism and speciation in a lowland flora from the Cape Floristic Region. Biol J Linn Soc 47:367–383

Cowling RM, Holmes PM (1992b) Flora and vegetation. In: Cowling RM (ed) The ecology of fynbos. Nutrients, fire and diversity. Oxford University Press, Cape Town, pp 23–61

Cowling RM, Samways MJ (1995) Endemism and biodiversity. In: Heywood VH (ed) Global biodiversity assessment. Cambridge University Press, Cambridge, pp 174–191

Cowling RM, Campbell BM, Mustart PJ, McDonald DJ, Jarman ML, Moll EJ (1988) Vegetation classification in a floristically complex zone: the Agulhas Plain. S Afr J Bot 54:290–300

Cowling RM, Holmes PM, Rebelo AG (1992) Plant diversity and endemism. In: Cowling RM (ed) The ecology of fynbos. Nutrients, fire and diversity. Oxford University Press, Cape Town, pp 62–112

Cowling RM, Pierce SM, Stock WD, Cocks M (1994) Why are there so many myrmecochorous species in Cape fynbos? In: Arianoutsou M, Groves RM (eds) Plant-animal interactions in mediterranean-type ecosystems. Kluwer, Dordrecht, pp 156–168

Cowling RM, Witkowski ETF, Milewski AV, Newbey KR (1995) Taxonomic, edaphic and biological aspects of narrow plant endemism on matched sites in mediterranean South Africa and Australia. J Biogeogr 21:651–664

Cowling RM, Rundel PW, Lamont BB, Arroyo MK, Arianoutsou M (1996a) Plant diversity in mediterranean-climate regions. Trends Ecol Evol 11:362–366

Cowling RM, Macdonald IAW, Simmons MT (1996b) The Cape Peninsula, South Africa: physiographical, biological and historical background to an extraordinary hot-spot of biodiversity. Biodivers Conserv 5:551–573

Ellstrand NC, Elam DR (1993) Population genetic consequences of small population size: implications for plant conservation. Annu Rev Ecol Syst 34:217–242

Endler JA (1989) Conceptual and other problems in speciation. In: Otte D, Endler JA (eds) Speciation and its consequences. Sinauer, Sunderland, pp 625–648

Fiedler PL (1986) Concepts of rarity in vascular plants with special reference to the genus *Calochortus* Pursh (Liliaceae). Taxon 35:502–518

Gaston KJ (1994) Rarity. Chapman & Hall, London

Gaston KJ, Lawton JH (1990) Effects of scale and habitat on the relationship between regional distribution and local abundance. Oikos 58:329–335

Gentry AH (1986) Endemism in tropical versus temperate plant communities. In: Soule M (ed) Conservation biology. The science of scarcity and diversity. Sinauer, Sunderland, pp 153–181

Greuter W (1994) Extinction in Mediterranean areas. Philos Trans R Soc London Ser B 344:41–46

Herrera CM (1992) Historical effects and sorting processes as explanations for contemporary ecological patterns: character syndromes in Mediterranean woody plants. Am Nat 140:421–446

Hopper SD (1992) Patterns of plant diversity at the population and species level in south-west Australian mediterranean ecosystems. In: Hobbs RJ (ed) Biodiversity of mediterranean ecosystems in Australia., Surrey Beatty, Chipping Norton, New South Wales, pp 27–46

International Council for Bird Preservation (ICBC) (1992) Putting biodiversity on the map: priority areas for global conservation. ICBC, Cambridge

Johnson SD (1992) Plant-animal relationships. In: Cowling RM (ed) The ecology of fynbos. Nutrients, fire and diversity. Oxford University Press, Cape Town, pp 175–205

Johnson KG, Budd AF, Stemann TA (1995) Extinction selectivity and ecology of neogene Caribbean reef corals. Paleobiology 21:52–73

Keeley JE (1985) Resilience of mediterranean shrub communities to fire. In: Dell B, Hopkins AJM, Lamont BB (eds) Resilience in mediterranean-type ecosystems. Junk, The Hague, pp 95–112

Kruckeberg AR, Rabinowitz D (1985) Biological aspects of endemism in higher plants. Ann Rev Ecol Syst 16:447–479

Kunin WE, Gaston KJ (1993) The biology of rarity: patterns, causes and consequences. Trends Ecol Evol 8:298–301

Lamont BB, Markey A (1995) Biogeography of fire-killed and resprouting *Banksia* species in south-western Australia. Aust J Bot 43:283–303

Lande R (1993) Risks of population extinction from demographic and environmental stochasticity and random catastrophes. Am Nat 142:911–927

Lawton JH (1993) Range, population abundance and conservation. Trends Ecol Evol 8:409–413

Lewis H (1962) Catastrophic selection as a factor in speciation. Evolution 16:257–271

Linder HP (1985) Gene flow, speciation and species diversity patterns in a species-rich area: the Cape Flora. In: Vrba ES (ed) Species and speciation. Transvaal Museum, Pretoria, pp 53–57

Linder HP (1995) Setting conservation priorities: the importance of endemism and phylogeny in the southern African orchid genus *Herschelia*. Conserv Biol 9:585–595

Linder HP, Vlok J (1991) The morphology, taxonomy and evolution of *Rhodocoma* (Restionaceae). Plant Syst Evol 175:139–160

Linder HP, Meadows ME, Cowling RM (1992) History of the Cape flora. In: Cowling RM (ed) The ecology of fynbos. Nutrients, fire and diversity. Oxford University Press, Cape Town, pp 113–134

Lombard AT, Cowling RM, Pressey RL, Mustart PJ (1977) Reserve design on the Agulhas Plain, South Africa: a flexible tool for conservation in a species-rich and fragmented landscape. Conserv Biol 11:1101–1116

Mace GM (1994) Classifying threatened species: means and ends. Philos Trans R Soc London Ser B 344:91–97

Major J (1988) Endemism: a botanical perspective. In: Myers AA, Giller PS (eds) Analytical biogeography. An integrated approach to the study of animal and plant distributions. Chapman and Hall, New York, pp 117–146

Margules CR, Nicholls AO, Pressey RL (1988) Selecting networks of reserves to maximise biological diversity. Biol Conserv 43:63–76

McDonald DJ, Cowling RM (1995) Towards a profile of an endemic mountain fynbos flora: implications for conservation. Biol Conserv 72:1–12

McDonald DJ, Juritz JM, Cowling RM, Knottenbelt WJ (1995) Modelling the biological aspects of local endemism in South African fynbos. Plant Syst Evol 195:137–147

McIntyre S (1992) Risks associated with the setting of conservation priorities from rare plant species lists. Biol Conserv 60:31–37

Moll EJ, Gubb AA (1981) Aspects of the ecology of *Staavia dodii* in the south western Cape of South Africa. In: Synge H (ed) The biological aspects of rare plant conservation. Wiley, New York, pp 331–342

Myers N (1990) The biodiversity challenge: expanded hot-spots analysis. The Environmentalist 10:243–256.

Nott MP, Rogers E, Pimm S (1995) Modern extinctions in the kilo-death range. Curr Biol 5(1):14–17

Nunney L, Campbell KA (1993) Assessing minimum viable population size: demography meets population genetics. Trends Ecol Evol 8:234–239

Pimm SL, Russell GJ, Gittleman JL, Brooks TM (1995) The future of biodiversity. Science 269:347–350

Pressey RL, Logan VS (1994) Level of geographic subdivision and its effects on assessment of reserve coverage: a review of regional studies. Conserv Biol 8:1037–1046

Rabinowitz D, Cairns S, Dillon T (1986) Seven forms of rarity and their frequency in the flora of the British Isles. In: Soule M (ed) Conservation biology: the science of scarcity and diversity. Sinauer, Sunderland, pp 182–204

Raven PH (1964) Catastrophic selection and edaphic endemism. Evolution 18:336–338

Raven PH, Axelrod AI (1978) Origin and relationships of the California flora Univ Calif Publ Bot 72:1–134

Rebelo AG (1992a) Red Data Book species for the Cape Floristic Region: threats, priorities, and target species. Trans R Soc S Afr 48:55–86

Rebelo AG (1992b) Preservation of biotic diversity. In: Cowling RM (ed) The ecology of fynbos. Nutrients, fire and diversity. Oxford University Press, Cape Town, pp 309–344

Rebelo AG, Cowling RM, Campbell BM, Meadows ME (1991) Plant communities of the Riversdale Plain. S Afr J Bot 57:10–28
Richardson DM, van Wilgen BW, Higgins SI, Trinder-Smith TH, Cowling RM, McKelly D (1996) Current and future threats to plant biodiversity on the Cape Peninsula, South Africa. Biodivers Conserv 5:607–647
Ricklefs RE, Latham RE (1992) Intercontinental correlation of geographic ranges suggests stasis in ecological traits of relict genera of temperate perennial herbs. Am Nat 139:1305–1321
Slingsby P, Bond WJ (1985) The influence of ants on the dispersal distance and seedling recruitment of *Leucospermum conocarpodendron* (L.) Buek (Proteaceae). S Afr J Bot 51:30–35
Soule ME (ed) (1987) Viable populations for conservation. Cambridge University Press, Cambridge
Terborgh J, Winter B (1982) Evolutionary circumstances of species with small ranges. In: Prance GT (ed) Biological diversification in the topics. Columbia University Press, New York, pp 587–600
Trinder-Smith TH, Cowling RM, Linder HP (1996a) Profiling a besieged flora: endemic and threatened plants of the Cape Peninsula, South Africa. Biodivers Conserv 15:575–589
Trinder-Smith TH, Lombard AT, Picker M (1996b) Reserve scenarios for the Cape Peninsula: high, middle and low road options for conserving the remaining biodiversity. Biodivers Conserv 5:575–589
Vane-Wright RI, Humphries CJ, Williams PH (1991) What to protect? Systematics and the agony of choice. Biol Conserv 55:235–254
Vrba ES (1980) Evolution, species and fossils: how does life evolve? S Afr J Sci 76:61–84
Wells PV (1969) The relation between mode of reproduction and extent of speciation in woody genera of the California chaparral. Evolution 23:264–267
Williams PH, Humphries CJ (1994) Biodiversity, taxonomic relatedness, and endemism in conservation. In: Forey PI, Humphreys CJ, Vane-Wright RI (eds) Systematics and conservation evaluation. Clarendon, Oxford, pp 269–287
Willis CK, Cowling RM, Lombard AT (1996a) Patterns of endemism in the limestone flora of South African lowland fynbos. Biodiv Conserv 5:55–73
Willis CK, Lombard AT, Cowling RM, Heydenrych BJ, Burgers CJ (1996b) Reserve systems for limestone endemic flora of the Cape lowlands:iterative versus linear programming techniques. Biol Conserv 77:53–62
Willis JC (1922) Age and area: A study of geographical distribution and origin of species. Cambridge University Press, London

11 Managing Biodiversity on the Cape Peninsula, South Africa: A Hotspot Under Pressure

D.M. Richardson, C. Gelderblom, B.W. van Wilgen and T.H. Trinder-Smith

11.1 Introduction

The Cape Floristic Region (CFR) which covers about 90 000 km^2 at the southwestern tip of Africa is especially well known for its exceptional species richness and high levels of endemism in vascular plant taxa. The region has been flagged as one of the world's foremost "hotspots" of plant diversity and endemism (Myers 1990). Although levels of diversity in most other taxonomic groups are generally less impressive than for plants, the CFR also boasts high levels of endemism for some invertebrate groups, amphibia and fish (references in Macdonald and Richardson 1986; Rebelo 1992).

The CFR, which occupies less than 4% of the area of southern Africa (sensu Cowling et al. 1997), is home to an amazing 40% of the subcontinent's flora of over 22 000 species. About 70% of the roughly 8600 plant species in the CFR are endemic, and many of these taxa reside in the fynbos biome which covers 80% of the CFR and is home to about 7300 plant species, 80% of them endemic. Three major vegetation types (fynbos, renosterveld, and subtropical thicket), together with small islands of Afromontane forest in some areas, form the natural vegetation matrix of the topographically complex fynbos biome. As has been the case in other mediterranean-climate regions, there has been extensive transformation of a large part of the region as a result of pastoral, agricultural and urban development. This has been particularly severe in the low-lying areas. In addition, the spread of alien trees and shrubs has been a particularly severe problem in the Cape (Richardson et al. 1992).

Partly because of these factors, a large proportion of the CFR's biota survives in a precarious state. Similarly, a large proportion of southern Africa's *Red Data Book* (RDB) taxa (70% of plants, 57% of freshwater fish, 43% of amphibians, 38% of butterflies and 35% of reptiles) occur within the CFR (Rebelo 1992). Inherent features of the endemic plants [e.g. very small population sizes (Cowling and McDonald, Chapter 9 in this volume;

Ecological Studies, Vol. 136
Rundel et al. (eds.) Landscape Degradation and Biodiversity in Mediterranean-Type Ecosystems

Richardson et al. 1996a)] and probably also of endemics in other groups also makes them highly vulnerable to extinction. In addition, as a result of the high beta and gamma diversity of the flora particularly, it is impossible to conserve all diversity in one contiguous area. Management for the conservation of this biodiversity is thus extremely complex (van Wilgen et al. 1992). A major problem is that data on the distribution of biodiversity components and the factors that threaten these resources are not available at a scale suitable for use in decision making.

By far the greatest concentration of *Red Data Book* species of plants, butterflies, reptiles and amphibians in the CFR occur in the southern part of the region, and in particular on the Cape Peninsula and environs in the southwestern corner of the CFR. As Rebelo (1992) has shown, with 15.1 RDB plant species per km^2, the Cape Peninsula ranks as one of the highest conservation priorities in the world. Furthermore, with 2285 plant species in an area of only 470 km^2, the Peninsula's regional species density is unparalleled anywhere else in the CFR or the world (Cowling et al. 1992). With 90 plant species endemic to the region, it also has the highest level of local endemism recorded thus far in the CFR (Trinder-Smith et al. 1996a). Cowling (1995) has termed this area "the jewel in the crown" of the CFR. These statistics, together with the growing threats to the region's biodiversity, clearly make the Cape Peninsula an important area for conservation action. Because of the large human population in the area, and the long history of human influences, the threats to the Peninsula's biodiversity are particularly severe and varied. It is both necessary and particularly informative to focus on this area when searching for new insights for managing biodiversity in the CFR. Recent developments, including moves to establish a National Park (discussed later), and the aim of applying for World Heritage Site status for the area, has focused the attention of researchers and planners on documenting the area's biodiversity and formulating objective methods for its effective management.

Fortunately, the biota of the Peninsula have received sustained attention from researchers for several decades. Consequently, the determinants of diversity for most taxa are probably better understood for this area than for any other part of southern Africa. This Chapter reviews recent research conducted to assess the status of biodiversity in the Cape Peninsula, to quantify the magnitude of current threats to this resource, and to formulate likely scenarios for future threats and measures that should be taken to alleviate such threats. These insights allow us to formulate guidelines useful for practical management. We use these insights to define environmentally sensitive areas on the Peninsula, and discuss the implications for zoning to ensure adequate management.

11.2 The Cape Peninsula – Physical Features, Biodiversity and Current Conservation Status

The Cape Peninsula, 470 km^2 in extent, is located at the southwestern tip of Africa. In a recent analysis of features influencing regional plant diversity in southern Africa, Cowling et al. (1997) found the Peninsula to have the highest topographical diversity (computed as the coefficient of variation of altitude values gridded at intervals of 1 minute of a degree latitude and longitude) of similarly-sized areas in the subcontinent. This, they argued, contributed profoundly to its greater plant species diversity when compared with other regions such as the Agulhas Plain, which has similar (or greater) edaphic diversity, but much lower topographic diversity. The impressive topography, which has undoubtedly contributed to the richness of endemics in other groups as well (Picker and Samways 1996), is dominated by the Peninsula Mountain Chain. This range includes the world-famous landmark of Table Mountain (maximum altitude 1085 m) which is now virtually encircled by urban developments, giving the area its distinctive "mountain in a city" attraction. Further details on geology, topography, soils and climate are given by Cowling et al. (1996).

Features of the biodiversity of the Peninsula are described in detail by Cowling et al. (1996); Picker and Samways (1996); Simmons and Cowling (1996) and Trinder-Smith et al. (1996a). Salient features for major taxonomic groups are given in Table 11.1. A striking feature of the Peninsula's flora is the large number of range-restricted, uncommon species. Such rarity seems to be attributable as much to specialization to the wide range of habitats provided by the rough topography as to the complex mosaic of distribution and abundance patterns arising from the effects of stochastic fire regimes on regeneration and persistence. Endemic and threatened

Table 11.1. Components of biodiversity on the Cape Peninsula, South Africa (471 km^2 in extent). Data for plants from Trinder-Smith et al. (1996a); data for other groups from Picker and Samways (1996)

Taxon	Number of species	Number of endemic species	Number of threatened species
Vascular plants	2285	90	141
Mammals	41	0	0
Birds	155	0	8
Reptiles	48	0	2
Freshwater fish	6	0	?
Amphibians	18	0	4
Invertebrates	?	112	?

plant species, which are concentrated in a few families (notably the Ericaceae), and which are mostly low-growing shrubs with ant-dispersed seeds are not evenly distributed across major vegetation types (Trinder-Smith et al. 1996a). Clearly identifiable hotspots of endemism are apparent in the landscape, most of which are associated with areas of high topographic relief; these are depicted in Fig. 11.1.

Endemic animal groups are also not randomly distributed, but are clustered in several nodes, most of which are also at high altitudes. This concurrence of hotspots between flora and fauna is probably a result of a common response to historical isolation and topographic variety (Picker and Samways 1996).

The dominant vegetation formations on the Peninsula are fynbos, renosterveld and subtropical thicket, with closed forests covering less than 3%. Cowling et al. (1996) described 15 major vegetation types (12 fynbos and 3 non-fynbos categories) on the basis of floristics, dominant taxa and structure. Although defined solely on the basis of dominant plant growth forms, Richardson et al. (1996b) argued that this delineation provides a reasonable basis for classifying habitats for other taxa, and therefore serves as a crude index for assessing habitat diversity.

An exceptional database of biological records and physical features exists on a Geographic Information System (GIS). Included in this database are: mapped natural plant communities; localities for all 141 plant taxa that are classified as threatened according to IUCN (World Conservation Union) criteria; precise locality data for all 45 native Proteaceae taxa, localities of many of the endemic animals, detailed maps of the extent and density of all alien woody plants (mapped in 1994); boundaries of plant and animal hotspots; boundaries of all conservation areas and of areas under the jurisdiction of all major landowners; details of the occurrence of fires over the last 20–30 years; and an impressive set of topographic and climatic coverages (Richardson et al. 1996b; van Wilgen 1996).

Cowling et al. (1996) review some milestones in the history of human settlement and impacts on biodiversity in the area; see also Wood et al. (1994) for discussion relating to the whole Cape Metropolitan Area (of which the Cape Peninsula forms about 10%). To generalize, the most significant impacts have been: the elimination of most large mammal species soon after the arrival of European settlers in 1652; the large-scale transformation of lowland vegetation, mainly by urbanization (the extent of which is depicted in Fig. 11.1); the transformation of mountain vegetation by the spread of alien trees and shrubs (Fig. 11.2); and inappropriate fire regimes (see Richardson et al. 1996b for discussion of other impacts). At least 13 plant taxa have become extinct on the Cape Peninsula in the 20th century (Trinder-Smith et al. 1996a), and degradation of water quality

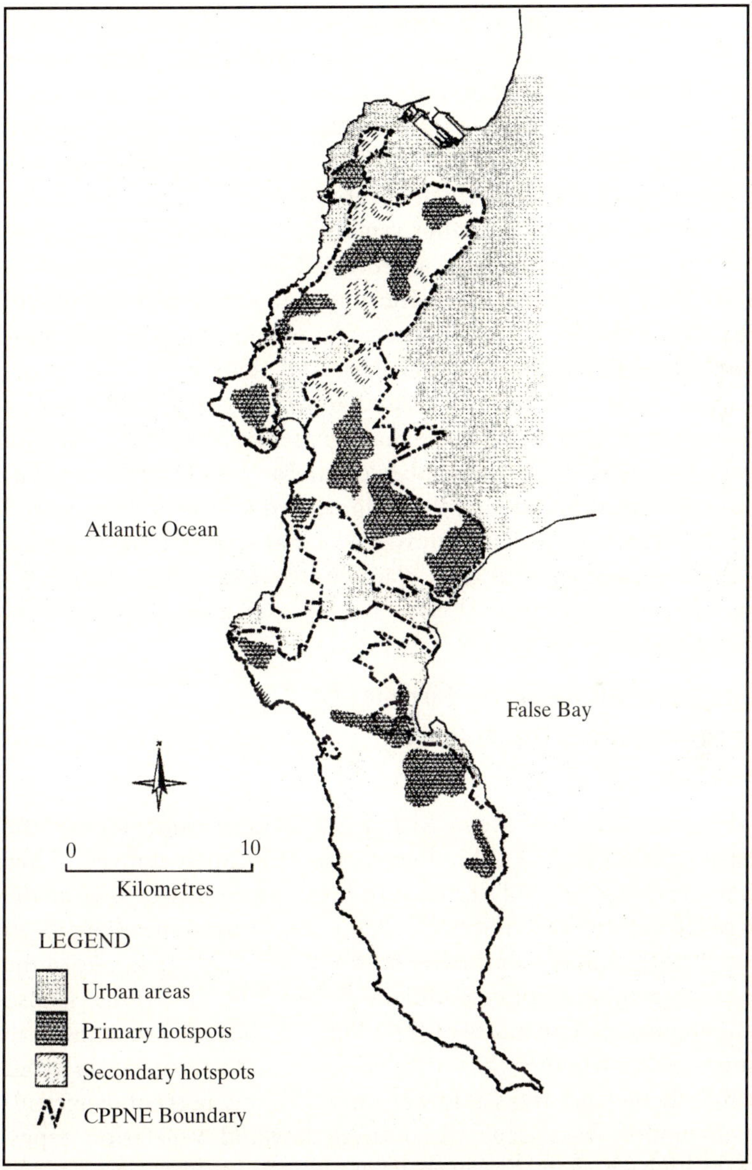

Fig. 11.1. Distribution of hotspots of plant endemism and the extent of urbanization on the Cape Peninsula (see text for details on the derivation of primary and secondary hotspots)

has already lead to local extinctions of endemic fauna (Picker and Samways 1996).

There are four isolated nature reserves which together preserve 29% of the area of the Peninsula (Cape of Good Hope Nature Reserve 8104 ha; Table Mountain Nature Reserve 3058 ha; Silvermine Nature Reserve

2166 ha; Rondevlei Bird Sanctuary 208 ha). Through accident, rather than careful planning, these reserves conserve a much greater proportion of the Peninsula's biodiversity than might be predicted from their relatively small area – this is because the reserves span steep and long environmental gradients (Trinder-Smith et al. 1996b). The Cape Peninsula Protected Natural Environment (CPPNE), which covers 291 km^2 and includes all the nature reserves and most of the other remaining natural areas, was proclaimed a Nature Area in 1983 and a Protected Natural Environment in 1989. It is for the CPPNE that World Heritage Site listing is being sought (CSIR 1995). This area also forms the basis for medium-term planning for the proposed Cape Peninsula National Park and the formulation of the scenarios and zoning proposals discussed in this chapter. Until now a major problems thwarting the effective management of the natural areas of the Peninsula has been the large number of landowners (14 different authorities control significant areas); this fragmented efforts and hampered attempts to initiate a standardized approach to management, notably with respect to fire management and alien plant control (van Wilgen 1996).

11.3 Assessing the Current Threats to the Cape Peninsula's Biodiversity

Using the database described above, Richardson et al. (1996b) assessed the status of various components of biodiversity on the Cape Peninsula. They showed that urbanization and agriculture have transformed 37% of the original area of natural vegetation. Lowland vegetation types have been worst affected, with almost half the transformation occurring in one of the 15 recognized vegetation types, sandplain proteoid fynbos, which is also severely threatened in other parts of the biome (Rebelo 1995). Almost all the remaining transformation has affected another five vegetation types that occur mainly on level areas at low altitudes. Ninety percent of agricultural transformation has occurred in three lowland vegetation types. Vegetation at high altitudes has been little affected by urbanization and agriculture (90% of its original area remains more or less intact), but alien trees and shrubs are now threatening biodiversity in these areas. Of the area not affected by urbanization and agriculture, 10.7% is currently under dense stands (>25% canopy cover) of alien plants and another 32.9% is lightly invaded. Dense stands of *Acacia cyclops*, the most widespread invader, cover 76% of the total area under dense alien stands.

All records for 6.6% of the 161 "special" plant taxa (endemic and/or threatened) occurred in 1-km^2 squares that are already affected by urbanization (see Richardson et al. 1996b for details), and 5.3% of Proteaceae taxa were similarly affected. These taxa are seriously threatened. For 42.5% of special taxa and 34.2% of Proteaceae, no mapped localities are currently affected by urbanization. For 86% of the special taxa more than half the localities are in 1-km^2 squares that are not currently affected by urbanization. All but two of the Proteaceae taxa also have half or more of their original range unaffected by urbanization. The 3313 ha of dense alien stands in 1994 affect almost a third of the 1170 known localities of special taxa, and 8.4% of these taxa currently occur only in areas already affected by aliens. Dense alien stands affect 30 of the 38 Proteaceae taxa in part of their range; for two species the entire known range is already affected. However, more than half the (recent) records for 35 of the 38 Proteaceae taxa are in areas that are not currently affected by dense alien stands. Table 11.2 provides a summary of the major threats to plants, vertebrates and invertebrates.

11.4 Considering Future Threats to the Cape Peninsula's Biodiversity

Richardson et al. (1996b) assessed the possible future impacts on biodiversity by considering the effects of several scenarios (taking biological and socio-political factors into account) involving increased urbanization and changes to alien plant control strategies and further spread over the next 50–100 years (Table 11.3). Increased urbanization is predicted to cause a steady, more-or-less linear, loss of biodiversity. The scenario of full urbanization outside the CPPNE (U1 in Table 11.3) results in an 11.4% reduction in the extent of natural vegetation, with sandplain proteoid fynbos (21.5%) and wetlands (36.5%) being worst affected. Almost a third of the localities of special taxa (2.9% for Proteaceae) are affected; for 7.8% of special taxa and 5.3% of Proteaceae all known localities are urbanized. The percentage of special taxa with more than half their known localities affected increases from 13.8% to 17.4%; for the Proteaceae the number of taxa thus affected remains unchanged. Scenario U2 (complete urbanization on suitable areas outside public land) leads to further deterioration, reducing the extent of natural vegetation to 24681ha, or 79.7% of the area remaining in 1994. Under this scenario, sandplain proteoid fynbos occupies only 842 ha (8.0% of its extent before human settlement). Just less than half the special taxa lose 50% or more of their known localities to urbanization, with 8.4% of them now occurring *only* in squares affected by urbanization. The additional habitat loss caused by scenario U2 affects 78

Table 11.2. Examples of the threats experienced by various taxa on the Cape Peninsula (the list is by no means exhaustive)

Taxon	Threat				
	Invasive alien plants	Fire	Habitat Degradation	Habitat Fragmentation	Habitat Loss
Plants	Alien trees and shrubs have invaded large areas, threatening many taxa with extinction	Where dense stands of alien trees occur, fires are very intense; this is detrimental to plant regeneration and to soil stability	Unknown due to lack of studies in this field (air pollution probably important)	The main effect is to make the maintenance of appropriate fire regimes difficult. The disruption of gene flow and other important ecosystems processes may also be important (but are poorly studied)	Large areas of some habitats have been transformed, e.g. urbanization threatens the only known localities of some endemic taxa
Vertebrates	Dense stands of alien trees affect the availability of suitable habitat for some vertebrates, particularly those endemic to the fynbos	Too-frequent fires are thought to have played a role in the decline of several reptiles such as the Cape Dwarf Chameleon (*Bradypodion pumilum*)	The fish and amphibians, many of which are near-endemics of the Peninsula, are particularly sensitive to the decline in water quality which is widespread in fresh water bodies	Loss of connectivity has serious implications for mobile species such as baboons as it disrupts gene flow and the ability of the animals to move between diverse habitats to track resource availability	Habitat loss has reduced the capacity of the area to support species with large range requirements, but probably the most important impact on vertebrates is disturbance by humans, including the active eradication of many of the large mammals
Invertebrates	Invasion of woody alien plants can be predicted to have a serious impact on the numerous endemic invertebrates which are specialized for survival on the fynbos vegetation	Too-frequent fires threaten endemic taxa such as the butterfly *Thestor yildizae*	Habitat degradation is also a particularly important problem for the freshwater invertebrates. Cavedwelling species which are limited to small areas are particularly vulnerable to the disturbance which is already occurring	The primary effect of habitat fragmentation is the loss of connectivity which may impede the usual rapid colonization of burnt areas.	Habitat loss is likely to be a problem for many species, particularly the decline in ravine forest which contains many endemics. Many of the endemics, like the plants, are highly localized and can be threatened by individual developments.

Table 11.3. Seven scenarios of possible changes in the extent of urbanization and invasion by alien trees and shrubs on the Cape Peninsula. The CPPNE is the Cape Peninsula Protected Natural Environment (see text). (Modified from Richardson et al. 1996b)

Scenario		Assumptions
Urbanization		
U1	No further urbanization within the boundaries of the CPPNE, but all remaining areas suitable for urbanization are developed	CPPNE legislation maintained in the face of increasing demand for land; large-scale development elsewhere
U2	All areas suitable for urbanization outside public land are developed	Relaxation of regulations restricting development in the CPPNE; only public land remains for conservation
U3	Development of all suitable areas, irrespective of land ownership	Total breakdown of legislation governing development on protected sites, or the modification of legislation to permit development over a larger area
Alien plants		
A1	All dense stands of alien plants are cleared	Major investment of funds towards integrated control; effective liaison between landowners, conservation authorities and volunteer groups to compile and implement effective strategies; biological control is effective and prevents re-establishment of dense stands after mechanical control
A2	All dense alien stands within the CPPNE are cleared, but all other potential sites become covered with dense stands	Substantial investment of funds, but efforts (see A1) are confined to the CPPNE; control efforts outside CPPNE poorly co-ordinated and ineffective; biological control moderately effective, but little effect on dense stands
A3	All dense alien stands on public land are cleared, but all other potential sites become covered with dense stands	Available funds restrict co-ordinated integrated control to public land; biological control moderately effective, but little effect on dense stands
A4	Alien plants spread to form dense stands in all potentially invadable areas	Reduction, in real terms, of funding available for control; existing dense stands persist and become even denser after each fire; dense stands serve as foci for further spread; stand density increases in lightly invaded areas; efforts by volunteer groups are ineffective; net effect of biological control and harvesting insufficient to contribute to overall control

Proteaceae localities, but still only two taxa have 100% of their range affected. Of the 161 special taxa, 57.4% of known localities and 40.1% of Proteaceae localities, are affected by urbanization under scenario U3 which involves development of all suitable areas, irrespective of land ownership; this scenario sees the natural vegetation reduced to only 39.3% of its extent in 1994, with dune asteraceous fynbos, wet restiod fynbos, sandplain proteoid fynbos (vegetation types which house many endemic and threatened taxa) and wetlands being almost totally transformed (Richardson *et al.* 1996). Under this scenario almost a quarter of special taxa are confined totally to urban areas or areas within less than 1 km of such developments - 24 more taxa than under scenario U2. All mapped localities for four Proteaceae taxa are urbanized. All but four of the 15 vegetation types will lose a large part of their area under the three urbanization scenarios described here. Another four types only loose major parts of their area under the worst-case scenario (U3).

Future impacts of alien plants on biodiversity were also assessed by Richardson et al. (1996b). Clearing all dense alien stands (scenario A1) would result in 62.9% of special taxa having less than half their known localities affected by dense stands (49.1% at present). Under this scenario, 92% of Proteaceae taxa have more than 75% of their localities unaffected by aliens. The vegetation types that would benefit most from this scenario are coastal scree asteraceous fynbos and dune asteraceous fynbos; realization of this scenario would result in 96% and 49% increases in the extent of uninvaded areas of these types, and a 12% increase in the total extent of natural vegetation not under dense stands of aliens. If funds allow alien control only within the CPPNE (but invasion continues outside this area; scenario A2), the total area of vegetation not under dense stands would remain virtually the same as it is at present, with the area of uninvaded vegetation increasing for some types and shrinking for others. Scenario A3, whereby control efforts are confined to publicly-owned land, sees the shrinkage of the total area under uninvaded vegetation to 82.4% of its current extent, with marked reductions in the extent of most vegetation types (only two types would benefit under this scenario). Scenario A4, which sees the collapse of control programmes and the rampant spread of the aliens to all potential sites results in the annihilation of the natural vegetation, with only 407 ha (1.5% of the current area) remaining.

The worst-case scenario (A4) results in about a quarter of special taxa and more than 15% of Proteaceae species being confined entirely to densely invaded sites, with most other taxa having large parts of their ranges affected. Species that are currently common and widespread will lose large parts of their range. These assessments clearly show the importance of initiating a major programme to control alien plants.

11.5 Facing the Future

The importance of the Peninsula as an area of great biological richness has long been recognised and there has been a long history of attempts to protect this heritage (reviewed in CSIR 1995). These endeavours recently gained momentum with the acceptance of the principle of a National Park by the Minister of Environmental Affairs and Tourism in 1996. When established, the National Park will afford the Peninsula the highest possible protection under South African legislation, as a Schedule 1 National Park. The plan is to transfer all state-owned land to the National Parks Board in 1998; the Parks Board will then negotiate with the remaining landowners in the area to ensure the effective management of the whole area by declaring these areas a Schedule 2 National Park.

With unified control of this important area now virtually in place, effective management of its biodiversity must now be implemented. Although previous management plans did exist for smaller areas, effective control of aliens and fire management can best be achieved by incorporating the whole landscape in planning. Furthermore, effective protection of all species and habitats can only be achieved if the entire peninsula is managed as a unit (which was not practical in the past under numerous authorities).

The potential for effective management of this important area has been enhanced by the development of the data base and management tools described earlier in the chapter. For example, the availability on GIS of localities of all "special" plant taxa enables managers planning a prescribed burn to consider the special requirements of such taxa known to occur in the given area. The available data also allow for the objective prioritization of areas for clearing alien plants on the basis of biodiversity criteria (e.g. to relieve pressure on "special" plant taxa or particular habitats or vegetation categories).

In the following section we describe an example of how the various data can be implemented to contribute to effective long-term management – by planning the zoning by determining the location of ecologically sensitive areas.

The new National Park on the Cape Peninsula faces daunting challenges. It has to ensure the survival of biodiversity in one of the world's principal hotspots of biodiversity, while at the same time catering for the needs of a large and growing human population. The problems of population pressure being experienced here are typical of mediterranean-climate areas. To ensure that conservation aims could be met, the Peninsula was zoned to ensure that hotspots of biodiversity would not be adversely affected by future developments. The zonation (Table 11.4) was based on levels of

Table 11.4. Proposed zones for the management of the Cape Peninsula National Park

Zone	Definition and function
Special protection	To give special protection to unique features
Remote	To ensure large areas are managed to ensure a sense of remoteness. The primary objective of management being the maintenance of ecological processes and the protection of threatened species and habitats
Quiet	To act as a buffer taking some of the pressure off the remote areas
Recreational and high use zones	To act as a focus for human recreational activities.

permitted use and included the "Remote" zone where the primary objective is conservation. The allocation of land to these zones took into consideration the occurrence of concentrations of endemic and/or threatened species, as described below. It also attempted to maintain large, contiguous "Remote" areas to facilitate the maintenance of ecological processes.

Hotspots of endemism and threatened plants and animals were located as a first step in this process. Each endemic, threatened and rare plant species was scored according to its level of endemism and threat. Endemic (confined to the Cape Peninsula), threatened (listed in the *Red Data Book* or in T. Trinder-Smith's unpublished revised list); and rare taxa (with less than five records on the peninsula, even if they occur elsewhere) were scored as follows: threatened = 1; threatened and rare = 2; endemic = 3; endemic and threatened = 4; and endemic, rare and threatened = 5. The distributions of all endemic and threatened species were then mapped on 1:50000 maps in consultation with local experts. The species tended to be clumped, and the outer boundaries of all species within each clump were used to delimit the hotspot boundaries. For each hotspot, a score was calculated by summing the individual scores of all the endemic, rare and threatened species it contained. Primary plant hotspots (areas with at least 10 threatened and/or endemic species and with a score of >20) and secondary plant hotspots (areas with 3–10 special taxa and a score of between 9 and 19) were identified and the localities of individual species were also flagged. The distributions of all species which were not incorporated in the hotspots were also mapped. The primary and secondary hotspots of endemism were then superimposed on the coverages of urbanization (Fig. 11.1) and dense alien tree and shrub cover (Fig. 11.2) to show major areas of concern. In addition to mapping the localities of all species at the scale of 1:50000, a database was created to record the presence/absence of all species in each of the 591 1-km^2 grid squares. Species were assigned scores on the same basis as for scoring plant hotspots. The cumu-

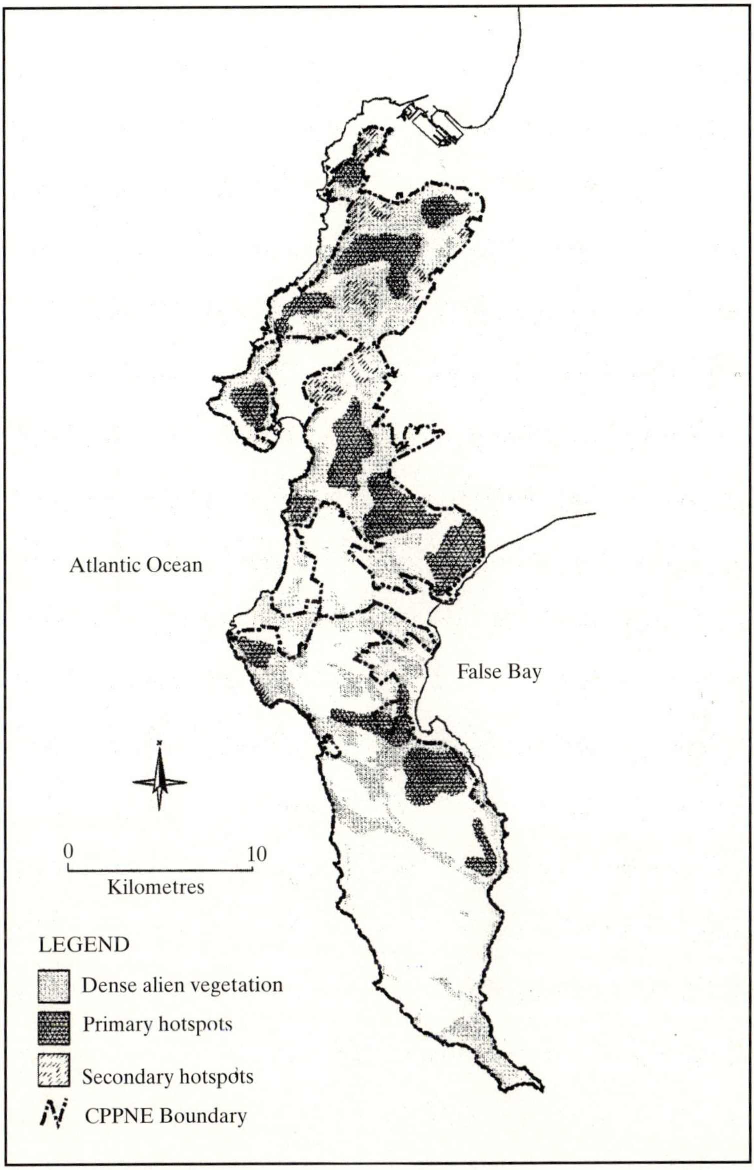

Fig. 11.2. Distribution of hotspots of plant endemism and the extent of dense stands of invasive alien trees and shrubs (>50% canopy cover) on the Cape Peninsula (see text for details on the derivation of primary and secondary hotspots)

lative score of each square was calculated by summing the scores of all the endemic, rare and threatened species in that square.

This information, together with information on the distributions of sensitive habitats such as wetlands, was used to identify ecologically sensitive areas (Figure 11.3). These areas were mapped and were used as input

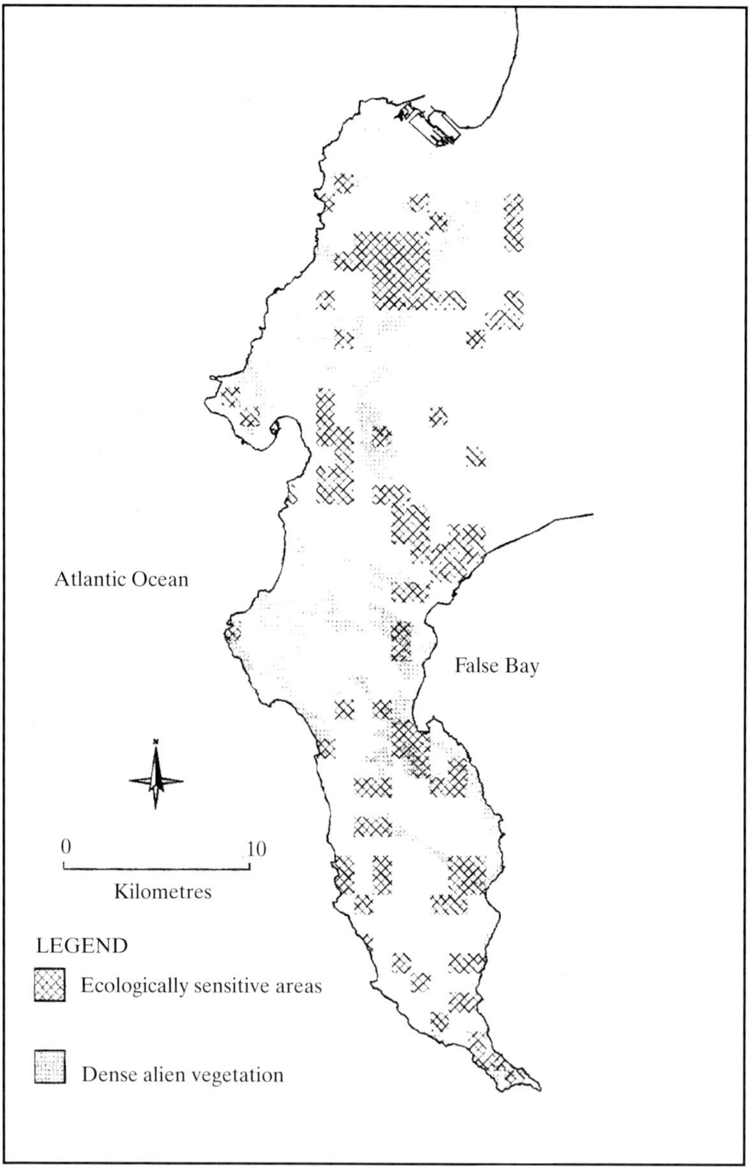

Fig. 11.3. The distribution of ecologically sensitive areas (see text) and dense stands of invasive alien trees and shrubs (>50% canopy cover) on the Cape Peninsula. Areas where the two overlap can be regarded as areas of high priority for clearing of aliens for the conservation of biodiversity

into the zoning of the National Park. Extremely sensitive areas such as some caves containing highly localized endemic invertebrates were zoned as areas for special protection. Less sensitive areas such as the plant hotspots were incorporated into the "Remote" and "Quiet Zones" which are to be managed primarily to maintain ecological processes and species as opposed to intensive recreation. Where an ecologically sensitive area abuts a current or future urban development it can be flagged as an area of special concern and appropriate action can be taken. Thus knowledge of ecologically sensitive areas greatly facilitates appropriate zoning and management. In addition, the extensive knowledge of species localities enables us to ensure that all species are adequately protected (Trinder-Smith et al. 1996b). The same applies to the vegetation.

11.6 Conclusions

Managers of areas with very rich biotas and multiple threats often need to take urgent decisions without enough information on all pertinent factors. They must make the best use of what data is available for making objective decisions. Explicit criteria are frequently difficult to formulate when information is sketchy. We believe the approach outlined in this Chapter has great potential for facilitating better management by facilitating the efficient management of spatial data.

Acknowledgements. Much of the data and interpretations reported on here came from an intensive exercise conducted between 1994 and 1996 which aimed to provide an objective basis for the scientific management of the Cape Peninsula. Results of these studies were published in a special issue of *Biodiversity & Conservation* (Volume 5, no. 5; May 1996). We thank all the participants in this venture for their inputs. Other work incorporated in this paper formed part of the planning for the new National Park and has been sponsored by the Western Cape Regional Services Council and WWF (South Africa)'s Table Mountain Fund. We also thank the CSIR for providing support for the participation of Caroline Gelderblom and Brian van Wilgen in the production of this Chapter. David McKelly is thanked for his help with GIS analyses.

References

Cowling RM (1995) Table Mountain: a world heritage site? Inaugural lecture. University of Cape Town New Series 191, 19 pp

Cowling RM, Holmes PM, Rebelo AG (1992) Plant diversity and endemism. In: Cowling RM (ed) The ecology of fynbos: nutrients, fire and diversity, Oxford University Press, Cape Town, pp 62–112

Cowling RM, Macdonald IAW, Simmons MT (1996) The Cape Peninsula, South Africa: physiographical, biological and historical background to an extraordinary hot-spot of biodiversity. Biodiv Conserv 5:527–550

Cowling RM, Richardson DM, Schulze RE, Hoffman MT, Midgely JJ, Hilton-Taylor C (1997) Species diversity at the regional scale. In: Cowling RM, Richardson DM, Pierce SM (eds) Vegetation of southern Africa. Cambridge University Press, Cambridge, pp 447–473

CSIR (1995) Nomination proposal for the Cape Peninsula Protected Natural Environment to be listed as a World Heritage Site. Prepared by the CSIR in collaboration with the Institute for Plant Conservation, Stellenbosch

Macdonald IAW, Richardson DM (1986) Alien species in terrestrial ecosystems of the fynbos biome. In: Macdonald IAW, Kruger FJ, Ferrar AA (eds) The ecology and management of biological invasions in southern Africa. Oxford University Press, Cape Town, pp 77–91

Myers N (1990) The biodiversity challenge: expanded hot-spots analysis. Environmentalist 10:243–256

Picker MD, Samways MJ (1996) Faunal diversity and endemicity of the Cape Peninsula, South Africa – A first assessment. Biodiv Conserv 5:591–606

Rebelo AG (1992) Red data book species in the Cape Floristic Region: threats, priorities and target species. Trans R Soc S Afr 48:55–86

Rebelo AG (1995) Sand plain fynbos. In: Low AB, Rebelo AG (eds) Vegetation of South Africa, Lesotho and Swaziland. Dept of Environment Aff Tour, Pretoria, p 74

Richardson DM, Macdonald IAW, Holmes PM, Cowling RM (1992) Plant and animal invasions. In: Cowling RM (ed) The ecology of fynbos: nutrients, fire and diversity, Oxford University Press, Cape Town, pp 271–308

Richardson DM, Cowling RM, Lamont BB (1996a) Non-linearities, synergisms and plant extinctions in South African fynbos and Australian kwongan. Biodiv Conserv 5:1035–1046

Richardson DM, van Wilgen BW, Higgins SI, Trinder-Smith TH, Cowling RM, McKelly DH (1996b) Current and future threats to plant biodiversity on the Cape Peninsula, South Africa. Biodiv Conserv 5:607–647

Simmons M, Cowling RM (1996) Why is the Cape Peninsula so rich in plant species? An analysis of the independent diversity components. Biodiv Conserv 5:551–573

Trinder-Smith TH, Cowling RM, Linder HP (1996a) Profiling a besieged flora: rare and endemic plants of the Cape Peninsula, South Africa. Biodiv Conserv 5:649–669

Trinder-Smith TH, Lombard AT, Picker MD (1996b) Reserve scenarios for the Cape Peninsula: high, middle and low road options for conserving the remaining biodiversity. Biodiv Conserv 5:671–684

van Wilgen BW (1996) Management of the natural ecosystems of the Cape Peninsula: Current status and future prospects. Biodiv Conserv 5:671–684

van Wilgen BW, Bond WJ, Richardson DM (1992) Ecosystem management. In: Cowling RM (ed) The ecology of fynbos: nutrients, fire and diversity, Oxford University Press, Cape Town, pp 344–371

Wood J, Low AB, Donaldson JS, Rebelo AG (1994) Threats to plant species diversity through urbanization and habitat fragmentation in the Cape Metropolitan Area, South Africa. Strelitzia 1:259–274

12 Biodiversity and Conservation Biology of Coastal Transition Zones from Mediterranean to Desert Ecosystems: An Intercontinental Comparison

K.J. Esler, P.W. Rundel and R.M. Cowling

12.1 Introduction

Comparative research on Mediterranean-type ecosystems over the past several decades has largely focused on evergreen, sclerophyllous vegetation. Little attention has been given to comparative research on ecological structure and community diversity in the coastal transitions from mediterranean climate zones to winter rainfall deserts. This has been a particularly important oversight by the MEDECOS research community, since the Mediterranean desert transition zones are potentially as diverse as their related "true" Mediterranean counterparts. These ecotonal areas appear to support a large diversity of plant taxa, many of which are endemic (e.g. Raven and Axelrod 1978; Shmida 1985; Cowling and Hilton-Taylor, in press).

It is clear that the vegetation in each of these zones is under increasing pressure as population numbers rise, and the demand for economic security increases. Threats are diverse and include those associated with the mining industry, as well as overgrazing by small livestock (goats and sheep), agriculture and urban development. Lack of research in these arid areas has not encouraged good land use practices and all three areas discussed here are poorly conserved. As pressure on these marginal farming areas increases, so should the efforts to improve, sustain and restore them. These efforts should be based on a scientific understanding of the dynamics of the vegetation, an understanding of the processes leading to degradation as well as those involved in the rehabilitation and restoration of vegetation. This requires both theoretical investigations into vegetation dynamics as well as the development of practical procedures by which to evaluate such theory.

The comparison of the characteristics of different systems is one of the oldest approaches in ecology, and one which is very familiar to researchers in mediterranean-type ecosystems. Comparisons between Californian chaparral and Chilean matorral (e.g. Mooney 1977; Miller 1981), and

Ecological Studies, Vol. 136
Rundel et al. (eds.) Landscape Degradation and Biodiversity in Mediterranean-Type Ecosystems

Australian kwongan and South African fynbos (Mileski 1983; Cowling and Witkowski 1994) have provided a large body of support for the theory of convergence. They have also highlighted the unique features of each of these systems, thus providing a focus for research questions and variety of research opportunities. More recently, comparative studies have contributed towards an understanding of the potential effects of global change. In order to predict patterns of vegetation response to global change, it is useful to examine areas with outwardly similar climatic conditions to see where the differences lie (Cowling and Midgley 1996). These differences can provide important insights into how global change might influence species composition over evolutionary time. Finally, a comparison of the effects of landscape degradation in similar systems can provide us with a convincing tool to lobby for pressing conservation issues. Monitoring the levels of landscape degradation is essential to determine the significance of the potential loss of biodiversity in these systems and to gain a perspective for future research and conservation.

In this Chapter, we focus on the biodiversity and conservation of the coastal transition zones from mediterranean to desert ecosystems in South Africa, southern California and northwestern Baja California, and Chile (Fig. 12.1). First, we provide a comparative analysis of the selective regimes for each area. Next we briefly review patterns of community structure, plant diversity, landscape degradation and conservation status of each of the areas. Finally, we discuss the importance of these areas as significant repositories of global plant diversity; the opportunities they provide for comparative research; and the need to improve their conservation status in the face of escalating threats.

12.2 Selective Regimes

12.2.1 Climate

In each of the mediterranean-desert transition zones, climatic characteristics are outwardly very similar, with parallel relationships between mean annual rainfall and latitude on all three continents (Fig. 12.2). Mean temperature variations along the latitudinal gradients are also not greatly divergent between continents. The close proximity of oceanic influences in all three regions greatly moderates seasonal changes in temperature.

The mediterranean climate (in the broad sense; di Castri 1973) of the west coast of South Africa is influenced by circumpolar westerly cyclonic fronts (cold fronts in winter associated with northwesterly winds), subtropical high pressure anticyclones (associated with southeasterly winds in

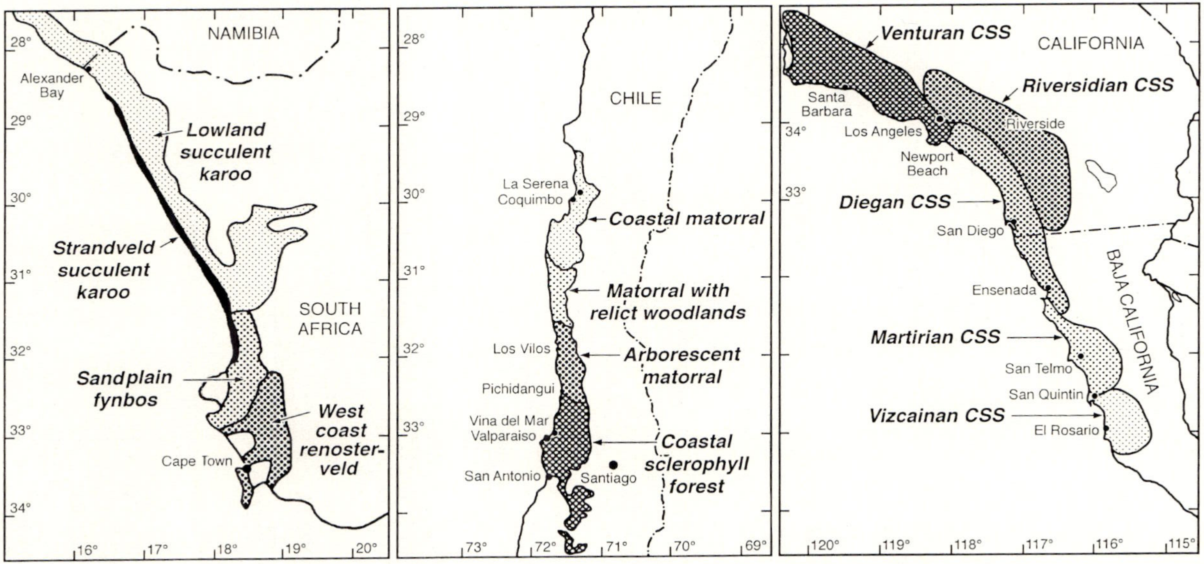

Fig. 12.1. Vegetation zonation along the three coastal mediterranean-desert transition zones in South Africa, California/ Baja California, and Chile. Vegetation units were taken from Low and Rebelo (1986), Westman (1983 a,b), and Gajardo (1994)

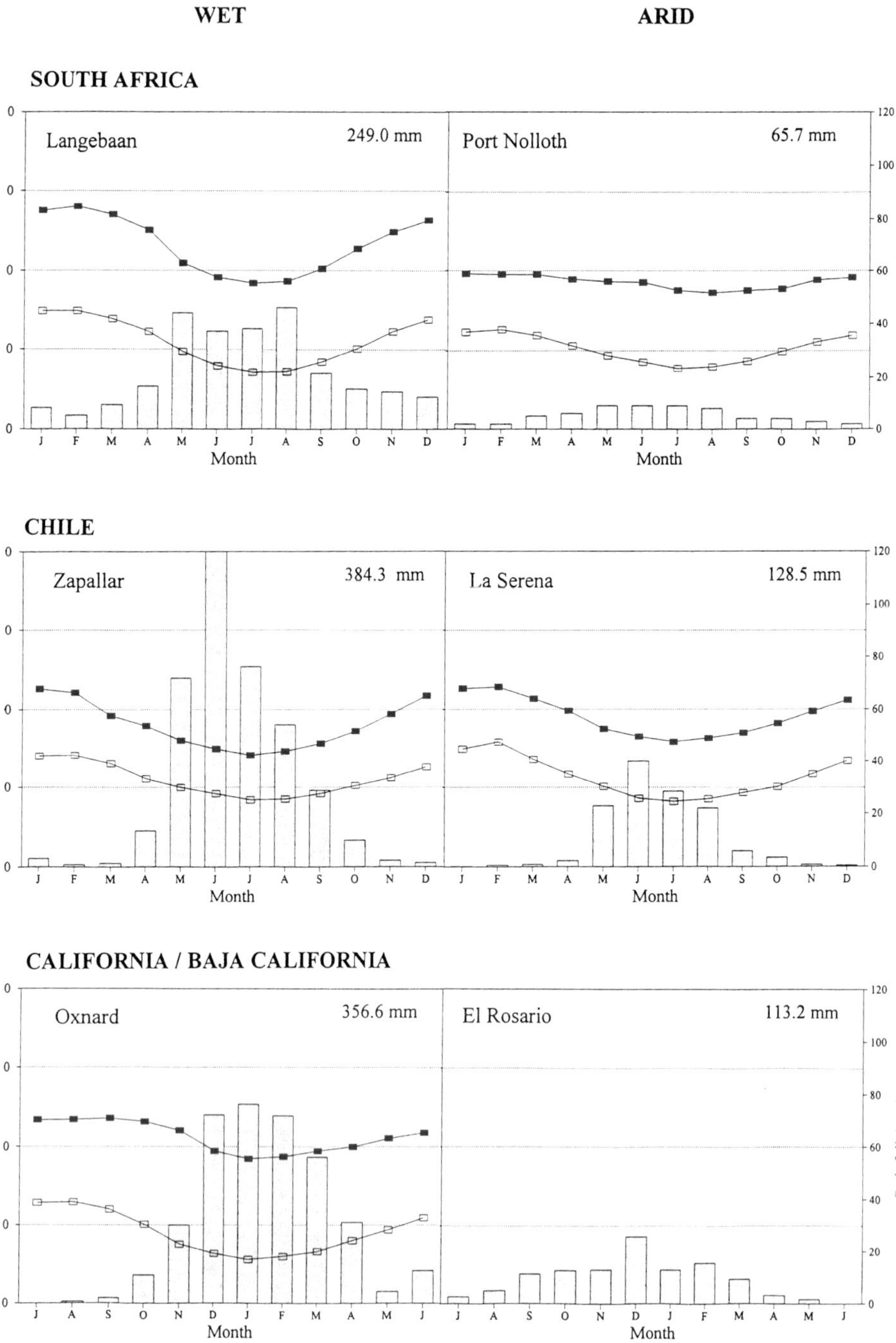

Fig. 12.2. Climate diagrams from the arid and higher rainfall ends of coastal transition zones between winter rainfall desert and mediterranean ecosystems for South Africa, California/Baja California, and Chile. Data shown are mean monthly maximum and minimum temperatures for and mean monthly rainfall (Weather Bureau 1996, National Climatic Data Center, Asheville, Miranda Reyes et al. 1991, Reyes Coca 1990, and Hajek and di Castri 1975).

summer) and a strong marine influence (the cold Benguela current of the Atlantic Ocean) (Tyson 1986). Mean annual rainfall decreases from about 400 to 500 mm in the south to less than 100 mm in the north. Over the same latitudinal gradient, low-layer fog frequently forms during summer, its incidence increasing from south to north. Due to the maritime influence, frost is absent.

The mediterranean climate along the California/Baja California transect is characterised by severe summer drought and mild, wet winters. Rainfall in winter is a characteristic, with a strong influence of a subtropical high pressure center eliminating summer precipitation in all but rare years. Mean annual rainfall decreases from about 400 to 500 mm in the north of the transect to less than 100 mm in the south. In summer temperatures are ameliorated when fog forms, a phenomenon particularly evident in parts of the Baja California coast, due to the effects of the cool offshore California current.

The mediterranean-type ecosystems in the central parts of Chile have been extensively studied. Rainfall in winter, as in California, predominates, with summer rainfall a rare event (Hajek and di Castri 1975; di Castri and Hajek 1976). Gradients of mean annual rainfall with latitude closely follow those of California/Baja California and South Africa.

Despite similarities in total annual rainfall, there are marked differences with respect to monthly and interannual patterns of rainfall events between each transect. The relatively high monthly variation in rainfall of California/Baja California and Chile results from the strong seasonality of rainfall pattern, while lower variation in South Africa indicates the greater spread of rainfall throughout the year (Esler et al. 1998). South Africa receives more summer rain since this region lacks high mountains to block westward movement of moist, subtropical air which enters the region in late summer, bringing associated thunderstorms. Interannual coefficient of variation in rainfall is an indication of the predictability of precipitation as a resource between years. For this character, climatic data thus strongly demonstrate that South Africa has the most predictable rainfall regime of the three regions, while Chile is the least predictable.

12.2.2 Geomorphology and Soils

Landscape age and the resultant geomorphic weathering through geologic time presents a fundamental difference between South Africa and the two comparative regions in the Western Hemisphere. South Africa is an ancient and relatively subdued landscape (coastal plain) on a sheared continental margin. The resulting coastal plain along the Atlantic coast in the mediterranean/desert transition zone is relatively wide and gently sloping.

In contrast, the landscapes of California/Baja California and Chile are in continental subduction zones and are consequently geologically young and topographically complex. Major mountain building has occurred as recently as the late Tertiary and Quaternary.

Soils of the coast plain of South Africa are dominated by windblown sands of marine origin and of varied ages. The predominant surface is a Late Tertiary/early Pleistocene sand mantle, mainly base-rich but with patches of acid sand throughout and especially in the wetter south. These sands are underlain by hardpans of ferruginized dorbank, calcrete, and ancient silcrete. Outcrops of ancient granites (Namaqualand Complex, Cape Granite Suite) occur sporadically. Consequently, with the exception of Quaternary sands along the coast and the acid plumes, soils are relatively shallow (0.1–0.5 m) and universally sandy (Boucher 1987; Desmet 1996). The coastline along this mediterranean-desert transition is fairly linear and is interrupted by relatively few half-heart and arcuate bays (Heydorn and Tinley 1980). The major part of the coastline is sandy (dune coast), although cliff faces up to 30 to 90 m in height and rocky platforms can also be found (Boucher and Le Roux 1993).

The mediterranean-climate region of southern California continues to the south into northwestern Baja California along the Pacific Coast to the west of two mountain ranges, the Sierra Juarez and Sierra San Pedro Martir. The coasts of northern Baja California and southern California are contiguous and therefore the topography and climate are very similar. Both coasts are indented by numerous bays and coves. In contrast to the South African west coast, much of this coastline is uplifted, with steep seacliffs and multiple rocky beach terraces evident along much of the coast. Sandy areas and dunes are present, but they are not a major landscape feature.

The Chilean coast differs physiographically from comparable coastlines in South Africa and California/Baja California in that a coastal range (Cordillera de la Costa) runs parallel and very close to the coast. The topographic result is a coastal plain that is generally narrow or even virtually absent at times. Where the coastal plain is present, it consists of a series of marine terraces that extend up to several hundred meters in elevation.

12.3 South Africa

12.3.1 Phytogeography, Community Structure and Biodiversity

The South African coastal transition zone between mediterranean and desert environments largely falls within the Succulent Karoo Region, a floristic region whose strong floristic affinities with the Cape Region in the true mediterranean-climate zone to the south, is now recognised as forming part of the greater Cape Flora (Hilton-Taylor 1987; Jurgens 1991; Fig. 12.1). This western, and strongly winter-rainfall part of the Succulent Karoo forms a distinct phytogeographical zone, the Namaqualand–Namib Domain (Jurgens 1991, see also Cowling and Hilton-Taylor, in press). Throughout this zone, the vegetation is predominantly a succulent shrubland that is especially rich in Mesembryanthemaceae. The southern, wetter part of the area falls within the Cape Region: the predominant vegetation comprises various sclerophyllous shrubland types.

The vegetation of this area is everywhere under strong edaphic control, and compositional change along edaphic gradients is very high (Boucher and Jarman 1977; Boucher 1987; Cowling and Hilton-Taylor, in press; P.G. Desmet and R.M. Cowling, unpubl. data). Starting in the south, dune thicket is associated with deep, calcareous sands; sandplain fynbos with acid sands; and west coast renosterveld with clay-rich soils derived from granite (Boucher 1987; Low and Rebelo 1996). Dune thicket is a dense, mainly evergreen and large-leaved shrubland with strong subtropical affinities. The component plants are mainly long-lived shrubs with bird-dispersed propagules e.g. *Euclea racemosa, Pterocelastrus tricuspidatus, Rhus* spp., and *Olea exasperata.* Patches of dune Thicket extend almost to the Orange River on plumes of recent calcareous sand. Sandplain fynbos has a structure typical of fynbos throughout the Cape Region, being dominated by restioids (e.g. *Willdenowia incurvata*), proteoids (e.g. *Leucospemum tomentosum* in the south and its sister species, *L. rodolentum* in the north), and numeous ericoid shrubs. Fynbos extends well into the desert zone on patches and plumes of acid sand (P.G. Desmet and R.M. Cowling, unpubl. data). West Coast Renosterveld is an ericoid shrubland dominated by Asteraceae (e.g. *Elytropappus rhinocerotis, Relhania* spp), grasses (e.g. *Ehrharta calycina, Pentaschistis* spp.) and numerous seasonal geophytes in the Iridaceae, Amaryllidaceae and Liliaceae. Renosterveld does not extend beyond the Berg River.

North of the Berg River, and especially north of the Olifants River, the Cape Region elements are largely replaced by those from the Succulent Karoo. The dominant matrix vegetation is termed Succulent Karoo strandveld (Low and Rebelo 1996) and this extends almost to the banks of the Orange River where the annual rainfall is little more 50 mm. Soils are deep, grey sands of marine origin (Quarternary to Recent) and shallower (but always >0.3 m), red sands overlying various hardpans. Strandveld is an open, semi-succulent to semi-deciduous shrubland varying in height from 0.5–2.0 m. Although there are considerable compositional differences along gradients of soil depth, pH and colour (Desmet 1996), especially in the dwarf leaf succulent component (mainly Mesembryanthemaceae), there are a number of species which typify this vegetation throughout its range. These include *Stoeberia utilis, Othonna cylindrica, Zygophyllum morgsana, Osteospermum oppositifolium* and *Pteronia onobromoides*. Spring-flowering annuals, mainly Asteraceae, provide spectacular displays, especially in disturbed sites. Surprisingly, a tropical C_4 grass, *Odyssea paucinervis*, forms a conspicuous component between the Spoeg and Olifants Rivers. Broom-like succulent *Euphorbia* species are commonly the dominant component on the inland margin of the strandveld. Different species are associated with soils of different depth. For example, east of Port Nolloth, *E. mauritanica* is exclusively associated with the deepest soils; *E. dregeana* on soils of intermediate depth; and *E. ephedroides* on the shallowest soils. These replacement patterns are visible across distances of a few hundred meters.

On the isolated areas of shallow sands (overlying hardpans or on bed-rock outcrops and inselbergs), and on the relatively extensive gravel plains in the Orange River area, strandveld is replaced by lowland Succulent Karoo (Desmet 1996; Low and Rebelo 1996). This vegetation is a dwarf, open shrubland dominated almost entirely by leaf succulent members of the Mesembrayanthemaceae, Crassulaceae and Asteraceae. The geophytic flora is very rich whereas annuals are not especially conspicuous. These communities are very species-rich (Cowling et al. 1994) and show marked compositional change along almost any conceivable environmental gradient (Jurgens 1987; Cowling and Hilton-Taylor, in press). They defy any attempt at overall floristic characterisation. Some important genera include *Ruschia, Antimima, Cephalophyllum, Cheirodopsis, Mesembryanthemum, Stoeberia, Meyerophytum, Monilaria, Phyllobolus, Leipoldtia, Conophytum, Lampranthus, Drosanthemum, Mitrophyllum, Dracophilus, Enarganthe, Wooleya*, and *Vanzijlia* (Mesembryanthemaceae), *Crassula* and *Tylecodon*, (Crassulaceae), and *Othonna* and *Senecio* (Asteraceae). Many species are relatively short-lived and community compositions can show pronounced compostional change over time (for details see Cowling and Hilton-Taylor, in press).

The flora of the South African region considered here is very rich, probably in excess of 3000 species. Desmet (1996) recorded 300 species in an area of 716 km^2 in the extreme north, where annual rainfall is about 50 mm. The mediterranean-climate zone of the south includes some of the most species-rich landscapes in the world (Cowling et al. 1992). This diversity is largely a function of high turnover along environmental and geographic gradients (Cowling and Hilton-Taylor in press, P.G. Desmet and R.M. Cowling, unpubl. data) of species-rich communities (up to 100 species in 0.1 hectare plots in the rocky areas of the transition zone) (Cowling et al. 1989).

The Succulent Karoo is also an area of pronounced local endemism (Cowling and Hilton-Taylor, in press). Some 28% of Desmet's (1996) 300 species in a desert area are local endemics. Most local endemics and otherwise rare species are dwarf leaf succulent shrubs (principally Mesembryanthemaceae) associated with Succulent Karoo on shallow soils and hard substrata such as rocky outcrops and pebble plains (Desmet 1996; Cowling and Hilton-Taylor, in press). Towards the south, fynbos shrubs contribute substantially to the endemic flora (Cowling and McDonald, Chapter 9 in this Volume).

12.3.2 Landscape Disturbance

The sclerophyllous shrubland vegetation of the Cape Region (southern, wetter part of the gradient) and the shrublands of the Namaqualand-Namib Domain of the Succulent Karoo region have been a focus for human activities for upwards of half a million years (Deacon 1992). Humans, therefore, have a very long history in the region, and no part of it could be considered to be absolutely pristine. About 30000 years ago, a dry period isolated the then inhabitants of the west coast from their original decendants further north. This led to the evolution of the Khoi-San peoples. The San were hunter-gatherers, who did not have a major impact on the vegetation, but from 2000 to 1500 years ago, the Khoi-Khoi started hearding sheep.

With the arrival of the Dutch settlers in 1652, there were some 17000–18000 Khoi sheep herders between Cape Town and the Berg river (Truter 1996). The Dutch established Saldahna Bay as a supply port as early as 1658 (Axelson 1977). In the 18th century, white settlers began to be established along the coast, but because of poor, sandy soils, intensive farming has not been a major concern. The discovery of copper and diamonds along the coast up to Port Nolloth greatly enhanced development. Today, the mining concerns are isolated and the major part of the coastal zone is relatively little affected by development. The region is sparsely populated, except for the extreme south where there is urban expansion and industrial development around Saldanha, and resort developments between Saldanha and Cape Town.

Disturbance regimes and conservation status are very dependent on land tenure. South of the Olifants River, most of the land is privately owned and used for agriculture (extensive grazing, pastures and various crops) and, to a lesser extent, resort development. Cultivation associated with centre-pivot irrigation is on the increase in this area, threatening rare plants, causing soil erosion and "mining" an already dwindling water resource (Heydendrych 1993). Studies investigating the impact of this marginal agricultural activity have recommended that the South African Conservation of Agricultural Resources act of 1983 be amended to halt this problem (Moll et al. 1989, Heydenrych 1993). Overgrazing by small-stock, mainly sheep, is apparent on some farms, whereas others are relatively well-managed. In the southern margin of this region, fynbos vegetation has been widely destroyed by strip-plowing and alien invasive plants, and thus very little of this vegetation type remains. Similarly, more than 90% of west coast renosterveld has been replaced by crops (Rebelo 1992) with remnants of this community confined to steep, granitic hillslopes (Boucher 1987).

Most of the strandveld Succulent Karoo (north of the Olifants River; Fig. 12.1), is owned by mining companies, especially De Beers Consolidated Mines Ltd. The chief mining activity is open-cast retrieval of diamonds from subterranean and near-shore gravels, derived from the paleo-Orange River System. Extraction of heavy metals from strandveld sands is a new enterprise that will have an impact on an extensive area in the southern part of this region. In the past, mining operations appear to have been conducted with minimal regard for landscape degradation. However, all operations post-1980 are subject to Environmental Management Reports which must stipulate budgets, methods and time frames for rehabilitation.

The primary mining impact is from prospecting through trench excavation (almost the entire coastline is scarred by parallel trenches, 10 m wide and 100 m to 3 km long), and overburden stripping in larger areas to reach the diamond-bearing gravels (Talkenberg 1982). Secondary influences, such as wind erosion, the creation of many access roads, tracks and borrow pits to service the mined areas, and the establishment of processing plants and slimes dams magnify the direct effects of mining (Jackelman and Moll 1989). However, direct mining impacts are restricted to a relatively small part of the area. For example, in the Alexcor concession, comprising about 800 km^2 in the area between Port Nolloth and Alexander Bay, only about 10% of the land is thus affected. The remainder is generally in a very good condition (Desmet 1996). Furthermore, mining is not economically feasible on the vast majority of the mine-owned land, including extensive stretches of the coastal plain. The condition of the vegetation in these areas is generally excellent, being only lightly grazed by sheep or indigenous ungulates. Much of this land is potentially available for formal conservation.

A major problem associated with rehabilitation of mined sites relates to the antiquity of the strandveld soils (Desmet 1996). The weathered, stratified subsoils, which are usually exposed on the surface of overburden dumps, have excessively high levels of various minerals, which retards plant establishment (Scott et al. 1994). Natural rehabilitation on overburden derived from younger (Quarternary) sands, which lack toxic subsoils, is extremely rapid, even under the low rainfall conditions of the north (Le Roux and Odendaal, no date). In the very dry area of the north, wind erosion from overburden dumps create mobile plumes which devastate large areas of relatively intact vegetation (Desmet 1996). Much more research is required to provide guidelines for rehabilitation and restoration.

Privately-owned farms are mainly resticted to the inland margin of the coastal plain. Vegetation condition varies greatly but is generally moderate to good. Communally-owned land occurs in the northern and central parts of the eastern fringe of the area. Here vegetation is very poor, with clear evidence of desertification.

Nature-based tourism (or ecotourism), based largely on the mass floral displays on disturbed lands, is a growing industry in the area, and certainly provides some economic opportunities. With the predicted downscaling of the diamond mining industry over the next few decades, appropriately developed ecotourism has the potential to become one of the a significant economic input to this region. However, much more needs to be done in order to realise this potential. In particular, the reserve network must be expanded; the perennial succulent flora must be more effectively promoted as an ecotourist resource; and the areas assets must be more vigorously marketed at home and abroad.

12.3.3 Protected Areas

The reserve network along the west coast of South Africa is inadequate. There are only two formal reserves, the West Coast National Park (180 000 ha) and Rocher Pan Nature Reserve (39 000 ha), both located south of the Olifants River. Of the 38 170 km^2 of strandveld Succulent Karoo, only 0.4% is conserved (Low and Rebelo 1996). Almost no sandplain fynbos and no west coast renosterveld in this area is currently protected. However, there are initiatives to expand the reserve system. The West Coast National Park is expanding to include more fynbos habitat (Desmet and Cowling 1996) and there are plans for a Biosphere Reserve in the extreme south of the area. The National Parks Board, De Beers mining company and the local community negotiating the establishment of a national park on mine-owned land along a 60-km-long coastal strip between the Groen and the Spoeg rivers. This will protect a wide range of habitats and species typical of the central strandveld region (P.G. Desmet and R.M. Cowling, unpubl. data). It is important that these initiatives are representative (ensure the inclusion of all major habitats, endemic/rare taxa, scenic features and geomorphic processes in the reserve system) and efficient (ensure that these features are protected on the smallest possible area, thereby reducing the financial burden of purchasing land). Finally, efforts to improve the conservation status of land outside the protected areas must be strengthened (through, for example, improved extension facilities), since it will not be possible to include all of the unique and unusual features of this spectacularly diverse area in the reserve system. It is hoped that ecotourism will provide a strong economic incentive for conservation efforts outside reserves.

12.4 California/Baja California

12.4.1 Phytogeography, Community Structure and Biodiversity

Coastal vegetation in Southern California and northwestern Baja California is characterized by an association termed coastal sage scrub, as well as soft chaparral in the older literature. Coastal sage scrub differs notably from chaparral vegetation by the dominance of drought-deciduous semi-woody shrubs which lose all or most of their leaves as summer drought reduces soil moisture availability. The widespread and characteristic coastal sage scrub dominants include species of *Salvia, Encelia californica, Artemisia california*, and *Eriogonum fasciculatum*. The extreme summer drought restricts most evergreen chaparral shrubs from coastal environments below about 300 m elevation except where soil moisture is available. Some chaparral species with deep root systems are present, however, most notably *Malosma laurina, Rhus integrifolia*, and *Heteromeles arbutifolia*.

The coastal transition zone between mediterranean and desert communities in southern California and Baja California has been divided into four floristic associations along a gradient of evapotranspirational stress (Westman 1983a,b). Coastal sage communities from north of Santa Barbara south to Los Angeles have been termed the Venturan association (Fig. 12.1). South of Los Angeles and extending into northern Baja California as far as Ensenada, coastal vegetation has been termed Diegan coastal sage scrub. Ecologically, these two associations are very similar, and their differentiation has been based on quantitative analyses of floristic sampling data.

South of Ensenada, however, there is a marked change in the ecological dominance and floristic composition of these coastal communities. While a matrix of widespread coastal sage species is present, there is an addition of many transition zone endemics as co-dominants, and an increasingly significant ecological role played by succulent species. This form of coastal sage scrub has been termed coastal sage succulent scrub (Westman

1983a,b) and maritime succulent scrub (Beauchamp 1986), but more commonly coastal succulent scrub (Westman 1983a). Similar communities on the California Channel Islands have been termed maritime desert scrub or shrub (Dunkle 1950, Thorne 1967), and maritime cactus scrub (Philbrick and Haller 1977). The coastal region from Ensenada to San Quintín has been classified as the Martirian association of coastal succulent scrub, and from San Quintín to south of El Rosario is separated as the Vizcainan association (Westman 1983a).

A second approach to the classification of these coastal succulent scrub communities has been proposed by Peinado et al. (1994, 1995) who have focused on a phytosociological approach that highlights the dominance of transition zone perennials in characterizing a large number of community types. These authors classify what has been termed the Vizcainan association of coastal succulent scrub as a desert community because of the importance of species with transition zone/desert affinities.

The floristics of the mediterranean/desert transition zone has been characterized in considerable detail. Regional coastal floras are available for Santa Barbara County, the Santa Monica Mountains, San Diego County, the California Channel Islands in California (Smith 1976; Wallace 1985; Beauchamp 1986; Raven et al. 1986; Junak et al. 1995). Although there is a general flora for Baja California (Wiggins 1980), it provides only crude data on biogeographic distributions. Local floras for northwestern Baja California exist only for the Punta Banda Peninsula near Ensenada (Mulroy et al. 1979).

With the exception of a limited degree of edaphic endemism on clay, sand, or serpentine-like gabbro soils, coastal areas of southern California have relatively low amounts of local endemism. This situation changes with the Martirian and Vizcainan associations of northwestern Baja California where there is a relatively high degree of endemism in this mediterranean/desert transition region. Here, endemism is centered most heavily in perennial species of subshrubs and shrubs (Mulroy et al. 1979; Oberbauer 1992). Important endemic shrubs include *Aesculus parryi, Ptelea aptera, Hazardia berberidis, H. ferrisiae, H. orcuttii, H. rosaricus,* and *Salvia munzii.* There is also high endemism in succulent species including *Bergerocactus emoryi, Echinocereus maritimus, Ferocactus fordii, F. viridescens, Mammillaria brandegeei,* and many species of *Dudleya.*

A strong floristic connection can also be seen in species with distribution patterns which link northwestern Baja California with the California Channel Islands to the north. Good examples of such distributions can be seen with *Pinus muricata, Salvia brandegeei, Rhamnus insula, Ribes viburnifolium,* and *Galvezia juncea.* The presence of many relict species and endemics that are relatively unspecialized compared to congeners outside this region have led Raven and Axelrod (1978) to suggest that the present transition conditions of rainfall and temperature may be similar to the climate in which these species once evolved. Notable paleoendemics in northwestern Baja California include *Adenothamnus validus, Harfordia macroptera, Ptelea aptera, Aesculus parryi*, and *Fraxinus trifoliata*. The first two of these are endemic genera.

12.4.2 Landscape Disturbance

Significant human impacts on landscapes along the coastal area of Southern California and northwestern Baja California are a relatively recent phenomenon. Native Indian populations entered this area about 6000–10000 years ago, but their numbers were relatively small. Although the Spanish exploration of this area began as early as 1542 with the expedition of Portola, permanent Spanish colonies were not established for two and half centuries. European impacts began in the late 18th century with the establishment of a series of Dominican missions along the coast. In Baja California, Missions Nuestra Señora de Rosário, Santo Domingo, San Vicénte de Ferrer, and Santo Tomas were established from 1774–1794, with each mission a day's ride north of the previous one. Indian communities were gathered into these missions and wheat, corn and beans were grown. These missions closed in 1832–33, however, as disease decimated Indian populations and political instability compromised government support for this development. Construction of similar missions up the coast of California begin in 1769, and these remained more active well into the 19th century.

Economic development of the Baja California coast was limited during the past century and first half of this century. While small towns along the coast continued to exist, poor road access and limited economic opportunities kept population densities low. An agricultural settlement of English colonists was established near San Quintín at the end of the 19th century, but failed.

The recreational development of the Baja California coast began to expand after World War II when military concerns had led to the establishment of the port of Ensenada and construction of new highway access from Tijuana. This development has accelerated in the past two decades to the point that very little undisturbed coastal lands exist between these two cities.

South of Ensenada there has not been this same development of homes and recreational sites along the coast. About the same time as development was expanding to the north, a 1970 law signed by Mexican President Lazardo Cardenas established government support for cooperative farms (*ejidos*) to encourage economic development of agricultural lands. This law promoted development in many of the fertile valleys around existing towns such as Santo Tomás, San Vicénte, Colonia Guerrero, and San Quintín. Using water supplies from the Sierra San Pedro Martír to the east, these areas now support large scale farming of grapes and olives around San Quintín and vegetable and grain crops such as tomatoes, chilies, beans, wheat, and barley. Extensive areas of coastal succulent scrub communities in these valleys and along coastal beach terraces have now been converted to agricultural land.

Economic development of the southern California coastal area occurred much earlier, with major population centers beginning to grow and prosper as California joined the United States in 1850 and immigration from the eastern United States increased rapidly. Disturbance to coastal areas was somewhat limited, however, by an absence of any major cultural use of coastal resources for grazing animals, and by an absence of important mineral deposits. Heavy grazing by goats has never occurred along the California coast, although it has been a factor on some of the California Channel Islands.

By the 1880s, large areas of the southern California coast were already being developed as urbanization increased, and such development of these areas for housing and recreational use has continued up until the present time. As a result, relatively little of the coast along the mediterranean-desert transition from Ventura to San Diego has remained undisturbed.

12.4.3 Protected Areas

The degree of preservation and protection of coastal plant communities in southern California is uneven, but generally poor. The Santa Monica Mountains, which mark the northern end of the mediterranean/desert transition zone, have been heavily impacted by expanding urbanization, particularly along the coast and adjacent to the urbanized corridor of the San Fernando Valley to the interior. Large areas of protected lands do exist, however, with the Santa Monica Mountains National Recreation Area and a series of lands managed by the California Department of Parks and Recreation. These include Leo Carrillo, Malibu Creek, and Topanga State Parks. Other significant natural areas in these mountains are managed by the Santa Monica Mountains Conservancy, the Mountains Restoration Trust, the University of California Natural Reserve System, and the County of Los Angeles.

The most successful protection now exists with Channel Islands National Park, established in 1980 and increased in size in 1986, which has had major accomplishments in preserving and restoring natural ecological communities on Santa Cruz, Santa Rosa, San Miguel, Santa Barbara and Anacapa Islands. Particularly important has been success in removing goats and reducing populations of feral rabbits and the effect that this removal has had on native plant species. Similar efforts at removing exotic species have now been carried out on Santa Cruz Island which enjoys protected status under a joint management by The Nature Conservancy, National Park Service and the University of California Natural Reserve System. Santa Catalina Island, although privately owned, has large areas managed to protect natural resources.

South of Los Angeles, there are few natural areas of protected coastal lands. Beach areas owned by the California Department of Parks and Recreation are extensive, but these are managed for recreation and generally have only poor remnants of native communities. The El Segundo dunes once covered more than 90 km^2 on the ocean margin of Los Angeles, but less than 1% of this habitat remains today (California Coastal Commission 1987). The large Camp Pendleton Marine Corps Base along the coast between Los Angeles and San Diego provides a "greenbelt" between urban areas, but only limited areas of this base remain in undisturbed condition near the coast. The most significant conservation area between these two cities is the Torrey Pines State Reserve north of San Diego which provides protection for one of two habitats for the rare *Pinus torreyana*, and other significant flora. Other small reserves of coastal sage scrub and estuarine communities in this area are preserved and managed by the California Department of Parks and Recreation and the University of California Natural Reserve System.

South of the American border, the conservation status of coastal ecosystems is very poor. No natural reserves, parks, or protected areas exist along any part of the coastal region of northwestern Baja California. Even popular sites such as the Punta Banda peninsula, which promote tourism to see the natural landscapes and biota of this area, enjoy no special protected status. The result of little or no efforts at resource management along this coastal area has thus unfortunately led to a serious decline in biodiversity and an increasing invasion of exotic species.

12.5 Chile

12.5.1 Phytogeography, Community Structure and Biodiversity

The coastal transition zone between mediterranean and desert communities in central Chile runs from the area south of Valparaíso, with a mean annual rainfall of about 440 mm, and continues north to include the coastal area north of La Serena where mean annual rainfall drops to about 100 mm. Four vegetation types have been identified along this coastal zone by Gajardo (1994) in his recent overview of Chilean vegetation (Fig. 12.1). The southernmost of these units has been termed coastal sclerophyll forest (*bosque esclerófilo costero*), a zone once dominated by low sclerophyll forests with such species as *Cryptocarya alba, Lithraea caustica, Peumus boldus, Schinus latifolius, Escallonia pulverulenta*, and *Maytenus boaria*. This community becomes lower and more shrubby on drier slopes, and mostly arboreal in shaded quebradas where moisture levels are highest. This vegetation has been heavily altered by centuries of human impact, and little remains in relatively undisturbed condition. A secondary vegetation of drought deciduous and evergreen matorral shrubs dominates much of this area today. The coastal sclerophyll forest occurs inland to the south of Valparaíso where it is replaced along the coast by a dry coastal espinal. The northern limit of this association is at Los Molles where a unique community of coastal sclerophyll forest is dominated by *Lucuma valparadisiaca* (Mooney and Schlegel 1966), with stands of the *Eulychnia castanea*, an unusual shrubby cactus of more northern affinity. Relict communities of sclerophyll forest with associations of tree species more characteristic of the forest regions of southern Chile are scattered in favorable sites along the coast within this region. Good examples of such forested ravines can be seen near Quintero and around Cachagua and Zapallar. In addition to these pocket forests, the coastal sclerophyll forest association also includes notable azonal communites on extensive coastal dunes that occur near Reñaca and at Concon to the north of Viña del Mar (Serey et al. 1976). Ecological resources along this coast extend not just to vascular plant vegetation, but also to communities of breeding sea birds and marine mammals on small islands adjacent to the mainland. Penguin rookeries can be readily seen on the Isla de los Pinguinos near Cachagua.

From Pichidangui northward past Los Vilos, Gajardo (1994) classified the vegetation as open arborescent matorral (*matorral estepario arborescente*) which is characterized by a woody matorral with both mixed dominance of taller and low shrubs. This community includes both evergreen shrubs such as *Lithraea caustica* and the less sclerophyllous *Baccharis concava* and *Flourensia thurifera*, but greater dominance is found today by drought deciduous shrubs such as *Eupatorium glechenophyllum, Bahia ambrosioides*, and *Lepichinea salviae* (Weisser and Rundel 1980). The unusual *Fuchsia lycioides* is a characteristic associate (Atsatt and Rundel 1982). Areas along the coastal terraces that have been heavily affected by grazing typically support a sparse community of unpalatable species such as *Puya chilensis, Baccharis linearis*, and *Haplopappus polyphyllus*.

Moving northward from the open arborescent matorral, the coastal vegetation is termed the open matorral with woods (*matorral esteperio boscoso*) by Gajardo (1994). This coastal vegetation type is dominated by a low open coastal matorral dominated by such shrubs as *Heliotropium stenophyllum, Lithraea caustica, Bahia ambrosioides, Baccharis concava*, and *Fuchsia lycioides*. Heavy grazing pressure has virtually eliminated vegetation cover over extensive areas, producing open assemblages of the same unpalatable species described above. Both *Puya chilensis* and *P. berteroniana* are widespread and abundant, as is the shrubby *Echinopsis coquimbana* along the coast. Small pockets of sclerophyll woodlands remain in sheltered canyons along the coast. The northern limit of this zone lies near the mouth of the Río Limari with Fray Jorge National Park. This park contains a remarkable fog forest community dominated by *Aextoxicon punctatum*, and represents a relict outlier of Valdivian forest vegetation, as described in more detail below. Also present in the park area is a unique arid thornscrub community dominated by *Porlieria chilensis* (Muñoz Pizarro and Pisano 1947; Gutiérrez et al. 1993) and riverine communities along the Río Limari.

The northern transition from mediterranean to desert ecosystems in central Chile begins just northward of 31°N, extending up past La Serena to near Los Hornos where the Panamerican Highway turns inland from the coast. This association has been termed open coastal matorral (*matorral estepario costero*; Gajardo 1994). This area, again severely impacted by grazing pressure over centuries, once supported an open community of low evergreen and drought deciduous shrubs. The dominants today are such species as *Heliotropium stenophyllum, Baccharis concava, Cassia coquimbensis, Bahia ambrosioides, Puya chilensis, Oxalis gigantea, Fuchsia lycioides*, and the local endemic *Myrcianthes coquimbensis* (Ramírez et al. 1989). Rocky areas of coastal terraces have extensive development of several trailing subarborescent cacti, most notably *Echinopsis coquimbana* and *Eulychnia procumbens*.

The mediterranean desert transition zones in central Chile, as with the other regions discussed in this chapter, are rich in both species diversity and endemism. Although no floristic analysis of this coastal region has been made, the native flora likely exceeds 1000 species. Local floras are available for the Zapallar region (Johow 1948) and for Fray Jorge National Park (Muñoz Pizarro and Pisano 1947; Skottsberg 1948; CONAF 1992), with the latter having more than 400 species. Prominent endemics can be see in the coastal succulent flora, including two unusual *Puya* species, *P. venusta* and *P. coquimbensis*, and a diverse assemblage of Cactaceae such as *Eulychnia procumbens, E. castanea, Echinopsis litoralis, Neoporteria subgibbosa, N. chilensis, N. clavata, N. horrida*, and *Copiapoa coquimbana*. Regional shrub endemics notably include the monotypic *Myrcianthes coquimbensis, Myrceugenia rufa, Lucuma valparadisiaca, Fuchsia lycioides*, and *Carica chilensis*. All of these with the exception of *F. lycioides* are considered rare or endangered (Benoit 1989).

12.5.2 Landscape Disturbance

The central and semi-arid coasts of Chile have had a long history of heavy impact by man. The city of Valparaíso was founded in 1542 when the Spanish conquistador Pedro de Valdivia declared it a seaport, and La Serena near the northern end of the mediterranean/desert transition zone dates to 1544. While these ports grew slowly in their first two centuries, Chilean independence early in the 19th century opened these ports to international trade and commerce. It was at this same time that mining discoveries of copper and silver in the coast ranges brought new forms of development and adverse ecological impacts. Extensive copper mines were opened at La Higuera and Brillador north of La Serena from 1820–30. Los Villos along the central transition coast began as a mining port in 1855.

Increased wealth in Chile by the end of the past century led to the opening of coastal areas north of Valparaíso as sites for vacation homes among the upper class. Viña del Mar, Cachagua, Zapallar and Papudo all grew in this manner. With all of this development and population growth along the coast came heavy pressures on coastal plant resources. Trees were felled for construction materials and firewood, and matorral shrubs were cut and burned for charcoal production. The history of human impacts on coastal environments along the northern mediterranean/desert transition zone near Coquimbo has been described in considerable detail (Bahre 1979; Etienne et al. 1986). The most significant aspect of this impact in this semi-arid region has been the dramatic effect of overgrazing by goats, and the consequent devegetation and desertification of extensive areas of coastal lands. This destruction of normal vegetation cover has had alarming effects on species diversity over broad regions, limiting plant growth to a few unpalatable species and promoting severe erosion (Fuentes and Hajek 1979, 1980; Castro and Vicuña 1990). Large area of coastal hills between Los Villos and Coquimbo have now been planted with *Atriplex nummularia*, a native of Australia, to restore plant cover and provide forage for animals. This project has met with poor success, however.

contrast with the large number of paleoendemic shrubs and large succulents which characterize the transition zones of California/Baja California and Chile. Endemic annuals in these latter two regions are principally edaphic specialists of relatively limited distribution.

With a global focus on tropical biodiversity, it is easy to overlook the remarkable levels of plant species diversity that occur in mediterranean-type ecosystems of the world (Cowling et al. 1996). As with the core mediterranean-type regions, these transition regions share the characteristic of a high number of rare and locally endemic taxa that survive in small populations. Habitat disturbance and transformation, nevertheless, are rapidly and permanently altering the habitats in which many of these species occur, and threatening their survival.

Conservation efforts to protect the coastal transition zones between mediterranean and desert have been slow to develop and remain inadequate. New scientific and popular attention to the significance of species diversity and the unique biological communities of the coastal transition zone in South Africa are a hopeful sign that more areas can be preserved in the future in this region. Programs of long-term planning for development and preservation in southern California also give some hope to greater protection of coastal sage ecosystems. In Baja California, however, destruction of coastal sage communities continues at an alarming rate with little or no active programs to help preserve elements of these coastal transition ecosystems.

Landscape disturbance and negative impacts on species diversity and the integrity of natural communities is nowhere more severe than in the coastal areas of the Chilean transition zone from mediterranean to desert communities. Widespread problems of overgrazing have continued for many years, and expanding urbanization is now an increasing problem. Without popular appreciation of this problem and pressure for a political solution, little or no examples of natural coastal matorral ecosystems will exist in the future.

A critical need exists in all three regions to develop conservation management plans to ensure the future stability and biodiversity of coastal transition zone ecosystems. Protection of these natural communities will not occur, however, until there is a greater and more widespread appreciation of both the scientific and potential economic value of these resourses.

Acknowledgments. KJE acknowledges the support of the FRD for providing her with a postdoctoral bursary in 1994. Support of PWR was made possible by research grants from the National Science Foundation and the University of California Pacific Rim Research Program. Thanks are due to S. Bullock for providing climate data for Mexico and to T. Mulroy, D. Erasmus, B. Palma, and R. Villaseñor for assistance with fieldwork.

References

Atsatt PR, Rundel PW (1982) Pollinator maintenance vs fruit production: partitioned and facultative reproduction effort in subdioecious *Fuschia lycioides*. Ann Mo Bot Gard 69:199–208

Axelson E (1977) A summary of the history of human settlement at Saldanha Bay. Trans R Soc S Afr 42:215–221

Bahre C (1979) Destruction of the natural vegetation of north-central Chile. Univ Calif Publ Geogr 23:1–117

Beauchamp RM (1986) A flora of San Diego County, California. Sweetwater River Press, National City

Benoit I, (ed) (1989). Red list of Chilean terrestrial flora. Corporación Nacional Forestal, Santiago

Boucher C (1987) A phytosociological study of transects through the western Cape coastal foreland, South Africa. PhD Thesis, University of Stellenbosch

Boucher C, Jarman ML (1977) The vegetation of the Langebaan area, South Africa. Trans R Soc S Afr 42:241–272

Boucher C, Le Roux A (1993) Dry coastal ecosystems of the South African west coast. In: van der Maarel (ed) Dry coastal ecosystems: Africa, America, Asia and Oceana. Ecosystems of the world 2b. Elsevier, Amsterdam, pp 75–88

California Coastal Commission (1987) California coastal resource guide. University of California Press, Berkeley

Castro C, Vicuña P (1986) Man's impact on coastal dunes in central Chile (32°–34° S). Thalassas (Chile) 4:17–21

Castro C, Vicuña P (1990) Caracterización de la erosión lineal en planicies costeras de Chile Central. Rev Geogr Norte Grande 17:37–45

CONAF (1992) Plan de manejo: Parque Nacional Fray Jorge. Documento de Trabajo No. 161, Corporación Nacional Forestal, Santiago

Cowling RM, Gibbs Russell BE, Hoffman MT, Hilton-Taylor C (1989) Patterns of plant species diversity in southern Africa. In: Huntley BJ (ed) Biotic diversity in southern Africa: concepts and conservation. Oxford University Press, Cape Town, pp 19–50

Cowling RM, Holmes PM, Rebelo AG (1992) Plant diversity and endemism. In: Cowling RM (ed). The ecology of fynbos: nutrients, fire and diversity. Oxford University Press, Oxford, pp 62–112

Cowling RM, Esler KJ, Midgley GF, Honig MA (1994) Plant functional diversity, species diversity and climate in arid and semi-arid southern Africa. J Arid Environ 27:141–158

Cowling RM, Witkowski E (1994) Convergence and non-convergence of plant traits in climatically and edaphically matched sites in mediterranean Australia and South Africa. Aust J Ecol 19:220–232

Cowling RM, Hilton-Taylor C (1998) Plant biogeography, endemism and diversity. In: Milton S, Dean WR (eds) The ecology of the Karoo. Cambridge University Press, Cambridge (in press)

Cowling RM, Rundel PW, Lamont BB, Arroyo MK, Arianoutsou M (1996) Plant diversity in Mediterranean-climate regions. Trends Ecol Evol 11:352–360

Deacon HJ (1992) Human settlement. In: Cowling RM (ed). The ecology of fynbos: nutrients, fire and diversity. Oxford University Press, Oxford, pp 260–270

Desmet PG (1996) Vegetation and restoration potential of the arid coastal belt between Port Nolloth and Alexander Bay, Namaqualand, South Africa. MSc thesis, University of Cape Town

Desmet PG, Cowling RM (1996) Assessment of the farm Van Niekerk's Hoop (300) for inclusion into the West Coast National Park. Institute for Plant Conservation, University of Cape Town

di Castri F (1973) Climatographical comparisons between Chile and the western coast of North America. In: di Castri F, Mooney HA (eds) Mediterranean type ecosystems: origin and structure. Springer, Berlin Heidelberg New York, pp 21–36

di Castri F, Hajek E (1976) Bioclimatológia de Chile. Vicerrectoria Academica, Universidad Católica de Chile, Santiago

Dunkle MB (1950) Plant ecology of the Channel Islands of California. Publ Allan Hancock Pac Expedit 13:247–386

Esler KJ, Rundel PW, Cowling RM (1998) The Succulent Karoo in a global context: plant structural and functional comparisons with North American winter rainfall deserts. In: Miltons S, Dean R (eds) The Karoo: ecological patterns and processes. Cambridge University Press (in press)

Etienne M, Faúndez, L, Valdés J (1986) Evaluación de la tasa de desertificacián en la zona árida de Chile Central. Ambiente y Desarrollo 2:105–112

Fuentes E, Hajek E (1979) Patterns of landscape modification in relation to agricultural practice in central Chile. Environ Conserv 6:265–271

Gajardo R (1994) La vegetación natural de Chile: clasificación y distribución geografica. Editorial Universitaria, Santiago

Gutiérrez JR, Meserve PL, Jaksic FM, Contreras LC, Herrera S, Vasquez H (1993) Structure and dynamics of vegetation in a Chilean arid thornscrub community. Acta Oecol 14:271–285

Hajek E, di Castri F (1975) Bioclimatográfia de Chile. Publ 45, Inst Geogr, Univ Católica de Chile, Santiago

Heydenrych B (1993) Farming practices in the Clanwilliam, Lamberts Bay and Elandsbaai areas: an investigation of selected environmental impacts of centre-pivot irrigation. Unpublished report of the Botanical Society of South Africa. FCC Report No 93/3, Flora Conservation Committee, Botanical Society of South Africa, Kirstenbosch

Heydorn AEF, Tinley KL (1980) Synopsis of the Cape coast – natural features, dynamics and utilization. Part 1. Estuaries of the Cape series. CSIR Res Rep No 380, National research Institute for Oceanography, Stellenbosch

Hilton-Taylor C (1987) Phytogeography and origins of the karoo flora. In: Cowling RM and Roux PW (eds) The karoo biome: a preliminary synthesis. Part 2: vegetation and history. South African National Scientific Programmes Rep No. 142, Foundation for Research and Development, Pretoria

Jackelman J, Moll EJ (1989) Langebaan to Alexander Bay: a coastal survey. Unpublished report of the Botanical Society of South Africa. FCC Rep No 89/3, Flora Conservation Committee, Botanical Society of South Africa, Kirstenbosch

Johow F (1948) Flora de las plantas vasculares de Zapallar. Rev Chilena Hist Nat 49:8–566

Junak S, Ayers T, Scott R, Wilkin D, Young D (1995) A flora of Santa Cruz Island. Santa Barbara Botanic Garden, Santa Barbara

Jürgens N (1987) Untersuchungen zur ökologie sukkulenter pflanzen des südlichen Afrika. Mitteil Inst Allgemeine Bot Hamburg 21:139–365

Jürgens N (1991) A new approach to the Namib Region. 1. Phytogeographic subdivision. Vegetatio 97:21–38

Le Roux A, Odendaal FJ (no date) Report on the study of natural recovery of overburden dumps on DBCM Properties in Namaqualand. Unpublished Report, Geology Dept, De Beers Namaqualand Mines, Kleinsee

Low AB, Rebelo AB (1996) Vegetation of South Africa, Lesotho and Swaziland. Department of Environmental Affairs and Tourism, Pretoria

Milewski A (1983) A comparison of ecosystems in mediterranean Australia and southern Africa. Ann Rev Ecol Syst 14:57–76

Miller PC (ed) (1981) Resource use by chaparral and matorral: a comparison of vegetation function in two mediterranean-type ecosystems. Springer, Berlin Heidelberg New York

Miranda Reyes F, Reyes Coca S, Espinosa I, Espinosa JG, Garcia Lopez J (1991) Climatológia de la region noroeste de México (Baja California, Baja California Sur, Sonora y Sinaloa). Parte II. Temperatura: series de tiempo del valor total mensual y estadísticas del año climatológico. Reporte Técnico CI0FT9018, Centro de Investigacion Científica y de Educación de Ensenada, Baja California, Mexico

Moll EJ, Moll NG and Savage D (1989) A preliminary report on agricultural land use in the Clanwilliam and Lambert's Bay areas. Unpublished report of the Botanical Society of South Africa. FCC Report No 89/2, Flora Conservation Committee, Botanical Society of South Africa, Kirstenbosch

Mooney HA (ed) (1977) Convergent evolution in Chile and California. Dowden, Hutchinson and Ross, Stroudsberg, Penn

Mooney HA, Schlegel F (1966) La vegetación costera del Cabo Los Molles en la província de Aconcagua. Bol Univ Chile 75:27–32

Mulroy TW, Rundel PW, Bowler PA (1979) The vascular flora of Punta Banda, Baja California Norte, Mexico. Madroño 26:69–90

Muñoz Pizarro C, Pisano E (1947) Estudio de la vegetación y flora de los parques nacionales Fray Jorge y Talinay. Agric Téc (Chile) 7:71–190

Oberbauer TA (1992) Vegetation of northwestern Baja California. Fremontia 20:3–10

Peinado M, Alcaraz F, Aguirre JL, Alvarez J (1994) Vegetation formations and associations of the zonobiomes along the North American Pacific Coast. Vegetatio 114:123–135

Peinado M, Alcaraz F, Aguirre JL, Delgadillo J, Aguado I (1995) Shrubland formations and associations in mediterranean-desert transitional zones of northwestern Baja California. Vegetatio 117:165–179

Peréz C, Villagrán C (1985) Distribución de abundáncias de especies en bosques relictos de la zona meditierránea de Chile. Rev Chilena Hist Nat 58:157–170

Philbrick RN, Haller JR (1977) The Southern California Islands. In: Barbour MG, Major J (eds) Terrestrial vegetation of California, Wiley Interscience, New York, pp 893–906

Ramírez C, San Martín C, San Martín A, Hauenstein E, Contreras D (1989) Descripción del *Myrcianthes-Bahietum* una nueva asociación vegetal arbustal del litoral de Coquimbo, Chile. Idesia (Chile) 11:19–28

Raven PH, Axelrod DI (1978) Origin and relationships of the California flora. University Calif Publ Bot 72:1–134

Raven PH, Thompson HJ, Prigge BA (1986) Flora of the Santa Monica Mountains, California 2nd edn. South Calif Bot Spec Publ No 2, Los Angeles

Rebelo AG (1992) Red data book on species in the Cape Floristic Region: threats, priorities and target species. Trans R Soc S Afr 48:55–86

Reyes Coca S, Miranda Reyes F, Garcia Lopez J (1990) Climatológia de la region noroeste de México (Baja California, Baja California Sur, Sonora y Sinaloa). Parte 1. Precipitation: series de tiempo del valor total mensual y estadísticas del ano climatológico. Reporte Técnico CI0FT9001, Centro de Investigación Científica y de Educación de Ensenada, Baja California, Mexico

Scott DS, Du Toit B, Johnson MA, Burns MER (1994) The effects of mining on soil properties and the ability of soil to support vegetation after mining at Kleinsee and Khoingnaas on the west coast of South Africa. Rep No FOR-C, Proj No FL719. Forestek:CSIR, Pretoria

Serey I, Bustamente R, Guerrero I (1976) Diversidad de la vegetación de las dunas de Con-Con. An Mus Hist Nat Valparaíso 9:23–27

Shmida A (1985) Biogeography of the desert flora. In: Evenari M, Noy-Meir I, Goodall DW (eds) Hot deserts and arid shrublands. Elsevier, Amsterdam

Skottsberg C (1948) Apuntes sobre la flora y vegetación de Fray Jorge (Coquimbo, Chile). Medd Göteborgs Botaniska Trädgard 18:91–184

Smith CF (1976) A flora of the Santa Barbara County region, California. Santa Barbara Botanic Garden, Santa Barbara

Talkenberg WFM (1982) An investigation of the environmental impact of surface diamond mining along the arid coast of South Africa. Msc Thesis, University of Cape Town

Thorne RF (1967) A flora of Santa Catalina Island, California. Aliso 6:1–77

Truter C (1996) West Coast: a tourist guide. Hirt and Carter, Cape Town

Tyson PD (1986) Climatic change and variability in Southern Africa. Oxford University Press, London

Villagrán C, Armesto JJ (1980) Relaciones florísticas entre las comunidades relictuales del Norte Chico y la zona central con el bosque del sur de Chile. Bol Mus Nat Hist Chile 37:87–101

Villagrán C, Riveros M, Villaseñor R (1980) Estructura florística y fisionómica de la vegetación de la Quebrada de Córdoba (El Tabo), Chile Central. An Mus Hist Nat Valparaíso 13:71–91

Wallace GD (1985) Vascular plants of the Channel Islands of southern California and Guadaloupe Island, Baja California, Mexico. Nat Hist Mus Los Ang, Contrib Sci 365:1–136

Weather Bureau (1986) Climate of South Africa: Climate statistics up to 1984. Rep No WB40, Government Printer, Pretoria

Weisser P, Rundel PW (1980) Estudio cuantitativo de un matorral costero en Pichidangui (Prov. Coquimbo, Chile). Anal Mus Hist Nat Valparaíso 13:47–57

Westman WE (1983a) Xeric mediterranean-type shrubland associations of Alta and Baja California and the community/continuum debate. Vegetatio 52:3–19

Westman WE (1983b) Island biogeography: studies on the xeric shrublands of the inner Channel Islands, California. J Biogeogr 10:97–118

Wiggins IL (1980) Flora of Baja California. Stanford University Press, Stanford

13 Distribution and Ecology of Geophytes in Chile. Conservation Threats to Geophytes in Mediterranean-Type Regions

A.J. Hoffmann, F. Liberona and A.E. Hoffmann

13.1 Introduction

In this chapter, we study the distribution of geophytes in Chile and review the literature on their biology and ecology. This may help in understanding the causes accounting for their distribution patterns, and the results may serve to determine the conservation status of geophytes in central Chile. To our knowledge, this is the first time that an assessment of the conservation status, based on objective data, has been attempted for geophytes. Finally, we analyze conservation problems for geophytes in other mediterranean-type regions. The design of conservation measures for geophytes is urgent in these regions, where anthropogenic intervention has occurred for many years. We refer briefly to the conservation status of geophytes in Chile, California, the Iberian Peninsula and South Africa. Rather than discussing particular problems in each country, we analyze what is known, what can be done, and what has been done toward the conservation of geophytes in mediterranean regions. On the other hand, many of these plants have attractive flowers (Hoffmann 1989), and knowledge of their biology and requirements is useful for cultivation, in turn helping to preserve some threatened species.

Geophyte is a generic term used for plants that survive both by seed and by diverse storage organs where the perennating buds or shoot apices are buried in the ground during the unfavorable season. The underground organs are commonly bulbs, corms, rhizomes or root tubers. Some bulbs are replaced annually, while others are perennial (Raunkiaer 1934; Rees 1989; De Hertogh and Le Nard 1993). Geophytes include both dicotyledonous and monocotyledonous species, but only the petaloid monocotyledons will be considered here. Monocotyledonous geophytes seem to have evolved in the tropics, from ancestors with a sympodial growth habit (Rees 1989). This growth form appears ideally suited for the development of resting organs that allow plants to cope with unfavorable periods when they spread to seasonal, unpredictable climates.

Ecological Studies, Vol. 136
Rundel et al. (eds.) Landscape Degradation and Biodiversity in Mediterranean-Type Ecosystems

Geophytes are important components of mediterranean-type and semi-desert climate vegetations (Raunkiaer 1934; Dafni et al. 1981a,b; Rees 1989; Dafni 1993), but are generally rare in extreme desert climates (Shmida 1981). In Chile, with a native vascular flora of around 5100 taxa (Marticorena and Quezada 1985; Marticorena 1990), there are about 250 geophyte monocotyledonous species, which represent ca. 5% of the flora. The geographic and climatic features of the country, with precipitation increasing and temperature decreasing toward higher latitudes, provides a good opportunity to study the distribution of geophytes in a north-south climatic gradient of about 4000 km, and to test the hypothesis that geophytes are especially frequent in the mediterranean regions (Raunkiaer 1934).

The natural distribution patterns of geophytes probably depend on the interactions between the environment and the biological characteristics of each species. Thus, local conditions may represent environmental constraints for the survival of some species. An influence of landform and soil type on the distribution of geophytes has been recently shown in Israel (Dafni 1993). Therefore, we gathered information on occurrence of species at diverse landforms; distribution on soil types; types of storage organ; breeding systems and flowering phenology; and proportion of geophytes that flower after leaf die-back (hysteranthous foliage) in relation to geophytes where leaf presence and flowering are simultaneous (synanthous foliage) according to Dafni et al. (1981a,b) and Rees (1989).

13.2 Latitudinal Distribution

The analysis of data available on the geographic range of 208 of the approximately 250 species mentioned for Chile (see Appendix) between Arica (18°29′ S) and Tierra del Fuego (54°20′ S) shows that geophytes exist throughout the country. Their diversity differs notoriously with latitude (Fig. 13.1). The number of species present at four degree-wide latitudinal belts were correlated to the latitudinal distribution of mean values of annual rainfall and temperature (Fig. 13.2). The following patterns were evident:

Few species exist in the latitudinal belts 18°–22° and 22°–26° S, at the northern end of the gradient. This coincides with an almost complete absence of precipitation and a mean annual temperature over 17 °C. An orchid of the genus *Aa* is one of the few geophytes found at these latitudes. Geophytes slowly increase in number toward the south: 24% of the species existing in the country occur in the 26°–30° S belt, with mean annual pre-

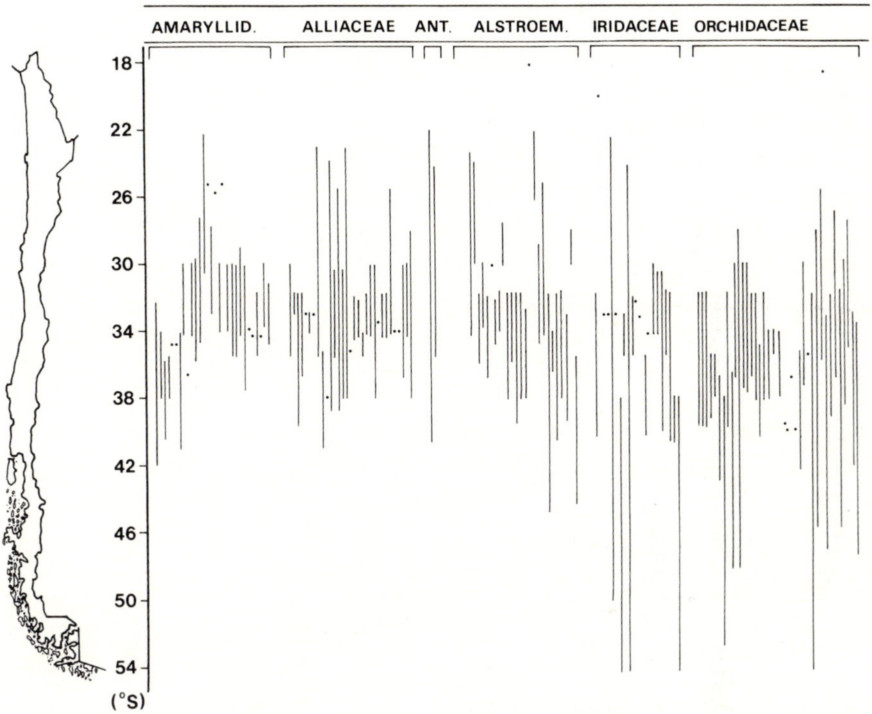

Fig. 13.1. Examples of the geographic ranges of geophyte species in Chile (not all species in each family are included). (·) indicates presence at a single location. *Amaryllid* Amaryllidaceae; *Ant* Anthericaceae; *Alstroem* Alstroemeriaceae

cipitation up to 100 mm and a mean annual temperature slightly over 16 °C. Between 30° and 34° S, and between 34° and 38° S, a dramatic increase in diversity occurs: 60% of the species are present at each of these belts. Together, they include 85% of the species existing in the country. Annual precipitation increases from 100 mm at 30° S to 1500 mm at 38° S, and temperature, which is 15.5 °C on average at 30° S, drops to 12 °C at 38° S. Between 38° and 42° S the number of geophytes decreases again, with less than 30% of the species present in this latitudinal belt. Here, annual precipitation exceeds 2000 mm and mean temperature is below 12 °C. At latitudes higher than 42° S, only 5 to 10% of the species occur in each latitudinal belt. Precipitation ranges from 2000 mm to over 4000 mm and the mean temperature is below 10 °C. At the southernmost belt (50°–54° S), a slight increase in the number of species occurs. Precipitation is scarce, less than 500 mm, and mean annual temperature is below 8 °C.

These results support the hypothesis that, along the climatic gradient of the country, the highest diversity of geophytes corresponds to the

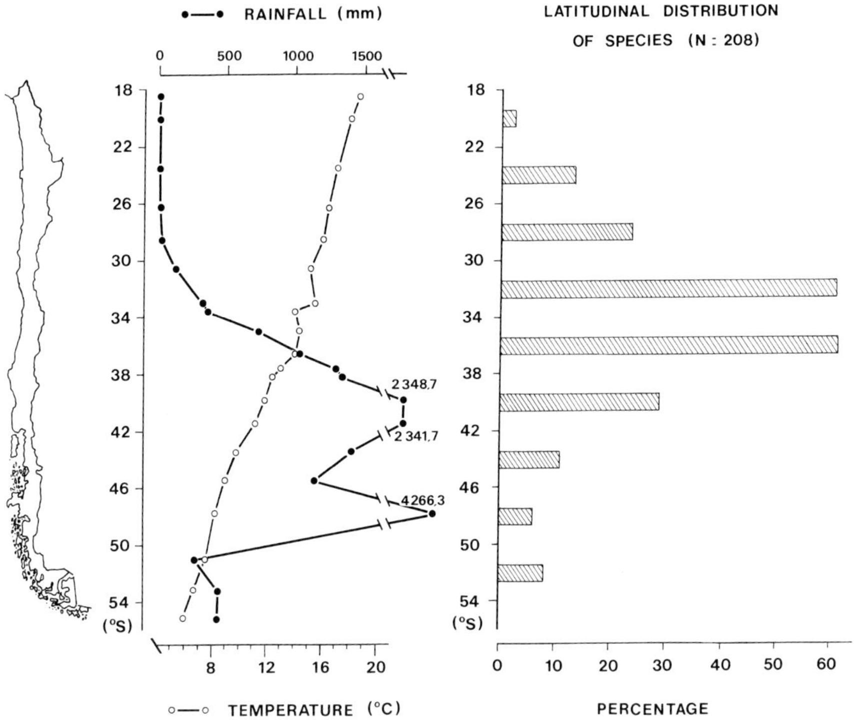

Fig. 13.2. Distribution of geophytes at 4° wide latitudinal belts in Chile, expressed as percentage of the total number of species considered (N = 208), together with mean temperature and rainfall along the gradient (climate data from di Castri and Hajek 1976)

mediterranean-type region. North of this region, the lack of precipitation and high temperature would prevent the survival of geophytes. And south of the mediterranean region, humidity appears to be excessive for most geophyte species.

Phylogenetic relatedness seems to influence the geographic distribution of geophytes. To investigate the replacement of species along the gradient, Jaccard's similarity values in floristic composition were calculated between pairs of latitudinal belts and subjected to a cluster analysis. The dendrogram indicates the following patterns of change in species composition (Fig. 13.3): Between 18° and 30° S, where low numbers of species occur, very low similarity was found between neighboring belts (as well as with the belts located at the higher latitudes), indicating quick species replacement at the northern part of the gradient. In contrast, the 30–34° and 34–38° belts, with high numbers of species, have a similarity value of ca. 60%. At latitudes higher than 46°, also with few species present, relatively high similarity values were found between neighboring belts. For instance, be-

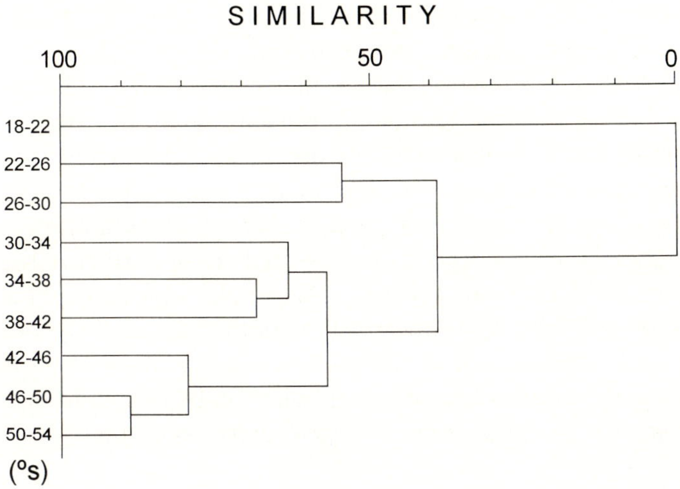

Fig. 13.3. Cluster analysis of Jaccard's similarity index in floristic composition, computed between pairs of latitudinal belts in Chile. 0 similarity is the minimum similarity detected

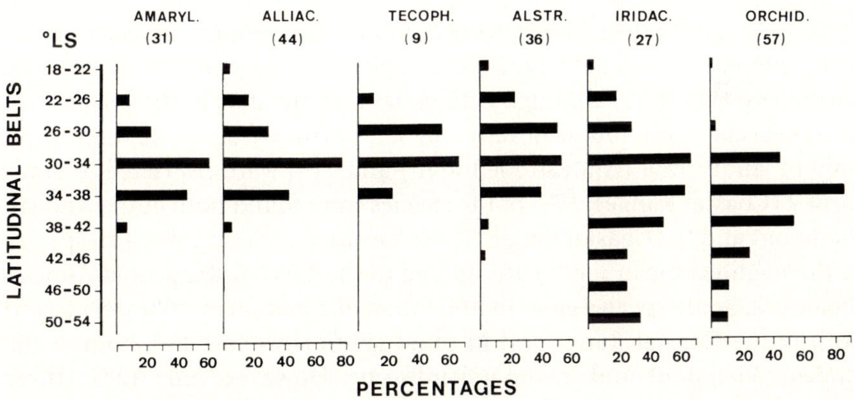

Fig. 13.4. Latitudinal distribution of geophytes in Chile, by family. *Amaryl* Amaryllidaceae; *Alliac* Alliaceae; *Tecoph* Tecophyllaceae; *Alstr* Alstroemeriaceae; *Iridac* Iridaceae; *Orchid* Orchidaceae. Number of species in *parentheses*

tween the 46°–50° and 50°–54° belts a similarity value of almost 70% was obtained. Also notable was the low similarity in species composition between southern belts (42°–46° to 50°–54°) and the central belts (30°–34° to 38°–42°).

When the geophytes were grouped according to family, differences also became apparent in the latitudinal distribution ranges of some families (Fig. 13.4). Species of Amaryllidaceae, Alliaceae, Tecophilaeaceae and Alstroemeriaceae were distributed between the northern end of the latitudinal range and the 38–42° belt, with highest frequency in the 30–34° belt (only one species of Alstroemeriaceae extended further south). The distributional range of 64% of the Amaryllidaceae and of 68% of the Alliaceae present in Chile was restricted to the mediterranean region. In contrast, species of Iridaceae and Orchidaceae were distributed throughout the country. Although the highest frequency of Iridaceae also occured in the 30–34° belt, several species were distributed toward the southernmost belts. In turn, Orchidaceae were scarcely represented at the northernmost end of the gradient (with only one species). Their highest frequency was found in the 34°38° belt, i.e., more toward the south than all the other families. In fact, some Orchidaceae were found even at the 50°–54° belt. It is likely that such differences in latitudinal distribution be related to biological characteristics (e.g. difference in storage organs: see below).

13.3 Landform Distribution

Data on landform distribution were found for 164 geophyte species occurring between 30° and 38°S in Chile. Geophytes grew in all landform types considered (Fig. 13.5) although with variable frequency. Forty-five percent of the species were found in only one landform. Of them, 22% occurred only in landform 4 (Andean Piedmont) and 11% were restricted to landform 2 (Coastal Range); 18% of the species were found both at the Andean Piedmont and the Coastal Range. These results show that 53% of all species in the mediterranean region are limited to the Coastal Range and Andean Piedmont. Some species grow in more than one landform: 10% were found both in the Coastal Range and in other landforms, and 25% both in the Andean Piedmont and in other landforms. However, only 1.9% (three species) grew in most landforms, and it is noteworthy that no species was present in all five landforms of the mediterranean region.

13.4 Soil Type

Data on soil types where geophytes grow were found for 81 species. About two thirds of these species grew in a specific soil type, and only 28% of species grew in more than one soil type (Table 13.1). Few species grew in clay or in swampy soil. Requirement of a specific soil type appeared to be related to phylogenetic relatedness (Table 13.1). Thus, 42% of the Amaryllidaceae and 40% of the Iridaceae grew mostly in sandy soils,

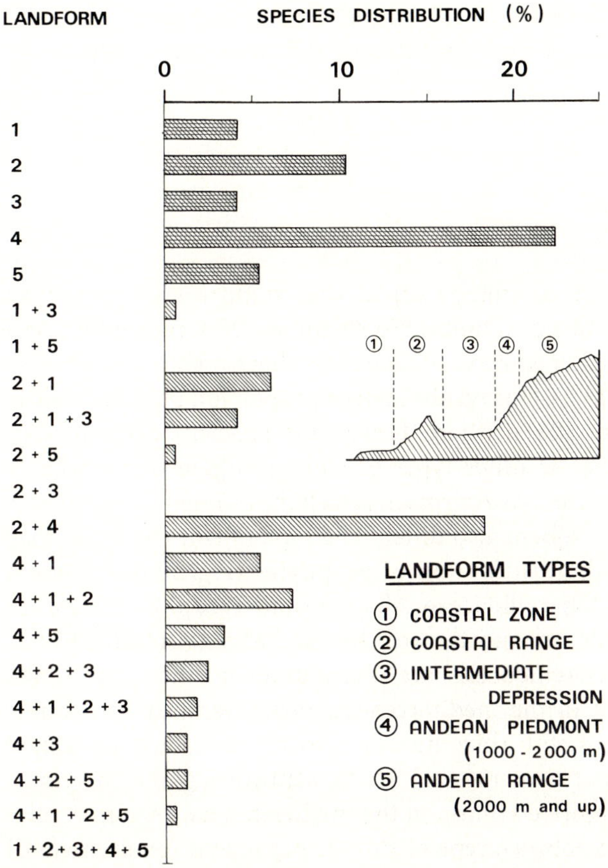

Fig. 13.5. Percentage of geophyte species present in five landform types in the mediterranean region of Chile

Table 13.1. Distribution of geophytes on different soil types. Only species of the main families were considered (data for 81 species)

Soil Type	Amaryll.		Alliac.		Alstroem		Iridac.		Orchid.		Total	
	N	%	N	%	N	%	N	%	N	%	N	%
Sand	8	42	4	27	1	6	4	40	2	9	19	23
Clay	0	0	0	0	0	0	1	10	1	5	2	3
Humus	3	16	5	33	3	19	1	10	9	43	21	26
Stones	2	10	2	13	5	31	2	20	3	14	14	17
Swamps	0	0	0	0	0	0	1	10	1	5	2	3
Varius soils	6	32	4	27	7	44	1	10	5	24	23	28
Total	19	10	15	10	16	10	10	10	21	10	81	10
	0		0		0		0		0		0	

whereas 33% of the Alliaceae and 43% of the Orchidaceae grew in humus. Species of Alstroemeriaceae grew in different soil types (44%) or in stony soils (31%).

13.5 Storage Organ

Information on the type of storage organ was found for 201 geophyte species. Among these, 42% had bulbs, 8% rhizomes, 28% root tubers and 22% rhizomes with thickened roots. Apparently there are no species with corms in Chile (Table 13.2). Our results show a proportion of bulbs similar to that in Israel (Dafni 1993), where 44% of the species have bulbs. In contrast, the frequencies of other types of storage organs are markedly different: 19% of the species have corms in Israel; the frequencies of species with rhizomes and root tubers also differ. The type of storage organ may have an influence on the distribution of geophytes (Fig. 13.6). Whereas geophytes with bulbs (Amaryllidaceae, Alliaceae and Tecophilaeaceae) are present in central Chile and northward, but are absent south of 38°S, geophytes with tubers (Orchidaceae) are absent in the northern part of the gradient, are abundant in the mediterranean region, and extend southward. Finally, geophytes with rhizomes, or with rhizomes and thickened roots (Iridaceae and Alstroemeriaceae) are present throughout the gradient although they are more common in the mediterranean region. A relationship may also exist between type of storage organ and requirement of soil type. However, further research is needed to clarify this point.

Table 13.2. Types of storage organs in geophytes of Chile (data for 201 species)

Family	Bulb	Rhizome	Root Tubers	Rhizome and Fleshy Roots	Species N
Amaryllidaceae	31	0	0	0	31
Alliaceae	43	0	0	0	43
Anthericaceae	0	0	0	2	2
Hyacinthaceae	1	0	0	0	1
Tecophilaceae	9	0	0	0	9
Alstroemeriaceae	0	0	0	36	36
Iridaceae	0	16	0	6	22
Orchidaceae	0	0	57	0	57
Total	84	16	57	44	201
%	42	8	28	22	100

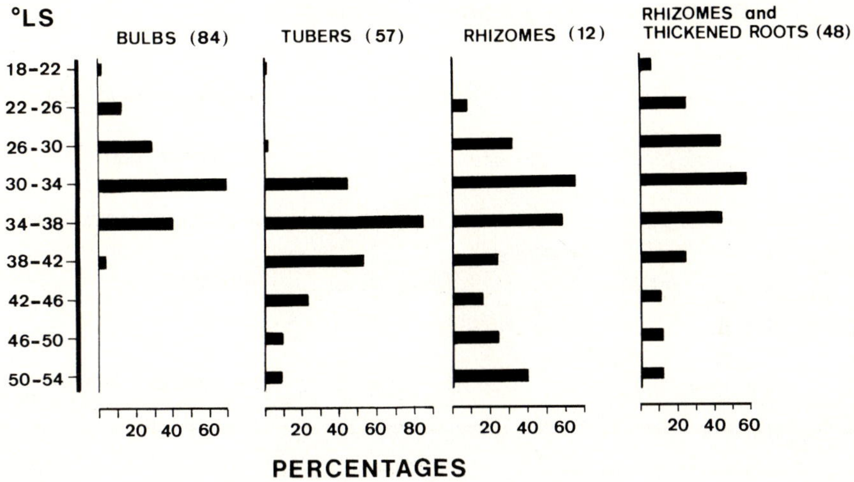

Fig. 13.6. Latitudinal distribution of geophytes in Chile according to their type of storage organ. Number of species in *parentheses*

13.6 Flowering Phenology

Data on flowering phenology were found for 157 geophyte species. Results show that at least one species is in bloom each month of the year, but blooming mostly occurs from early spring to early summer, with a peak in November (Fig. 13.7). No bimodal distribution was observed. This result is at variance with geophyte flowering in the mediterranean region of Israel, where a clear bimodal distribution is observed (Shmida and Dafni 1989), as well as with results described for several temperate angiosperm assemblages (Kochmer and Handel 1986).

A relationship between flowering phenology and phylogenetic relatedness is apparent (Fig. 13.7). While the overall flowering peak of monocotyledonous geophytes occurs in November, that of two families is skewed toward the austral spring with the Alliaceae peaking in September and the Amaryllidaceae in October. The flowering peak of the Alstroemeriaceae and Iridaceae is centered in November. Finally, the flowering peak of one family, the Orchidaceae, is skewed toward the austral summer (December), as also reported by Bernhardt (1995). These results thus concur with Kochmer and Handel (1986), who found that phylogenetic constraints are important in determining flowering phenology.

Differences in timing of flowering have been explained as natural selection to avoid competition for pollinators. In Chile, knowledge on pollination vectors is very limited for geophytes. Data are available only for *Hippeastrum bicolor* and *H. advenum* (Saavedra et al. 1996). These two

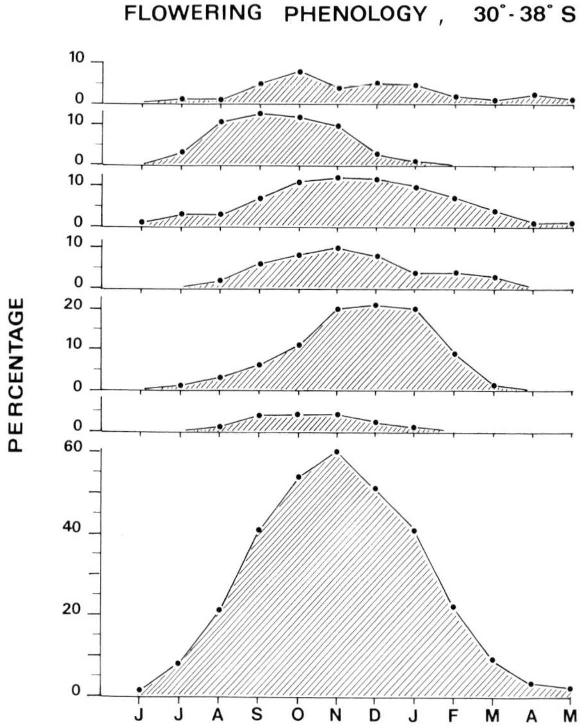

Fig. 13.7. Percentages of geophytes in bloom each month in the mediterranean region of Chile. Number of species per family in *parentheses*

species grow in the Andean piedmont of central Chile, but in different microhabitats. Flowering is markedly segregated in time. While *H. bicolor* blooms in spring and is visited by a hummingbird (*Patagona gigas*) and several large Hymenoptera, Coleoptera and Lepidoptera, *H. advena* blooms in late summer and is visited by small Hymenoptera, Diptera and Coleoptera. Therefore, the latter geophyte, with smaller and less showy flowers than the large red-colored *H. bicolor* may have evolved to flowering displacement and utilization of less specific pollinators (Saavedra et al. 1996; Hoffmann and Saavedra unpubl. data).

Data on flowering after foliage die-back were found for 101 species. Results indicate that 20% of the species have hysteranthous foliage with this characteristic in the timing of flowering. Scarce information is available in Chile on the latitudinal distribution of species with synanthous vs hysteranthous foliage, although some species, such as *Chloraea galeata* and *C. virescens*, appear to have hysteranthous foliage in the drier end of their distribution and synanthous foliage in moister habitats. In the mediterranean flora of Israel a somewhat higher proportion of species is hysteranthous (Dafni et al. 1981b). In general, it may be expected that geophytes with hysteranthous foliage be more common in drier habitats, as has been described for Israel (Dafni et al. 1981a,b) and South Africa, where a transition from synanthous to hysteranthous forms occurs from the moister eastern Transvaal toward the South Cape (Dafni et al. 1981a).

Concerning the influence of elevation on the frequency of hysteranthous foliage, among nine species found by Arroyo et al. (1981) at Cordón El Cepo (2300–3400 m elevation), Cordillera de los Andes (33°10′ S), only two geophytes (*Alstroemeria pallida* and *Rhodophiala montana*) had hysteranthous foliage, suggesting that this habit is not adequate for the climatic conditions at high altitudes. Some phylogenetic constraints may also bear on leafless autumn-flowering, because species with hysteranthous leaves exist among all the geophyte families but are absent among the Iridaceae (our results).

13.7 Breeding Systems

Information on the breeding system of geophytes was found for 11 species. Among them, 73% were self-incompatible, 18% self-compatible and 9% partially self-compatible (Table 13.3). These proportions agree well with

Table 13.3. Breeding systems in Chilean geophyte families as compared with other regions (Saavedra et al. 1996)

Families	Species	SI[a]		SC[b]		SI/SC[c]	
		N°	%	N°	%	N°	%
Chile							
Amaryllidaceae	7	4		2		1	
Liliaceae	1	1		0		0	
Iridaceae	3	3		0		0	
Total	11	8	73	2	18	1	9
Other regions							
Amaryllidaceae	17	13		3		1	
Liliaceae	42	41		0		1	
Iridaceae	2	2		0		0	
Total	61	56	92	3	5	2	3

[a] Self-incompatible.
[b] Self-compatible.
[c] Partially self-compatible.

those reported for 65 species of geophytes from diverse geographic origins (Table 13.3). Of these latter species, 92% were self-incompatible (see Saavedra et al. 1996 for a review). These results stress a strong tendency for xenogamy among bulbous geophytes in comparison to other herbaceous species: out of 104 non-geophyte perennial herbs drawn from many genera, only 27% were self-incompatible (Arroyo and Squeo 1990). Nevertheless, some species self-pollinate via mechanical autogamy when pollinators are lacking, as reported for orchids at higher altitudes, where insect vectors are infrequent or sluggish (Arroyo and Squeo 1990).

The type of breeding system may be important for the survival capability of species growing in areas severely modified by anthropogenic activities. Although geophytes reproduce sexually by seeds, they also reproduce by means of vegetative propagation. In sites where population size is severely reduced, or where the activity of specific pollinators is impaired by human intervention, geophytes propagate only vegetatively. Hence, although they survive in the short term, their ability for outcrossing decreases, restricting the potential for variation, adaptation and evolution.

13.8 Conservation of Geophytes in Mediterranean-Type Regions

In countries where lists of threatened plants are available, the highest percentages of rare and endangered species usually are from areas with a mediterranean climate. The geographical diversity in these regions results in many types of habitats and different plants adapted to each one. Some of these plants are naturally rare, and as a result of the increasing human population with the resultant habitat alterations, endemic or rare species may be destroyed even before they are catalogued.

In Chile, over 70% of the geophytes are endemic (Table 13.4), and there are 13 endemic genera. As early as 1976, Muñoz mentioned eleven geophytes among seventy threatened or endangered plant species in the country: three Iridaceae (*Herbertia lahue*, *Calydorea xiphiodes*, *Tigridia philippiana*), one species of the Tecophilaeaceae (*Tecophilaea cyanocrocus*), two Alstroemeriaceae (*Leontochir ovallei*, *Alstroemeria gayana*), four Amaryllidaceae (*Gethyum atropurpureum*, *Ancrumia cuspidata*, *Garaventia graminifolia*, *Erinna gillesiodes*), and one species of the Arachnitaceae (*Arachnitis uniflora*) (Prance and Elias 1978). The number of endangered geophytes may be much higher: in a recent analysis of the conservation status of geophytes, Hoffmann (1989) considered that of a group of 136 Chilean geophyte species, 77 had conservation problems:

Table 13.4. Distributional ranges of geophytes

Family	Species	Endemics		Present only between 30°–38°		Present at one or two locations <50 km apart	
	N	N	%	N	%	N	%
Amaryllidaceae	31	26	84	20	65	15	48
Alliaceae	44	35	80	30	68	16	36
Anthericaceae	2	1	50	0	0	0	0
Hyacinthaceae	1	1	100	0	0	0	0
Tecophilaceae	9	9	100	4	44	1	11
Alstroemeriaceae	36	31	86	11	31	2	6
Iridaceae	27	14	52	6	22	2	7
Orchidaceae	57	32	56	23	40	6	10
Corsicaceae	1	1	100	0	0	0	0
Total	208	150	72	94	45	42	20

one was extinct, six were endangered, 40 were vulnerable, 31 were rare. Further, almost half of the 208 species of geophytes in our study were found exclusively in the mediterranean region of Chile, between 30 and 38°S. Among these species, 20% have been reported for one locality only, or for two localities separated by less than 50 km, i.e., they are narrow endemics. Therefore, at least 40 species may be considered as endangered, and if the endemics restricted to the mediterranean region are also included, their number increases to about 95 endangered geophyte species (Table 13.4).

In Spain and Portugal several geophyte species are threatened. Among them are daffodils (*Narcissus*). Although miniature garden varieties have been obtained from wild species, the demand for miniature daffodils presently exceeds supply. However, the ability to develop new kinds of miniature daffodils is hampered by the severe decrease in number of many of the small, wild species. One of them is *N. calcicola*, a tiny daffodil that is considered endangered because of overcollecting.

South Africa, in turn, has intense pride and awareness of its exceptionally rich native flora, estimated conservatively as 6000 species, but claims of more than 10000 species have also been heard. Unfortunately many of them are threatened: in the most recent update (1982), 1621 species were listed as either extinct, endangered or critically rare. South Africa harbors about 22% of the world's amaryllid species. Of these, 66 species in some 13 genera fall in different categories of endangerment. This amounts to 30% of southern African species (Koopowitz and Kaye 1986). But there is strict legislation to protect the flora, many plant societies are devoted to wildflowers, and plant nurseries tend to specialize in indigenous plants. A fair number of species may be expected to be saved thanks to these initiatives.

In California, many active groups and foundations are trying to promote the use and conservation of endemic plants. In the open oak woodlands, where scattered oak trees stand in open grassland, small bulbous and herbaceous plants thrive in late winter and early spring. This vegetation type is adapted to a natural fire ecology. Man has strenuously "protected" the area from natural blazes, altering the natural system. Only in the last few years it has been realized that burns are necessary to maintain the characteristic sclerophyllous vegetation of the coastal area.

From our study, we may also conclude that some biological and ecological features of geophytes render them even more vulnerable to human impact. Many species seem to have specific soil type requirements, consequently being restricted to specific habitats. It is likely that such restrictions be related to morphological characteristics, for instance structure of the storage organ. On the other hand, although knowledge on the breeding system exists for a few species in Chile, based on what is known on geophytes elsewhere, it may be assumed that many geophytes in the country are self-incompatible, i.e., other individuals of the same species are needed for successful fertilization. Therefore a severe reduction in number of individuals, or a decrease in insect pollinators due to excessive use of pesticides may result in a decrease in seed production, thus limiting the species to vegetative reproduction.

13.9 What Is Known?

Conservation problems have been studied in detail for a few geophyte families, and some case histories follow.

The Amaryllidaceae are a small, uniform group of petaloid monocotyledonous species that are allied to the Liliaceae (Koopowitz 1986). As a whole, the family has supplied many desirable ornamentals. Two major centers of distribution can be recognized: South America and South Africa. Certain genera, such as *Narcissus*, also have centers of diversity in the Iberian Peninsula.

The total number of amaryllids presently at risk in the world is not known, but the Amaryllidaceae possess several biological features that render them particularly vulnerable. They have a very long generation time. Time taken to produce flowers varies between 3 and 9 years. The longevity of many species is much longer, and clumps of some bulbs might be well over a century old. Many species and genera are self-incompatible, which makes pollinators indispensable. Species in the African genera *Booephane, Brunsvigia, Clivia* and *Cyrtanthus*, the American *Eucharis, Euchrosia, Hyppeastrum, Rhodophiala* and *Urceolina* (Koopowitz 1986) as well as those of the Chilean genus *Hippeastrum* (Saavedra et al. 1996) are self incompatible.

Among the Iridaceae, *Gladiolus guenzii*, which grows as evergreen clumps on sand dunes, also figures in the endangered species list of South Africa. It has been grown in culture for some years, and if good hybrids were obtained the plant could make a good perennial landscaping plant for warm climates. *Gladiolus citrinus* is now almost impossible to find in the wild. It occurred only along a four-mile stretch of the western coast of the Cape Peninsula. Out of 70 plants counted in 1975, only 18 were found 2 years later. These remaining plants lie between picnic grounds and a children's playground. Fortunately the plant has been brought into cultivation at Kirstenbosch, and seeds have been sent to interested gardeners around the world. Scientists believe that this species is very primitive and close to the ancestral species that evolved into the large gladiolus group (Koopowitz and Kaye 1986). If so, preservation of this species involves a special responsibility.

In the International Union for Conservation of Nature (IUCN) database, two geophyte families, Orchidaceae and Liliaceae, are mentioned among the nine families with most threatened species worldwide. Of them, the highest proportion of threatened species (14.1%) was found among the Liliaceae.

Among the Orchidaceae, *Disa uniflora*, one of the most fabulous orchids in the world, occurs on Table Mountain in South Africa. It was collected for flower sale, and was popular even as late as the 1960's, but later it almost disappeared. The species has been protected since 1920, but flower picking and export of plants continued. Fortunately, Swedish gardeners discovered a way to germinate and grow the plants and now it is also cultivated in the National Garden in South Africa. Many species have narrow ranges, so that even very localized modifications of the landscape may represent severe threats for the survival of some species. According to Davis et al. (1986), the following are among the major threats to wild bulb populations in mediterranean type regions:

Conversion of Natural Bulb Fields to Urbanization and Agriculture. This is particularly keen in central Chile, where the highest diversity of geophytes coincides with the highest density of population in the country (Fig. 13.8). Urban developments, road construction, forestry (mainly *Pinus* and *Eucalyptus* plantations), tourist resorts on the coastal zone, dam construction, fire, etc., have severe impacts on geophyte populations.

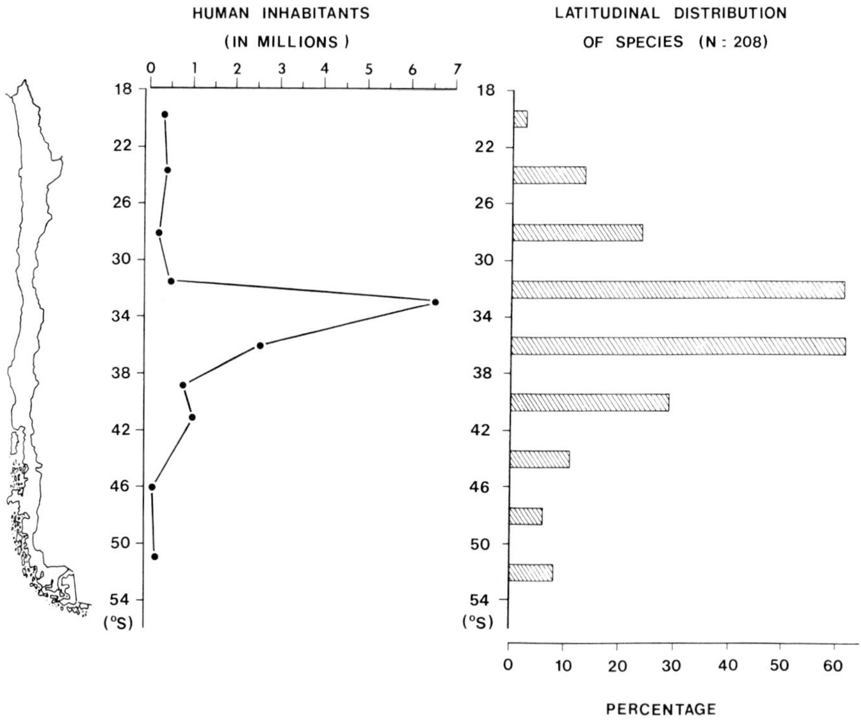

Fig. 13.8. Relationship between distribution of human inhabitants in Chile and latitudinal distribution of geophyte species. Number of species in *parentheses*

The same problems are found in the Iberian Peninsula (Koopowitz 1990), and apparently also in South Africa, where the greatest richness of the flora stretches along the same boundaries as the city of Cape Town, its suburbs, and surrounding farms. This area encompasses less than one percent of South Africa yet it contains 65% of the country's endangered species.

Overcollecting of Wild Bulbs and Seeds. Natural populations of geophytes are often affected by bulb and seed collecting or by cut-flower marketing. Dozens of wild bulb species are under threat of over-exploitation, and some species have been lost. Interest in wild bulbs has increased of late, which becomes evident in the greater emphasis on bulb species noticeable in an increasing number of catalogs during the past decade. Bulbs are collected and sent to intermediate destinations, mainly Dutch bulb companies that re-export them world-wide (Read 1989). Horticulturists, dealers and buyers are always searching for novelty. Novel species or varieties originate both from plant breeders' efforts and from Nature, but the discovery of species suitable for gardening has not always led to propagation of these species in nurseries because for many people gardening is also a business where short-term profit outweighs other considerations. Thus, if a given species is obtainable from the wild at lower prices than from artificial propagation, then the wild supply will be exhausted first (Read 1989).

In Chile, *Tecophilaea cyanocrocus* has disappeared in the wild due to bulb collectors, and cutting of wild flowers for commercial purposes may also affect other natural geophyte populations (Hoffmann 1989). Another cause for overcollecting is that in some locations, mainly in the arid North, bulbs and rhizomes are intensively collected for food (for instance *Aa* and *Conanthera*), to such extent that they have become very rare and difficult to find. Turkey appears to be the greatest source of wild collected bulbs, and seemingly at least 20 types of bulbs are exported annually in excess of 100 000 individuals per species. In Portugal, a million wild *Narcissus* bulbs are exported each year, and in Spain, a subspecies of *Narcissus* was collected to extinction. This type of exploitation also occurs in several other mediterranean-climate countries, such as Italy and South Africa.

Desertification. Increased desertification may be important for geophyte conservation, especially in the more xeric ends of mediterranean regions, where the unpredictability of rainfall increases. In Israel, diverse bulb species are affected by droughts. Soil water availability and soil temperature regulate the time of flowering in geophytes. Scarce rains result in smaller sized bulbs, thus decreasing flowering and seed production in the following season (Boeken 1989), and some geophytes may fail to emerge above ground during the growing season, as has been reported for *Tulipa salsola* (Boeken unpubl. data). In other species, the main factor controlling time of flowering is the drop in soil temperature after the summer (Boeken and Gutterman 1986; Boeken 1989). Therefore, in areas denuded of vegetation, sprouting and flowering of geophytes may be altered by modifications in the soil temperature cycle. This is important, because a displacement in flowering time may severely affect pollinators, with a consequent decrease in seed production.

Herbivory. Geophytes are often eaten by herbivores. In Chile and Turkey, goats severely browse on the foliage of geophytes; in Israel, porcupines are important bulb predators (Guttermann and Boeken 1988). Although some geophytes in Chile are heavily consumed by the fossorial rodent *Spalacopus*, it has been observed that, at least in the case of *Leucocoryne ixioides*, the number of plants does not decline in time, and actually the populations exhibit an overlapping distribution with the rodent (Contreras and Gutiérrez 1991). Total bulb biomass does not differ between areas with and without fossorial rodents, because even though the bulbs are smaller in rodent-present areas, they are more numerous.

Disturbance. Although most disturbances are threats to bulb populations, sometimes they result in increases in geophyte populations. Destruction of the natural vegetation may induce an increase of some geophytes. In Chile, sun-loving (heliophytic) species, such as *Alstroemeria* spp. numerically increase on grazed fields, at road sides or after fire. The same happens with *Rhodophiala advenum* in areas bared of vegetation (Hoffmann pers. obs). Similar effects have been described in Israel (Dafni 1993), where exposed phryganas that have been subjected to grazing and/or fire are rich in geophytes. Some species are even used as indicators of overgrazing. Superficial ploughing also seems to promote propagation of geophytes (Dafni 1993).

Extinction. The presence or absence of geophyte species is difficult to ascertain *in situ*, because some species may not bloom for many years. This explains why few cases of extinction are well documented. That is also the case in Chile. An important number of species was described during the last century, and some are very rare. A few of them may be already extinct, particularly in the Coastal Range and the Andean Piedmont in central Chile. *Tecophilaea cyanococcus* is one of the few species that have been carefully sought. However, plants have not been found over the last 50 years, hence this species is considered to be extinct in its natural habitat. Fortunately, it has been successfully cultivated in several places, and even some varieties have been obtained.

13.10 What Can Be Done, What Has Been Done?

Even though the first reaction is that little can be done, considering the extent of damage already inflicted to the vegetation, several measures that have been proposed to protect plants in general (Koopowitz and Kaye 1986) may be applied to geophytes.

Preserves. The most obvious solution to conservation problem is to set aside preserves where threatened species can be protected *in situ*. To be effective, however, conservation preserves must be carefully managed, which requires an good knowledge of the ecology of the area.

Bulb collections in Botanical Gardens. Collections of many species may be kept in botanical gardens for exhibition and scientific study.

Cultivation. Another approach is bulb cultivation, a conservation measure that appears more realistic at present. Cultivation is an important alternative tool to collecting. Cultivars present advantages over wild-collected specimens in several respects. They are likely to be healthier, and easier to establish, because they have been raised in homogeneous conditions and, further, they can be expected to flower at predictable times. Cultivation also ensures a renewable, reliable supply of stock. For practical reasons, new germoplasm is permanently needed for plant breeding purposes. There are, however, some conservation organizations that think that wild species should neither be cultivated nor kept in private hands (Koopowitz and Kaye 1986). They disregard the fact that all domesticated plants have their roots in wild species, and that it is in private collections where many rare varieties and species are maintained.

Biotechnology. Modern technology can save threatened species for future generations. Cryogenic gene banks are a promising tool. Most plant seeds and spores can be processed so that they can be kept frozen for a long time to be thawed and germinated in the future. A few banks already exist but the technique has hardly been utilized. In Chile, a project is under way in the Instituto de Investigaciones Agropecuarias (INIA) to keep seeds of endangered bulb species from the mediterranean-climate region.

Protection measures and legislation. Protection has been recognized as an important aspect for plant conservation. Many of the endangered species occur naturally in developing countries and are transferred to industrialized countries. Therefore, in many cases the ability to conserve a species depends on international cooperation. Recently, an increased readiness has developed to help preserve species, and diverse initiatives have arisen. First, the threatened plant status has to be accurately assessed. Then, actions to enhance plant conservation have to be formulated. One of these initiatives is the publication of Red Books on Endangered Species (Lucas and Synge 1978). The first *Red Data Book* on plants was published in 1970, and since then they have appeared in many countries. The first Chilean *Red Data Book* only included trees and shrubs, but the 1989 re-issue includes a few groups of herbaceous plants, and among them the geophytes (Benoit 1989). At least in Chile, the Red Data Book is being used by ecologists that participate in the environmental impact assessment of development projects.

The IUCN set up CITES (Convention on International Trade in Endangered Species of Wild Flora and Fauna) in 1973, and the TPC (Treatened Plant Committee) in 1974. CITES has identified some bulb species that need to be protected (Read 1989), and NRDC (Natural Resources Defense Council) entered negotiations with the Dutch bulb industry in an effort to regulate bulb trade. Correct identification of bulbs collected from the wild *vs* propagated bulbs is mandatory to help conserve wild stock, and also to support growers rather than diggers (Campbell 1989). Proper identification procedures started in 1990. As an example of practical consequences for the conservation of geophytes, in 1990 some genera, such as *Galanthus* and *Sternbergia* were placed in Appendix II of CITES, and since then international trade of these species requires a CITES license. For some threatened species a complete ban on trade should be established. In some cases, however, alternatives to a ban on trade of wild bulbs should also be considered, because if all trade is halted it becomes more likely that the land will be converted to agricultural use. Perhaps the most desirable alternative is sustainable harvest from the wild, because it produces incentives to leave areas relatively untouched.

An alternative is the trading of farmed species. This is often considered the best option, because many endangered bulbous plants are easy to cultivate. Examples of successful cultivation of threatened species are *Tecophilaea cyanocrocus, Gladiolus aureus, G. waterneyeri, G. citrinus, Morea loubeseri, M. atropunctata* and *Ixia maculata*. All of these species have been saved from the brink of extinction through cultivation.

Over the last years an increasing number of nature-minded people are buying pieces of more or less unaltered land in the aim of avoiding further anthropic intervention. These private endeavors may substantially contribute to the preservation of natural habitats in the country, hence this kind of initiatives should be encouraged and imitated. In the case of geophytes, such actions will be especially significant in Chile and other countries with a mediterranean-type climate, where the highest diversity of geophytes occurs. In these regions, only the adoption of diverse conservation measures can be effective for the survival of geophytes.

Acknowledgements. We appreciate the invitation of the editors to write this chapter. Facilities granted to the senior author by Facultad de Ciencias Biológicas, Pontificia Universidad Católica de Chile are gratefully acknowledged. We wish to thank J. Armesto, A. Dafni, and F. Jaksik their valuable comments on drafts of our paper, and also D. Aedo, I. Aguirre, M. Bobadilla, C. Marticorena and C. Yañez, who all contributed in different ways to its preparation.

References

Arroyo MTK, Armesto JJ, Villagrán C (1981) Plant phenological patterns in the high Andean Cordillera of central Chile. J Ecol 69:205–223

Arroyo MTK, Squeo F (1990) Genetic self-incompatibility in the endemic Patagonian genus *Benthamiella* (Solanaceae). Gayana, Bot 47:51–55

Benoit IL (ed) (1989) Libro Rojo de la flora terrestre de Chile. CONAF, Santiago de Chile

Bernhardt P (1995) Biogeography and floral evolution in the Geoblasteae (Orchidaceae). In: Arroyo MTK, Zedler PH, Fox MD (eds) Ecology and biogeography of mediterranean ecosystems in Chile, California, and Australia. Springer, Berlin Heidelberg New York, pp 116–134

Boeken B (1989) Life histories of desert geophytes – the demographic consequences of reproductive biomass partitioning patterns. Oecologia 80:278–283

Boeken B, Gutterman Y (1986) The effect of temperature in different habitats on flowering time of *Sternbergia clusiana* in the Negev Desert. In: Dubinsky Z, Steinberger Y (eds) Environmental quality and ecosystems stability, vol III. Bar-Ilan University Press, Ramat-Gan, Israel, pp 151–158

Campbell FT (1989) The problem of wild-collected bulbs from the perspective of the natural resources. Herbertia 45:1–5

Contreras LC, Gutiérrez JR (1991) Effects of the subterranean herbivorous rodent *Spalacopus cyanus* on herbaceous vegetation in arid coastal Chile. Oecologia 87:106–109

Dafni A (1993) The geophytes of Israel – distribution, ecology and reproductive cycles. Proc 5th Optima Meet, Istanbul, pp 83–90

Dafni A, Cohen D, Noy-Meir I (1981a) Life-cycle variation in geophytes. Ann Mo Bot Gard 68:652–660

Dafni A, Shmida A, Avishai M (1981b) Leafless autumnal-flowering geophytes in the mediterranean region. Phytogeographical, ecological and evolutionaty aspects. Plant Syst Evol 137:181–193

Davis SD, Gregerson SJM, Henson P, Leon L, Lamlein C, Villalobos J, Synge H, Zantowska J (1986) Plants in danger, what do we know? International Union for Conservation of Nature and Natural Resources. Gland, Switzerland

De Hertogh AA, Le Nard M (1993) Botanical aspects of flower bulbs. In: De Hertogh AA, Le Nard M (eds) The physiology of flower bulbs. Elsevier, Amsterdam, pp 7–19

Di Castri F, Hajek ER (1976) Bioclimatología de Chile. Vicerrectoría Académica de la Universidad Católica de Chile, Santiago

Gutterman Y, Boeken B (1988) Flowering affected by daylength and temperature in the leafless flowering desert geophyte *Colchicum tunicatum*, its annual life cycle and vegetative propagation. Bot. Gaz. 149:382–390

Hoffmann AE (1989) Chilean monocotyledonous geophytes. Taxonomic considerations and their state of conservation. Herbertia 45:13–28

Kochmer JP, Handel SN (1986) Constraints and competition in the evolution of flowering phenology. Ecol Monogr 56:303–325

Koopowitz H (1986) Conservation problems in the Iridaceae. Herbertia 42:21–25

Koopowitz H (1990) Conservation and bulbous plants. Herbertia 46:22

Koopowitz H, Kaye H (1986) Plant extinction: a global crisis. Stone Wall, Washington D.C.

Lucas G, Synge H (eds) (1978) The IUCN plant red data book. Morges, Switzerland

Marticorena C (1990) Contribución a la estadística de la flora vascular de Chile. Gayana Bot 47:85–113

Marticorena C, Quezada M (1985) Catálogo de la Flora Vascular de Chile. Gayana Bot 42:1–157

Prance GT, Elias TS (eds) (1978) Extinction is forever. New York Botanical Garden, NewYork
Raunkiaer C (1934) The life forms of plants and statistical plant geography. Clarendon, Oxford
Read M (1989) Overexploitation of wild bulbs by the horticultural trade. Herbertia 45:6–12
Rees AR (1989) Evolution of the geophytic habit and its physiological advantages. Herbertia 45:104–110
Saavedra MF, Arroyo MTK, Hoffmann AJ (1996) The breeding system of *Hippeastrum advenum* (Ker-Gawl.) Herb. (Amaryllidaceae), an endemic bulbous geophyte of the mediterranean-type climate region in central Chile. Bot Jahr Syst, Pflanzengesch Pflanzengeogr 118:1–8
Shmida A (1981) Mediterranean vegetation of Israel and California, similarities and differences. Israel J Bot 30:105–123
Shmida A, Dafni A (1989) Blooming strategies, flower size and advertising in the "Lily-group" geophytes in Israel. Herbertia 45:111–123

Note: A complete list of the taxonomic literature used for obtaining data used in the study on geographic distribution, biology and ecology of the geophyte species of Chile, will be sent by the authors on request.

Appendix 13.1. List of the genera present in Chile that were included in the study, and number of species of each that were considered

Family	Genus	N° of species
Amaryllidaceae (31)		
Tribu Hippeastreae	*Habranthus*	1
	Phycella	7
	Placea	5
	Rhodophiala	17
	Traubia	1
Alliaceae (44)		
Subfamily Allioideae (31)	*Ipheion*	1
	Nothoscordum	4
	Tristagma	3
	Brodiaea	9
	Zoellnerarium	1
	Leucocoryne	11
	Pabellonia	2
Subfamily Gilliesioideae (13)	*Ancrumia*	1
	Erinna	1
	Garaventia	1
	Gethyum	1
	Gillesia	5
	Miersia	1
	Solaria	1
	Speea	2

Appendix 13.1. cont.

Family	Genus	N° of species
Anthericaceae (2)	*Pasithea*	1
	Trichopetalum	1
Hyacinthaceae (1)	*Camassia*	1
Tecophilaeaceae (9)	*Conanthera*	7
	Tecophilaea	2
Alstroemeriaceae (36)	*Alstroemeria*	32
	Bomarea	3
	Leontochir	1
Iridaceae (27)		
Tribu Sisyrinchioideae (24)	*Chamelum*	2
	Sisyrinchium	13
	Libertia	4
	Phaiopleps	1
	Solenomelus	2
	Taipenia	2
Tribu Tigrideae (3)	*Calydorea*	1
	Herbertia	1
	Tigridia	1
Corsicaceae (1)	*Arachnytes*	1
Orchidaceae (57)	*Aa*	1
	Bipinnula	5
	Brachystele	1
	Chloraea	29
	Codonorchis	1
	Gavilea	19
	Habenaria	1

Part IV

Demography and Ecophysiology of Succession

14 Coupling Demography, Physiology and Evolution in Chaparral Shrubs

J.E. KEELEY

14.1 Introduction

Historically, since fire is a recurrent catastrophic disturbance, mediterranean-climate shrubs have been classified by their mode of postfire regeneration, i.e., obligate seeders, facultative seeders or obligate resprouters. While these terms are useful, they are too restrictive in that they only refer to a species' response to fire and do not adequately describe modes of reproduction for all taxa.

Two modes of seedling recruitment in mediterranean climate regions are (1) *disturbance-dependent* species that restrict recruitment to postfire conditions (gap species) and (2) *disturbance-free* species that successfully recruit only in the long-term absence of fire (gap-avoiding species). The former exploit post-fire environments for population expansion and the latter require unusually long fire-free conditions for population expansion.

Here I will contrast these two recruitment modes, provide evidence that they are closely tied to physiological patterns of water use, and speculate on the evolutionary scenario around their development.

14.2 Disturbance-Dependent (Gap) Recruitment

Disturbance-dependent species establish seedlings in the first year after fire, and rarely do so in later years. Demographically, these species are characterized by dense cohorts of even-aged shrubs that thin dramatically with time after fire. Mature shrubs produce fruits that are not animal dispersed but tend to be deposited locally; being dormant the seeds disperse more in time than in space (Keeley 1992c). Seedlings arise after fire from a long-lived "permanent" soil seed bank and germination is cued either by intense heat or by chemical stimulus from smoke or charred wood (Keeley 1991). Such species have been referred to as "seeders." In chaparral many species of

Ecological Studies, Vol. 136
Rundel et al. (eds.) Landscape Degradation and Biodiversity in Mediterranean-Type Ecosystems

Arctostaphylos and *Ceanothus* lack the ability to regenerate vegetatively from lignotubers and are often referred to as "obligate seeders." Other species in these genera, and the widespread *Adenostoma fasciculatum*, are referred to as "facultative seeders" because they are capable of regenerating vegetatively, as well as reproducing by seed after fire.

14.3 Disturbance-Free (Gap-Avoiding) Recruitment

In chaparral, species of *Prunus*, *Quercus*, *Rhamnus*, and *Rhus* seldom establish seedlings after fire, however, they resprout vigorously and thus, with respect to fire, are termed "obligate-resprouters." Seedling recruitment occurs in older, more mesic stands and dense cohorts of seedlings are rare (Keeley 1992a,b). Successful reproduction is generally restricted to beneath the shrub canopy and is absent from gaps; recruitment is correlated with, and seemingly facilitated by, more mesic conditions of low light and high litter depth. Mature shrubs produce large fruit crops that are widely dispersed by animals (Keeley 1992c). Soil seed banks are transitory because seeds are short-lived ($<$1 year) and germinate readily with adequate moisture (Keeley 1991, 1997); a dormant seed bank does not accumulate, accounting for the lack of postfire seedling recruitment.

14.4 Morphological and Physiological Correlates

14.4.1 Root Systems

In general, species with disturbance-dependent recruitment are shallow rooted whereas species with disturbance-free recruitment have substantially deeper root systems (Hellmers et al. 1955; Kummerow et al. 1977). While some of this difference is accounted for by species-specific differences in root development, much of it is likely to be accounted for by vast differences in age of rootstocks. Species with disturbance-dependent recruitment have root masses initiated after the most recent fire, whereas disturbance-free recruiters resprout after fire from root masses that are of considerable age and depth. These latter species have roots that may penetrate bedrock to 9 m or more, or penetrate rock fractures. During summer this weathered rock mantle is capable of holding more water within a matric potential range accessible to shrubs than the overlying soils (Jones and Graham 1993).

14.4.2 Water Relations

Correlated with rooting habit are marked differences in exposure to, and tolerance of, water stress. During summer drought shallow-rooted shrubs are subjected to extremely negative soil water potentials; shallow rooted *Ceanothus* and *Arctostaphylos* commonly have predawn stem xylem water potentials of −6.5 to −8 MPa (Parsons et al. 1981; Poole et at. 1981; Davis and Mooney 1986). These species are capable of tolerating even more extreme water stress, e.g., Schlesinger et al. (1982) reported pre-dawn water potential of −12 MPa for *Ceanothus*. *Ceanothus* and others capable of sustaining extreme soil water potentials have seasonal and diurnal osmotic adjustments (Bowman and Roberts 1985) and Davis (1989, Chapter 17; this volume) anatomical characteristics, which reduce embolism at high tensions, are an important factor in tolerance of high water stress.

In contrast, deep-rooted species of *Quercus*, *Prunus*, *Rhus*, and *Rhamnus* are generally subjected to substantially higher summer water potentials in the range of −3 MPa (Poole and Miller 1981; Oechel 1988; Davis 1989). When exposed to extremely low water potentials, usually only experienced at the seedling stage, they suffer high mortality. Oechel (1988) reported a minimum non-lethal water potential for *Quercus* seedlings of −7.5 MPa, far greater than that observed for a shallow-rooted obligate-seeding *Ceanothus*. Even at more moderate water stress, deep rooted species (e.g., *Rhus*) exhibit stomatal closure at far lower water stress than shallow-rooted species (e.g., *Ceanothus*) (Miller 1981).

The critical role rooting depth plays in tolerance of water stress is illustrated in a study by Thomas and Davis (1989). During the summer drought, deep-rooted *Rhus* (*Malosma*) *laurina* shrubs were exposed to very little water stress whereas shallow-rooted *Ceanothus megacarpus* were exposed to substantially higher water stress. On the other hand, seedlings of both species, having roughly similar root mass development, were exposed to remarkably similar summer water stress. However, the *Rhus* seedlings failed to survive, due to much greater susceptibility to water-stress-induced embolism. *Rhus* exhibits a 50% loss in hydraulic conductivity when branches reach a xylem pressure potential of only −1.6 MPa, whereas the obligate-seeding *Ceanothus* does not exhibit 50% embolism until −11 MPa (Kolb and Davis 1994; Jarbeau et al. 1995; Davis, Chapter 17, this Vol.). These patterns may be tied to differences in the size of pores in xylem pit membranes. In *Rhus*, larger vessels and wider pores contribute to greater xylem efficiency when adults have access to water that is available via deep roots, but contributes to greater vulnerability to embolism when water is limited by lack of sufficient root development, as is the case with seedlings (Jarbeau et al. 1995). Based on anatomical surveys (Carlquist and

Hoekmann 1985), this hypothesis predicts that deep-rooted species of *Quercus*, *Prunus*, *Rhamnus*, *Rhus* and others, produce seedlings that are more vulnerable to drought than shallow-rooted species such as *Ceanothus*.

14.5 Coupling Demography, Physiology and Evolution

Adenostoma, *Arctostaphylos*, and *Ceanothus* have adapted their reproductive biology to exploit wildfires for seedling recruitment and population expansion (disturbance-dependent recruitment). In contrast, *Quercus*, *Rhamnus*, *Prunus*, and *Rhus* have not, and these taxa require long fire-free intervals for seedling recruitment (disturbance-free recruitment). These demographic modes are characterized by character syndromes that reflect physiological and morphological divergence (Table 14.1).

Disturbance-dependent recruitment derives from the ready availability of resources in postfire environments, which have placed high selective value on delaying germination to postfire conditions. Summer droughts in these hot, high light environments, however, have imposed strong selection for physiological tolerance of water stress. As a consequence, these disturbance-dependent species have evolved vascular cells more resistant to embolism and greater osmotic tolerance to extremely low water potentials. Recruitment in postfire gaps has selected for rapid growth rates and may have selected against developmental patterns that generate adventitious buds and lignotubers, leading to the obligate-seeding mode. In turn, this obligate-seeding mode, with the increased frequency of sexual reproduction, may have allowed for a greater physiological and anatomical fine-tuning of adaptation to drought. Additionally, with enhanced drought tolerance, there is less selective value to resprouting, an adaptation which can be interpreted as a means of maintaining an established root system with access to year-round moisture. Safe sites for recruitment are rare in time but when they occur are spatially extensive, putting little premium on mechanisms that enhance spatial dispersal but a high premium on maintenance of deep dormancy, with germination cued to fire.

Disturbance-free recruitment restricts seedling establishment to cooler, low light, moister conditions under the shrub canopy. These species are highly susceptible to drought-induced embolism and thus avoid summer drought by maintaining year-round access to water with deep massive root systems. This drought avoidance strategy works well for adults but makes seedling recruitment in a drought-prone environment precarious, thus, there has been no effective selection for delaying seed germination to postfire conditions, and seeds are neither dormant nor long-lived (Keeley 1997). Additionally, safe sites for seedling recruitment are rare and thus these taxa all have highly attractive animal-dispersed propagules.

Convergent and parallel evolution is evident in the very similar pattern of disturbance-dependent and disturbance-free recruitment in the Mediterranean macchi or garrigue of Europe. Here, *Cistus*, *Cytisus*, and others

Table 14.1. Reproductive syndromes correlated with tolerator and avoider strategies of water stress

Examples:	*Ceanothus* *Arctostaphylos* *Adenostoma*	*Quercus* *Prunus* *Rhamnus*
Water stress mode:	Tolerators	Avoiders
Mechanism:	Physiological Anatomical	Morphological (deep roots)
Potential drought-induced mortality:		
-Adults:	Moderate	Very low
-Seedlings:	Moderate	Very high
Recruitment mode:	Disturbance-dependent	Disturbance-free
Gaps:	Gap species	Gap-avoider
Safe sites:	Burned sites	Under canopy
Availability		
-In time:	10–100 year	1 year
-In space:	Extensive	Limited
Seed dormancy:	Deep	Weak
Seed bank:	Persistent (10–100 year)	Transient (<1 year)
Germination cues:	Heat Smoke, char	(Cold)
Dispersal strategy:	Temporal	Spatial
-Mode:	Passive	Animal
-Shadow:	Narrow	Widespread
-Season:	Spring-summer	Fall-winter
Seed size:	Small	Large
Ecological pattern:		
-Niche width:	Narrow	Wide
Evolutionary pattern:		
-Speciation in chaparral:	Low – very high	Low – moderate
Biogeography Pattern:		
-Chaparral endemics:	>75%	<3%
-Outside North America:	<1%	>75%

are disturbance-dependent whereas many taxa such as *Quercus*, *Prunus*, *Rhamnus*, and others require fire-free conditions for recruitment. Particularly remarkable is the marked degree of similarity between California and Europe in the character syndromes tied to these demographic modes (Keeley 1994). The former species have passively dispersed seeds that accumulate in soil seed banks and are fire stimulated, whereas the latter taxa have animal dispersed fruits and transitory seed banks. As in chaparral, disturbance-dependent recruiters such as *Cistus* are physiologically more tolerant of drought stress than disturbance-free recruiters such as *Quercus* (Rundel 1995).

Herrera (1992), however, offers an alternative explanation for such character syndromes. Invoking historical effects, he argues that the regeneration mode in gap-avoiding species is ill-adapted to the mediterranean climate and represents evolutionary inertia. I agree with Herrera on the importance of phylogeny; indeed, it is clearly reflected in that many of the same genera are gap-avoiders in both California and the Mediterranean Basin. The fact that post-fire gap species are much more distantly related between these two regions suggests this is a more recent evolutionary step.

It is noteworthy that the character syndrome expressed by gap-avoiding species in both California and the Mediterranean Basin is widely shared outside of these regions and likely quite ancient. Kollmann (1995) describes the regeneration niche for "fleshy-fruited" shrubs in northern Europe as one in which seedling establishment is greatest under the canopy of mature scrub, largely because this vegetation is the focal point of dispersal. Successful recruitment of saplings, however, is hindered by low light (Kollman and Reiner 1996) and thus successful recruitment is more likely on intermediate "successional" sites that provide better light environments and are somewhat attractive to dispersers. As a consequence of these environmental constraints, Kollman (1995) suggests that there is a relatively narrow window of time during which successful recruitment may occur. If mediterranean climate gap-avoiders share such an evolutionary pathway, an alternative to the view that this syndrome is ill-adapted to the contemporary setting is that the window of opportunity for successful dispersal, seedling establishment, and sapling recruitment has narrowed.

In other words, in the current environment, disturbance-dependent gap species are adapted to recruitment in a postfire environment that is temporary but widespread. The mediterranean climate summer drought has dramatically increased the predictability of such safe-sites for recruitment. Thus, taxa in both California and the Mediterranean Basin, adapted to such conditions, are relatively recently evolved. In contrast, disturbance-free gap-avoiding species are adapted to recruiting in a "niche" that in this contemporary setting is less abundant and spatially disjunct. Prior to the onset of the mediterranean climate, this regeneration niche may have been more widespread and the same may have been true of these species. Today the window has simply narrowed, but it is still there.

14.6 Conclusions

Competition for limiting resources such as water, light and nutrients, selects for different adaptive options. In mediterranean-climate ecosystems some shrub species physiologically *tolerate* extreme summer drought whereas others *avoid* drought by maintenance of deep roots. Seedling access to resources differs from access by adults and thus mode of handling drought stress may greatly affect mode of regeneration (Table 14.1). *Drought tolerators* are capable of establishing on severe sites and thus capitalizing on the predictable availability of high light and nutrients after fire. Many of these species restrict seedling establishment to the immediate postfire environment and thus have disturbance-dependent recruitment. *Drought avoiders* depend upon deep root systems to avoid lethal drought stress, but this strategy is unavailable to their seedlings. Instead, successful reproduction must avoid burned sites and thus these species have disturbance-free recruitment that is generally restricted to more mesic sites under the shrub canopy.

References

Bowman WD, Roberts SW (1985) Seasonal and diurnal adjustments in the water relations of three evergreen chaparral shrubs. Ecology 66:738–742

Carlquist S, Hoekmann DA (1985) Ecological wood anatomy of the woody southern California flora. IAWA. Bull 6:319–347

Davis SD (1989) Patterns in mixed chaparral stands: differential water status and seedling survival during summer drought. In: Keeley SC (ed) The California chaparral: paradigms reexamined. Natural History Museum of Los Angeles County, Los Angeles, Sci Ser No 34, pp 97–105

Davis SD, Mooney HA (1986) Water use patterns of four co-occurring chaparral shrubs. Ecology 70:172–177

Hellmers H, Horton JS, Juhren G, O'Keefe J (1955) Root systems of some chaparral plants in southern California. Ecology 36:667–678

Herrera CM (1992) Historical effects and sorting processes as explanations for contemporary ecologicial patterns: character syndromes in Mediterranean woody plants. Am Nat 140:421–446

Jarbeau JA, Ewers FW, Davis SD (1995) The mechanism of water-stress-induced embolism in two species of chaparral shrubs. Plant Cell Environ 18:189–196

Jones DP, Graham RC (1993) Water-holding characteristics of weathered granitic rock in chaparral and forest ecosystems. Soil Sci Soc Am J 57:256–261

Keeley JE (1991) Seed germination and life history syndromes in the California chaparral. Bot Rev 57:81–116

Keeley JE (1992a) Demographic structure of California chaparral in the long-term absence of fire. J Veg Sci 3:79–90

Keeley JE (1992b) Recruitment of seedlings and vegetative sprouts in unburned chaparral. Ecology 73:1194–1208

Keeley JE (1992c) Temporal and spatial dispersal syndromes. In: Thanos CA (ed) Proc 6th Int Conf on Mediterranean climate ecosystems, plant-animal interactions in Mediterranean-type ecosystems. University of Athens, Athens, Greece, pp 251–256

Keeley JE (1994) Seed germination patterns in fire-prone mediterranean-climate regions, In: Arroyo MTK, Zedler PH, Fox MD (eds), Ecology and biogeography of Mediterranean ecosystems in Chile, California and Australia. Springer, Berlin Heidelberg New York, pp 239–273

Keeley JE (1997) Seed longevity of non-fire recruiting chaparral Shoubs. Four Seasons 10(3):36–42

Kolb KJ, Davis SD (1994) Drought tolerance and xylem embolism in co-occurring species of coastal sage and chaparral. Ecology 75:648–659

Kollmann J (1995) Regeneration window for fleshy-fruited plants during scrub development on abandoned grassland. Ecoscience 2:213–222

Kollmann J, Reiner SA (1996) Light demands of shrub seedlings and their establishment within scrublands. Flora 191:191–200

Kummerow J, Krause D, Jow W (1977) Root systems of chaparral shrubs. Oecologia 29:163–177

Miller, PC (ed) (1981) Resource use by chaparral and matorral. Springer, Berlin Heidelberg New York

Oechel WC (1988) Minimum non-lethal water potentials in Mediterranean shrub seedlings. In: di Castri F, Floret Ch, Rambal S, Roy J (eds), Time scales and water stress. Proc 5th International Conference on Mediterranean ecosystems (MEDECOS V). International Union of Biological Sciences, Paris, pp 125–13

Parsons DJ, Rundel PW, Hedlund R, Baker GA (1981) Survival of severe drought by a non-sprouting chaparral shrub. Am J Bot 68:215–220

Poole DK, Miller PC (1981) The distribution of plant water stress and vegetation characteristics in southern California chaparral. Amer Midl Nat 105:32–43

Poole DK, Roberts SW, Miller PC (1981) Water utilization, in Miller PC (ed), Resource use by chaparral and matorral. Springer, Berlin Heidelberg New York, pp 123–149

Rundel PW (1995) Adaptive significance of some morphological and physiological characteristics for the water relations in Mediterranean plants: facts and fallacies. In: Roy J, Aronson J, di Castri F (eds) Time scales of biological responses to water constraints, the case of mediterranean biota. SPB Academic, Amsterdam, pp 119–140

Schlesinger WH, Gray JT, Gill DS, Mahall BE (1982) *Ceanothus megacarpus* chaparral: a synthesis of ecosystem properties during development and annual growth. Bot Rev 48:71–117

Thomas CM, Davis SD (1989) Recovery patterns of three chaparral shrub species after wildfire. Ecology 80:309–320

15 Physiological Ecology of Mediterranean Seeds: Links with *Ex Situ* Conservation of Plants

C. VÁZQUEZ-YANES and A. OROZCO-SEGOVIA

15.1 Introduction

Seed biology studies in Mediterranean type ecosystems are relatively numerous and document interesting peculiarities in the seed germination behavior. Some peculiar germination syndromes will be described later in this chapter. Seed studies become most important because the Mediterranean ecosystems are among the most intensely utilized by man, consequently they are usually severely altered by anthropogenic disturbances. Many plant species populations have been currently reduced to dangerously low numbers, which may not insure self perpetuation of local ecotypes or entire species. Among the available germplasm conservation procedures, *ex situ* and *in vitro* preservation of seeds and explants are life insurance alternatives for valuable germplasm that must not be ignored. Seed storage has been undertaken by many botanic gardens and research institutions and it is now the most widely utilized *ex situ* conservation approach and is it currently in operation in some Mediterranean regions (Gómez-Campo 1985, 1987; Hernández-Bermejo and Heywood 1990).

15.2 Mediterranean Seed Traits

15.2.1 Soil and Canopy Seed Banks

Although few works document the presence of soil seed banks, the most persistent seeds in soil are mainly hard coated seeds that germinate massively after fires. Discontinuous banks may be found before the rainy season formed by the species that dispersed their seeds during late spring and summer (Keeley 1991; Bell et al. 1993; Thanos et al. 1995). Another kind of seed bank, that is typical of Mediterranean vegetation mainly in Australia but it has been reported elsewhere, is the canopy seed bank. Seeds remain

Ecological Studies, Vol. 136
Rundel et al. (eds.) Landscape Degradation and Biodiversity in Mediterranean-Type Ecosystems

in the fruits attached to the plant for long time after ripening until a fire induce seed dispersal by gravity, wind or the explosive dehiscence of the dry fruits (Thanos and Skordilis 1987; Thanos et al. 1989; Vogl et al. 1977; Brits 1987; Thanos and Marcou 1991; Bell et al. 1993).

15.2.2 Dormancy Mechanisms

Seed biology studies in Mediterranean document interesting peculiarities in the seed germination behavior like heat, pre-sown dryness, charred wood and smoke effects promoting germination in some species (Keeley 1991; Brits et al. 1993; Brown 1993). The most frequent germination impediment prolonging seed dormancy in Mediterranean plants seems to be an impermeable hard coat that becomes permeable with time or by the heat of a fire. The role of these coats regulating germination has been described in detail (Mohamed-Yasseen et al. 1994). It is a very unsophisticated dormancy mechanism removed by heat and abrasion, and do not involve any complex physiological features. The response of these seeds to smoke and charred wood reported in California, may indicated the existence of a chemical sensor in the seeds which has not been characterized yet. In some cases chemical inhibitors are present in hard coated seeds that are also destroyed by heat or by other factors associated with the breaking of the seed coat (Corral et al. 1990; Keeley 1991; Bell et al. 1993). These dormancy mechanisms interact with other environmental factors as light, temperature and moisture.

A short lasting dormancy, which is usually released by cold temperatures has been described in different Mediterranean areas. This trait has been related to germination in early spring, which is the period when a massive number of annual plants become established in many areas. Even some species not showing this requirement elsewhere, may have it in their Mediterranean populations (Ren and Abbot 1991). Whatever great variability has been found in this requirement even among populations of the same species growing in different areas of the Mediterranean Basin (Skordilis and Thanos 1995). The opposite has been described in Australia; some species of seeds require a period of elevated temperatures during late summer as an after-ripening factor that allows germination during autumn (Bell et al. 1993).

A generalization can be made for the temperature responses of Mediterranean species. They have a very low minimum and relatively low maximum for germination (0–25°C) in comparison to the adjacent areas (Thompson 1970, 1973). This is due to the relationship between lower temperatures and high humidity characteristic of the Mediterranean climate. However populations of the same species growing in the Mediterranean and areas outside it may germinate at relatively higher temperatures (Thompson 1973).

Light regulated dormancy seems to be common in several habitats. Light gap colonizers annuals and some perennials of the California chaparral and Greek forest have positive photoblastic seeds (light requiring) which germinate after fire due to the change in light quality produced by destruction of vegetation and litter (Thanos and Skordilis 1987; Thanos et al. 1989; Keeley 1991). The opposite case is common at the coastal sand dunes of Greece. Several species produce seeds that are light inhibited. They germinate when they are buried in the sand insuring favorable humidity and temperature conditions for establishment (Thanos et al. 1991).

All dormancy mechanisms interact with a variety of environmental factors producing a diversity of germination responses that may spread germination over different seasons. This could be the origin of the scant permanent seed bank formed by soft coated seeds often found in many areas, although very little information exist about it (Thanos and Georghiou 1988; Panneta and Randall 1993; Pérez-García 1993; Pérez-García et al. 1995).

15.3 Seed Longevity

The strong ecological constrains present in most Mediterranean ecosystems like the long summer drought, frequent vegetation fires and winter freezing temperatures are all conditions driving the natural selection of seed characteristics toward the development of traits that convey a long lasting longevity. Hard coated seeds of any habitat regularly retain the longest longevity at above zero temperatures in storages like herbaria, because of the protection that the impermeable coat provides against humidity changes. Many Mediterranean seeds survive in the soil or on plant canopies for long periods of extreme dryness, heat or cold temperatures. Due to this factors it can be predicted with a high degree of certitude that most plant species growing in Mediterranean climates must be easily storable (orthodox) for germplasm conservation purposes. However, some temperate tree genera like *Quercus*, present in some Mediterranean regions, have been found to produce recalcitrant seeds (Hendry et al. 1992).

According to their storage properties seeds have been divided in two main groups: storable seeds or "orthodox", and non-storable ones or "recalcitrant" (Roberts 1973). Recalcitrant and orthodox seeds diverge in their physiology and ecology. Recent studies on the development of the recalcitrant seeds of black mangrove indicate that there are differences in the production and circulation of plant growth regulators, like abscisic acid. This hormone drive the genetic responses that might in part be responsible for the dissimilarities in metabolic activity and dehydration tolerance found in these seeds with respect to the orthodox type (Ferrant et al. 1993).

15.4 *Ex Situ* Conservation of Seeds

To include more accurately the diversity found among the longevity of seeds of temperate plants, a subdivision of the two storage categories has been proposed (Bonner 1990):

True orthodox seeds can be stored for long periods at sub-freezing temperatures, if their moisture contents are reduced below 10%. For short term storage of 5 years or less, temperatures of 0–5°C are satisfactory. For longer storage, temperatures less than −15°C are commonly used. Many common temperate plant genera such as *Alnus, Betula, Carduus, Digitalis, Fraxinus, Lotus, Pinus, Plantago, Platanus, Prunus*, and possibly all the hard coated and seed bank forming seed species are included in this storage behavior.

Sub-orthodox seeds are those that can be stored under the same conditions as true orthodox seeds, but for shorter periods due to their high lipid contents or to fragile seeds with thin seed coats. Some tree seeds once ranked as recalcitrant, through careful drying are now known to share the characteristics of this group. This group includes genera such as *Carya, Juglans, Populus* and *Salix*.

Temperate recalcitrant seeds cannot be dried but they can be stored for several months at near freezing temperatures from −3° to 0°C. This seeds keep active metabolism and can sometimes germinate during storage. Includes genera such as *Acer* and *Quercus*.

Tropical recalcitrant seeds cannot be dried and are very sensitive to low temperatures even during short periods of time. Includes many genera of tropical moist forest trees.

Although predictions about seed storage behavior can be made based on ecological traits, life history patterns and the seed storage characteristics of taxonomically related plants, trials are required to classify seeds as orthodox or recalcitrant. Once a seed sample has been classified as orthodox, the arrangements for its storage can be undertaken by following a well known track which has been described in detail (Ellis et al. 1985; Cromarty et al.

1990). For a modest storage facility, the following procedure should be practical. To reduce moisture content seeds are placed in a cool environment of about 15° to 19°C at relative air humidity of about 18–20% for a month. A cool drying chamber provided with air conditioning and dehumidifiers is ideal, if available, but a glass desiccator provided with dry silica gel or activated alumina placed in a cool room can also be used. A seed storage facility of good size should be provided with a drying room to carry on continuous dehydration of seed samples. Seeds should never be dried with heat.

After the seeds are dried, they should be placed in secure, well sealed, precisely labeled containers like hermetic glass and metal vials or heat sealable plastic or plastic confined in aluminum foil envelopes. Many seeds can endure further deep drying, so storage can be made in closed vials containing a dehydrating non toxic substance like silica gel (Gómez-Campo 1985).

To insure long storage survival, orthodox seeds should be stored at the lowest temperature accessible such as those that can be obtained in a domestic deep freezer (around −18°C). A well-insulated thermos filled with liquid nitrogen may also be used for the storage of resistent seeds.

Several sophisticated freezing rooms for the storage of seeds are in operation in different botanic gardens and seed storage facilities around the world. e.g. the renown ones at Kew Gardens in West Sussex U. K. and at the National Seed Storage Laboratory of the U. S. Department of Agriculture in Fort Collins, Colorado. National seed storage facilities should exist in every country. For developing countries the ideal ones are subterranean, provided with thick concrete walls, because they can be well insulated from surface temperature changes, conditions inside may remain more stable, require less energy and may survive catastrophes. An example of such a facility is the Pine Ridge Forest Nursery at Smoky Lake, Alberta, Canada. It consists of a building with reinforced 43 cm thick concrete walls that, except for the entrance ramp and the top of the mechanical services room, is buried beneath a 2 m layer of soil (Pitel and Wang 1984). The advantages of underground seed storage facilities include protection against fire, significant saving in energy expenses due to insulation from outside temperatures, minimum impact of power or equipment failure and breakdown due to insulation effect of the soil, and savings in the external cost and maintenance of the building.

A seed storage facility built underground in areas with constant belowground freezing soil temperatures like the circumpolar permafrost or some permanent mountainous ice caps would allow the survival of valuable living seeds during very anarchic periods of war, revolutions, economic chaos or political unrest for the benefit of future, more civilized, human generations. In any case, a consistent and steady politics on germplasm conservation, a responsible supervisor and personnel and continuous financial support are a must to ensure the perpetuation of a well-managed storage facility.

15.5 Seed Recollection and Handling

Ex situ preservation of germplasm has been severely criticized as a valid conservation strategy because, among other things, the serious limitations to safeguard a dependable representation of the natural genetic variability of the populations of a species (Hamilton 1994). Therefore, any valuable collection of live seed samples should be based on an adequate set of rules, routines and precautions that should be applied during recollection and handling preceding storage. Common sense dictates many of these and some have been summarized in different publications (Bass 1979; Hanson 1985). In germplasm conservation schemes the recollection of high quality seeds which preserve a good representation of the genetic diversity of the plant populations is critical to create a worthwhile seed storage collection.

Most plants have genetic variability over their distribution range that should be surveyed and preserved when taking seed samples. These should be collected from wild sources if available, because the seeds collected from manipulated or cultivated population of plants may uncover traits which were deliberately or unintentionally selected by man and consequently genetic diversity is diminish. Seeds should be collected from various individual plants in order to preserve some of the variability found within a local population. Any increase in number and surface covered will allow an improvement in the representation of such variability.

The quality of a seed sample often varies widely depending on provenance, maturity level, degree of parasitism and predation, seasonal resource limitations for reproduction within a giving year and seed recollection and handling techniques employed. The quality of the seeds collected is of great importance.

Recalcitrant seed have always short survival in storage conditions. Fortunately there are other alternatives for *ex situ* conservation, which comprise explant conservation, using *in vitro* techniques that are being currently explored thoroughly by many plant physiologist and biotechnologist, they already offer promising applications for conservation purposes, which undoubtedly will become more comprehensive in the near future (Adams and Adams 1992).

15.6 Conclusions

Seed germination syndromes in Mediterranean-climate ecosystems closely resemble what could be expected according to the characteristics of the different seasons and prevailing environmental conditions. As a consequence, long term seed storage is an appropriate *ex situ* conservation practice for the great majority of the species. Soft-coated seeds seem to be priority for conservation due to their short survival in the soil seed bank. There is a great disparity on the amount of the available information on seed dormancy, viability and germination syndromes among the different Mediterranean regions.

Acknowledgements. This paper was written when the senior author was "Charles Bullard Fellow" (Harvard Forest) at Harvard University. Facilities to work provided by Dr. F. A. Bazzaz are deeply appreciated. Mariana Rojas Aréchiga provided helpful technical support.

References

Adams RP, Adams JE (ed) (1992) Conservation of plant genes: DNA banking and in vitro biotechnology. Academic Press, San Diego

Bass LN (1979) Physiological and other aspects of seed preservation. In: Rubinstein I, Phillips R, Green CE, Gengenbach GB (eds) The plant seed: development, preservation and germination. Academic Press, London, pp 145–170

Bell DT, Plummer JA, Taylor SK (1993) Seed germination ecology in southwestern Western Australia. Bot Rev 59:24–54

Bonner FT (1990) Storage of seeds: potential and limitations for germplasm conservation. For Ecol Manag 35:35–43

Brits GJ (1986) Influence of fluctuating temperatures and H_2O_2 treatment on the germination of *Leucospermum cordifolium* and *Serruria florida* (Proteaceae) seeds. S Afr J Bot 52:286–290

Brits GJ, Calitz FJ, Brown NAC, Manning JC (1993) Desiccation as the active principle in heat-stimulated seed germination of *Leucospermum* R. Br. (Proteaceae) in fynbos. New Phytol 125:397–403

Brown NAC (1993) Promotion of germination of fynbos seeds by plant-derived smoke. New Phytol 123:575–583

Corral R, Pita JM, Pérez-García F (1990) Some aspects of seed germination in four species of *Cistus* L. Seed Sci & Technol 18:321–325

Cromarty AS, Ellis RH, Roberts EH (1990) The design of seed storage facilities for genetic conservation. International Plant Genetic Resources Institute Publ, FAO, Rome

Ellis RH, Hong TD, Roberts EH (1985) Handbook of seed technology for genebanks, vol I and II. International Plant Genetic Resources Institute Publ, FAO, Rome

Ferrant JM, Pammenter NW, Berjak P (1993) Seed development in relation to desiccation tolerance: a comparison between desiccation-sensitive (recalcitrant) seeds of *Avicennia marina* and desiccation-tolerant types. Seed Sci Res 3:1–13

Gómez-Campo C. (1985) Seed banks as an emergency conservation strategy. In: Gómez-

Campo C (ed) Plant conservation in the Mediterranean area. Junk, Dordrecht, pp 237–247

Gómez-Campo C (1987) A strategy for seed banking in botanic gardens: some policy considerations. In: Bramwell D, Hamann O, Heywood V, Synge H (eds) Botanic gardens and the world conservation strategy. Academic Press, London, pp 151–160

Hamilton MB (1994) *Ex situ* conservation of wild plant species: time to reassess the genetic assumptions and implications of seed banks. Conserv Biol 8:39–49

Hanson J (1985) Procedures for handling seeds in seed banks. International Plant Resources Institute Publ, FAO, Rome

Hendry GAF, Finch-Savage WE, Thorpe PC, Atherton NM, Buckland SM, Nilsson KA, Seel WE (1992) Free radical processes and loss of seed viability during desiccation in the recalcitrant species *Quercus robur* L. New Phytol 122:273–279

Hernández-Bermejo JE, Heywood CV (eds) (1990) Conservation techniques in botanic gardens. Koeltz Scientific Books, Koenigstein

Keeley JE (1991) Seed germination and life history syndromes in the California chaparral. Bot Rev 57:81–116

Mohamed-Yasseen Y, Barringer SA, Splittstoesser WE, Costanza S (1994) The role of seed coats in seed viability. Bot Rev 60:426–439

Panetta FD, Randall RP (1993) Variation between *Emex australis* populations in seed dormancy/non-dormancy cycles. Aust J Ecol 18:275–280

Pérez-García F (1993) Effect of the origin of the cypsela on germination of *Onopordum acanthium* L. (Asteraceae). Seed Sci & Technol 21:187–195

Pérez-García F, Iriondo JM, Martínez-Laborde JB (1995) Germination behaviour in seeds of *Diplotaxis erucoides* and *D. virgata*. Weed Res 35:495–502

Pitel JA, Wang BSP (1984) A review of papers published in the proceedings of the IUFRO Int Symp Forest Tree Seed Storage. Commonw For Rev 63:55–66

Ren Z, Abbot JR (1991) Seed dormancy in Mediterranean *Senecio vulgaris* L. New Phytol 117:673–678

Roberts EH (1973) Predicting the storage life of seeds. Seed Sci & Technol 1:499–514

Roberts EH and Ellis RH (1989) Water and seed survival. Ann Bot 63:39–52

Skordilis A, Thanos CA (1995) Seed stratification and germination strategy in the Mediterranean pines *Pinus brutia* and *P. halepensis*. Seed Sci Res 5:151–160

Thanos CA, Georghiou K (1988) Ecophysiology of fire-stimulated seed germination in *Cistus incanus* ssp. *creticus* (L.) Haywood and *C. salvifolious* L. Plant Cell Environ 11:841–849

Thanos CA, Marcou S (1991) Post-fire regeneration in *Pinus brutia* forest ecosystems of Sanos island (Greece): 6 years after. Acta Oecol 12:633–642

Thanos CA, Skordilis A (1987) The effects of light, temperature and osmotic stress on the germination of *Pinus halepensis* and *P. brut*ia seeds. Seed Sci & Technol 15:163–174

Thanos CA, Kadis CC, Skarou F (1995) Ecophysiology in the aromatic plants thyme, savory and oregano (labiatae). Seed Sci Res 5:161–170

Thanos CA, Marcou S, Christodoulakis D, Yannitsaros A (1989) Early post-fire regeneration in *Pinus brutia* forest ecosystems of Samos island (Greece). Acta Oecologica/Oecol Plant 1:79–94

Thanos CA, Georghiou K, Douma DJ, Marangaki CJ (1991) Photoinhibition on seed germination in Mediterranean maritime plants. Ann Bot 68:469–475

Thompson PA (1970) Characterization of the germination response to temperature of species and ecotypes. Nature 225:827–831

Thompson PA (1973) Geographical adaptation of seeds. In: Heydecker W (ed) Seed ecology. Butterworths, London, pp 31–57

Vogel RJ, Armstrong WP, White LL, Cole KL (1977) The closed-cone pines and cypresses. In: Barbour MJ, Major J (eds) Terrestrial vegetation of California. Wiley Interscience, New York, pp 295–358

16 Aspects of Demography in Post-Fire Mediterranean Plant Communities of Greece

M. Arianoutsou

16.1 Introduction

Fire is a major ecological factor in many biomes of the world (Rundel 1981). In the Mediterranean ecosystems fire acts as an integral part of their evolutionary history, having shaped their adaptive traits (Naveh 1975). It is beyond any doubt that the specific characteristics of fire regimes have a profound effect on these ecosystems (Mooney and Conrad 1977), the most important of which is related to community and population structure. These are critical issues for any conservation attempt. In this perspective, the biological features that are mostly influenced by fire are reproductive and demographic characteristics which assure a rapid reestablishment by resprouting and/or seed germination (Keeley and Zedler 1978; Arianoutsou and Margaris 1981a). The specific regeneration behaviour of the plants is closely related to their physiological traits and is strongly influenced by fire regime (fire season, intensity, interval).

Resprouting occurs from dormant buds or lignotubers effectively protected by the soil. It starts almost immediately after fire in the evergreen sclerophyllous species, while it takes several months, until the onset of the rainy season for the seasonal dimorphic (phryganic) species. This difference has been attributed to the different depths of their root systems. The resprouting ability of the plants is determined by their developmental phase. Many shrubby taxa are successful resprouters, either obligate or facultative.

In several Mediterranean ecosystems, fire has been shown to trigger a flush of germination (Purdie 1977; Trabaud 1980; Arianoutsou and Margaris 1981b). Post-fire seed germination is an adaptation of the species which form permanent soil seed banks of hard-coated seeds (Cistaceae, Leguminosae) and of those with a canopy seed bank (Proteaceae, Pinaceae). Seed germination has been attributed to several factors such as mechanical rupture of the hard seed coat induced by the increased temperatures developing during fire (Arianoutsou and Margaris 1981b; review

Ecological Studies, Vol. 136
Rundel et al. (eds.) Landscape Degradation and Biodiversity in Mediterranean-Type Ecosystems

in Rundel 1981), inactivation of germination inhibitors present in the soil (Christensen and Muller 1975; Keeley and Keeley 1989), effect of charred woody stems (Keeley and Nitzberg 1984) and change in light quality (Roy and Arianoutsou 1985).

The temporal destruction of either the whole or the aboveground part of the mature plant by a fire incident triggers the restart of the demographic cycle of the community through the relevant cycle of the plant taxa present. Many studies have been done in California, Australia, and South Africa on plant community diversity and dynamics in relation to fire, and these are extensively reviewed by Kruger (1983). Not as much work, though, has been done in the Mediterranean Basin, apart from a few demographic studies (Naveh 1974; Papanastasis 1977; Arianoutsou and Margaris 1981a; Arianoutsou 1984; Trabaud and Oustric 1989; Ne'eman et al 1993). This chapter aims to review and synthesize information on the dynamics of post-fire Mediterranean communities in Greece, as part of a broader study which is evolving under the framework of two research projects sponsored by the European Union – Modelling vegetation dynamics and degradation in Mediterranean ecosystems (MODMED) and Management techniques for optimisation of suppression and minimisation of wildfire effects (PROMETHEUS).

16.2 Mediterranean Ecosystems of Greece

Mediterranean ecosystems constitute 40% of the terrestrial vegetation of Greece. They can be grouped into three major types according to the prevailing vegetation, and they cover a rainfall regime of very low precipitation (close to 100 mm) to adequately higher (700–900 mm). Analytically, the three types are:

1. Communities with seasonal dimorphic dwarf plants, known as phrygana. This community is found mainly in the complex of Cyclades islands, in the islands of Crete, Limnos, Chios, Kefalonia, Lefkada, in West Akarnania and in Southeast Peloponnese. Their vertical structure is restricted to one single layer, that of the dominant shrubs, which are seldom taller than 50 cm. At the mature stage, phrygana is a relatively closed community, leaving some open space between the individuals. Among the most common woody plants are *Coridothymus capitatus* (thyme), *Phlomis fruticosa* (Jerusalem sage), *Cistus* spp. (rockroses), *Sarcopoterium spinosum* (thorny burnet) and *Euphorbia acanthothamnos* (Greek spiny spurge). The majority of the woody species are

seasonal dimorphic, that is they change their leaves and type of branches twice a year, in order to face water stress. *Euphorbia acanthothamnos* is a summer drought plant. They are all shallow rooted and they grow on calcareous, relatively poor soils.

2. Communities with evergreen sclerophyllous shrubby plants, otherwise referred to as maquis. These cover part of the Halkidiki peninsula in Macedonia, part of continental central Greece and parts of the islands in the Ionian Sea. Deep-rooted plants taller than 1 m, such as *Quercus coccifera* (kermes oak), are the predominant species in these communities. Other common taxa are *Pistacia lentiscus* (lentisc), *Arbutus unedo* (strawberry tree), *Myrtus communis* (myrtle), *Juniperus phoenicea* (phoenician juniper) and *Juniperus oxycedrus* (prickly juniper). Their leaves are thick, larger than those of the phryganic species and remain the same throughout the year, having a life span of two years. They form very dense stands of one layer, leaving almost no space between individuals at the mature stage. They grow on more fertile soils.
3. Forest communities with pine trees. *Pinus halepensis* (Aleppo pine) forests are found in Attica, in Euboia, in western Peloponnese and in Halkidiki, while *Pinus brutia* (brutia pine) forests grow on the islands of the north and southeast Aegean, in Crete and its islets, (see Fig. 16.1a). The trees can be up to 15 m tall, and generally have a rather poor understory stratum at the middle of their life span (which can be as long as 100 years). However, species diversity is quite high immediately after fire and rich also at the mature stage. The understorey stratum consists of either phryganic species or species of the evergreen sclerophyllous formations.

Phrygana exploit the areas with low precipitation and higher summer temperatures, while evergreen sclerophyllous ecosystems occur in more humid and less warm environmental regimes. Mediterranean coniferous forest can be found in situations closer to those prevailing in the latter case (Fig. 16.1b).

16.3 Fire Frequency

Fire incidents in the Mediterranean Basin are very common and a high frequency of fires has extended back into the past. Evidence of fire can be identified in the Iron Age, 2600 years B.P., when shepherds and farmers set fires in order to open the forest and the ground for improving pasture and cultivated land (Barry 1960; Dugrand 1964).

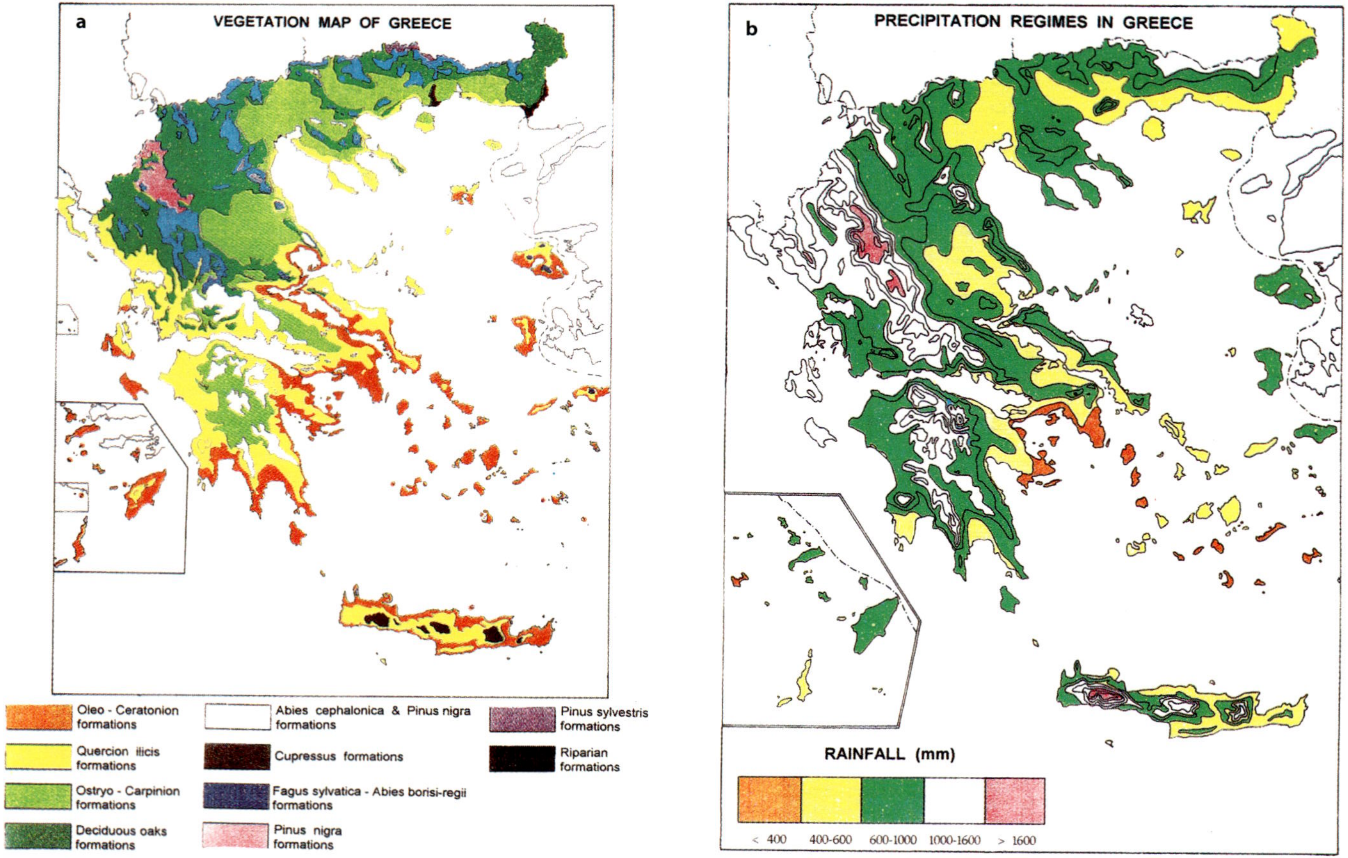

Fig. 16.1a,b. Vegetation map of Greece. *Brown* areas correspond to phrygana, *yellow* to Aleppo pine forests and *light green* to maquis (Modified from Mavromatis 1978) **b.** Precipitation regimes in Greece (Modified from Mariolopoulos and Karapiperis 1955)

In Greece, the first archaeological evidence of fire occurrence in the xerophytic landscapes goes back to the end of the Middle Stone Age: remnants of ashed wood in the Kastritsa cave, near Ioannina Lake (Higgs et al. 1967). Liakos (1973) has reviewed the early use of fire and wildfire in ancient Greece.

Most fire events occur in evergreen sclerophyllous ecosystems (Fig. 16.2), where they consume approximately 13024 ha $year^{-1}$ (Table 16.1), which corresponds to nearly 50% of the total area burned. Following in size are fire events occurring in phrygana and *P. halepensis* forests (Table 16.1; Fig. 16.2 and 16.3). Most of the fires are caused accidentally or by arson.

While there are extensive and detailed studies on fire frequency in California, South Africa and Australia, there are but a few data for the Mediterranean Basin. Le Houerou (1973) states that the so-called red belt of southern France is burned regularly every 25–30 years. Forest Service records in Greece are scarce and sometimes not very reliable. Moreover, these data do not reveal any clear pattern of fire frequency, presumably due to multiple reasons encountered as possible fire causes.

16.4 Plant Adaptive Traits

Table 16.2 summarizes the adaptive traits that taxa dominant in the Mediterranean ecosystems of Greece have evolved in order to cope with fire. Most of the phanerophytes are resprouters and generally obligate

Table 16.1. Vegetation types and five data. (Period of reference 1965–1990; data adapted from Kailidis 1992)

Vegetation type	Total area burned (ha)	Total number of fires	Mean area burned/year (ha)
Abies spp. forest	1304	147	50.15
Mixed coniferous forests	2534	11	97.46
Pinus nigra forests	9152	383	352.00
Pinus halepensis forests	122015	4385	4692.88
Pinus brutia forests	61684	1159	2372.46
Other coniferous forests	1288	139	49.54
Forests with deciduous *Quercus* spp.	42459	1514	1633.04
Forests with other deciduous species	13393	553	515.12
Evergreen sclerophyllous shrublands	338608	10801	13023.38
Phrygana	167084	4020	6426.31

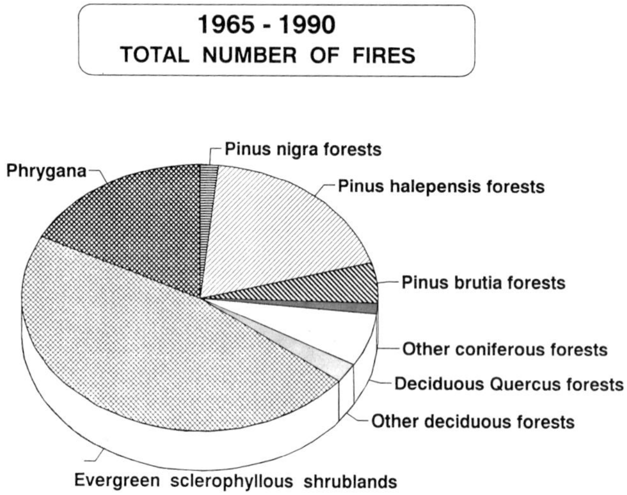

Fig. 16.2. Percentage of fire incidents occurring in several vegetation types in Greece between 1965–1990. (Data adapted from Kailidis 1992)

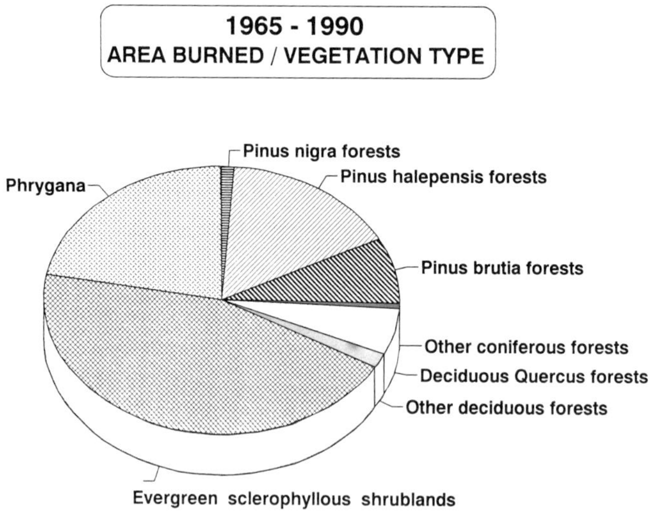

Fig. 16.3. Percentage of areas burned in several vegetation types in Greece between 1965–1990. (Data adapted from Kailidis 1992)

resprouters. The only exceptions are *Pinus halepensis* and *Pinus brutia* which are obligate seeding species. Among the chamaephyte subshrubs both regenerative modes are encountered, so as the taxa can be characterised as either facultative resprouters and/or facultative seeders. Members of the Cistaceae family are only obligate seed regenerators.

Table 16.2. Regeneration mode in some typical mediterranean plant taxa

Taxa	Growth form	Life form	Sprouting	Seeding
Pinus halepensis	Tree	Ph	−	+
Pinus brutia	Tree	Ph	−	+
Quercus coccifera	Tall Shrub	Ph	+	−
Pistacia lentiscus	Tall Shrub	Ph	+	−
Pistacia terebinthus	Tall Shrub	Ph	+	−
Arbutus unedo	Tall Shrub	Ph	+	−
Arbutus andrachne	Tall Shrub	Ph	+	−
Erica arborea	Tall Shrub	Ph	+	−
Erica manipuliflora	Tall Shrub	Ph	+	−
Olea europea	Tall Shrub	Ph	+	−
Phillyrea latifolia	Tall Shrub	Ph	+	−
Cistus creticus	Shrub	Ph	−	+
Cistus salvifolius	Shrub	Ph	−	+
Cistus monspeliensis	Shrub	Ph	−	+
Phlomis fruticosa	Shrub	Ph	+	+
Euphorbia acanthothamnos	Shrub	Ph	+	+
Thymelea tartonraira	Subshrub	Ch	+	+
Hypericum empetrifolium	Subshrub	Ch	+	+
Fumana thymifolia	Subshrub	Ch	−	+
Teucrium polium	Subshrub	Ch	+	+
Ballota acetabulosa	Subshrub	Ch	+	+
Coridothymus capitatus	Subshrub	Ch	−	+
Asparagus aphyllus	Subshrub	Ch	+	−
Phagnalon graecum	Subshrub	Ch	+	+
Vicia villosa	Herb	Th	−	+
Vicia tetrasperma	Herb	Th	−	+
Lotus ornithopodioides	Herb	Th	−	+
Cynosurus echinatus	Herb	Th	−	+
Asphodelus aestivus	Herb	Cr	+	?
Arisarum vulgare	Herb	Cr	+	?
Ophrys lutea	Herb	Cr	+	?
Brachyphodium ramosum	Herb	H	+	−
Scorzonera crocifolia	Herb	H	+	?

Ph = Phanerophytes.
Ch = Chamaephytes.
Th = Therophytes.
Cr = Cryptophytes.
H = Hemicryptophytes.

Hemicryptophytes show both adaptation traits, while cryptophytes (including geophytes) seem to regenerate preferably by resprouting. Most of the representatives of the herbaceous therophytes are annuals and recover from fire through seed germination. However, there is evidence that many perennial herbs are resprouters. An extensive and detailed survey of the regeneration strategies of the Mediterranean taxa of Greece is currently underway and is expected to give more accurate percentages.

Data from other Mediterranean countries of the world reveal that among about half or more of trees or shrubs in regional floras are resprouting species (Keeley 1977a; Kruger 1977; Gill and Groves 1981), although there are characteristic genera that tend to have more seeding species.

16.5 Post-Fire Succession

Plant succession after fire in Mediterranean communities consists largely of the regeneration and development through the respective life cycle of the species present before fire (Specht et al. 1958; Hanes 1971; Arianoutsou 1979; Kazanis and Arianoutsou 1996). It is a process of secondary autosuccession in which the burned stand, although initially appearing to be different from the pre-burned one, retains its floristic identity in time. The most obvious change is that of the composition and relative abundance of annual herbaceous plants, followed by that of herbaceous perennials (Kazanis and Arianoutsou 1996). An impressively rich flora of these two groups is recorded in all Mediterranean plant communities of Greece after fire, being promoted either by increased temperature regimes prevailing during and immediately after fire (such as the leguminous plants) or by other factors which are currently investigated (e.g. nitrogenous compounds, fluctuating temperatures, changes in the light quality). As succession proceeds and the canopy closes, herbaceous vegetation is either restricted to small openings, as in the case of phryganic ecosystems and to a lesser degree in coniferous forests, or remains dormant in the soil in the form of seeds waiting for the next fire to come (Arianoutsou 1979; Kazanis and Arianoutsou 1996). Similar results have been reported by Specht et al. (1958) for heath stands in Australia, Bond (1980) for fynbos stands in South Africa, Horton and Kraebel (1955) for chaparral, Espirito-Santo et al. (1993) for sclerophyllous formations in Portugal, Faraco et al. (1993) for pine woodlands and shrublands in Spain, Mazzoleni and Pizzolongo (1990), De Lillis and Testi (1990), Lucchesi and Giovannini (1993) for similar communities in Italy.

The qualitative information on plant life histories and field observations allows us to summarize some highlights of the pyric succession in the Mediterranean plant communities of Greece, which are shown in Table 16.3. Mature stands vary considerably in the relative contribution of seeders versus that of sprouters both between the different types of Mediterranean ecosystems in Greece and within these types in relation to their

Table 16.3. Highlights of early post-fire succession in Mediterranean ecosystems in Greece

	Phrygana	Evergreen sclerophyllous shrublands	Pine forests
Annuals and short-lived ephemerals	Abundant	Abundant	Abundant
Geophytes	Many	Many	Many
Herbaceous perennials	Many (mostly grasses)	Many (mostly grasses)	Many (mostly grasses)
Regeneration mode of dominant woody species	Sprouting/seeding	Sprouting	Overstorey: seeding Understorey: sprouting/seeding
Timing of plant recruitment	Immediately after fire (in the 1st rainy season for the seeding species)	Immediately after fire	In the first post-fire rainy season
Fire-stimulated germination	Yes	Not in shrubs	Yes
Development of canopy cover	100% in less than 10 years	100% in ~ 5 years	100% in more than 10 years
Diversity trends	- Initially increases - Decreases with age	- Initially increases - Decreases with age	- Initially increases - Decreases with age

successional stage (Table 16.4). Woody seeding species forming soil seed banks exist in all stages of phrygana, in relatively high numbers. Among these species rockroses are the most prominent. Resprouting woody species also occur throughout the successional cycle of phrygana, such as Jerusalem sage and Greek spiny spurge. In the evergreen sclerophyllous shrublands though, only sprouting shrubs like kermes oak, lentisc, strawberry tree etc., occur throughout the fire cycle. In pine forests sprouting and seeding shrubs can be found in either stages, but in very different numbers and cover. In the early successional stages, resprouters are competing with young tiny pine seedlings and seedlings of the other taxa. As the canopy closes, the understorey taxa are diminished in number and relative cover. Herbaceous species are present in the early successional stages of all three vegetation types, but they become less and less abundant as succession proceeds. Sprouting herbaceous perennials occur mainly at the early stages in the evergreen sclerophyllous shrublands and pine forests (Table 16.4).

Table 16.4. Plant groups of different regeneration modes at three distinctive post-fire successional stages of mediterranean ecosystems in Greece

	Phrygana			Evergreen sclerophyllous shrublands			Pine forests		
	Early	Middle	Late	Early	Middle	Late	Early	Middle	Late
Seeding trees/ shrubs	+[a]	+[a]	+[a]	–	–	–	+	+	+
- canopy seed bank	–	–	–	–	–	–	–	+[b]	+[b]
- soil seed bank	+	+	+	–	–	–	+[a]	+[a]	+[a]
Sprouting shrubs	+	+	+	+	+	+	+[c]	+[c]	+[c]
Herbaceous perennials									
- seeding	?	?	?	?	?	?	?	?	?
- sprouting	+	<+	<+	+	–	–	+	<+	<+
Annuals									
- seeding	+	<+	<+	+	<+	–	+	<+	<+

[a] Seeding shrubs.
[b] Seeding trees of the overstorey.
[c] Sprouting shrubs of the understorey.
<+ Less abundant.

Similar patterns are reported for the other mediterranean-climate regions of the world. In chaparral, stands may consist of resprouting *Adenostoma fasciculatum* (Hanes 1977), but pure stands of the seeder *Ceanothus megacarpus* have been also reported (Schlesinger and Gill 1978). In Australian heathlands, sprouters appear to predominate (Russell and Parsons 1978; Specht 1981).

Among the various attempts which have been made so far in order to explain this variation two are more worthy of mention. Keeley (1977b) hypothesized that the relative success of seeders depends on the availability of openings in the post-fire regenerating vegetation, since they can exploit these spots and they have greater possibility to establish themselves away from competition from the resprouters. Thinning of a stand of sprouters progressively occurs in a normal fire interval, while live and dead fuel also accumulates. Thus the seeders, being relatively long-living taxa, have the advantage of establishing in these openings, from the dormant seed bank that they have formed in the ground. This pattern is mostly realised in long fire intervals, while in the short ones the ratio of seeders versus resprouters is decreased. It is therefore expected that under a variable fire frequency both forms may coexist. On the other hand, sprouters pose a competitive pressure on the seeders during post-fire regeneration, which has to do with exploitation of available moisture. As emphasized by Specht (1981), the extent of the foliage of the sprouters is determined by the available moisture. Thus, the more humid the environment, the less gaps are available to seeders. Therefore, the relative prevalence of either regeneration mode is considered a consequence of the site productivity.

Site productivity is not independent from fire regime, however, since repeated fires may lead to great loss of nutrients; on the other hand, prolonged absence of fire might have an analogous result, as decomposition rates and subsequent nutrient release are slowed down. It is fire, therefore, that acts as the primary force inducing this or the other pattern, through several secondary factors.

Data from the South African fynbos agree rather with the latter argument. Fynbos stands, including those on moist sites, have a number of species of both reproductive types, though often dominated by seeding shrubs (Kruger 1979; Bond 1980). The balance between seeders and sprouters is attributed to both an increase and a decrease in fire frequency.

All these quite often contradictory hypotheses need more concrete explanations, which require a more detailed demographic approach that will reveal the mechanisms underlying these patterns and which will predict the relative recruitment and mortality of sprouters and seeders as defined by site history.

16.6 Plant Demography

Studies performed in California have revealed two patterns of community recruitment: one in coastal sage shrubs, where there is a continual seedling production resulting in mixed-aged stands (Westman 1981), while in the chaparral seeding species produce seedlings only the first two post-fire years (Horton and Kraebel 1955; Quick 1959). This temporal seedling recruitment is one of the potential causes for chaparral stand decline (Westman 1981). Production of new stems by the resprouting species has not drawn much attention.

Data inserted in Tables 16.2–16.4 show the physiological traits of the plants and the way they are expressed in post-fire succession. Since vegetative regeneration seems not to be affected by the stage of the succession, at least not as much as regeneration by seeds, I will deal mostly with the latter. Data from Arianoutsou and Margaris (1981a) show that seed germination in the field occurs after the first rains in most of the seeding woody species of the burned phryganic ecosystems (Fig. 16.4), during the first post-fire

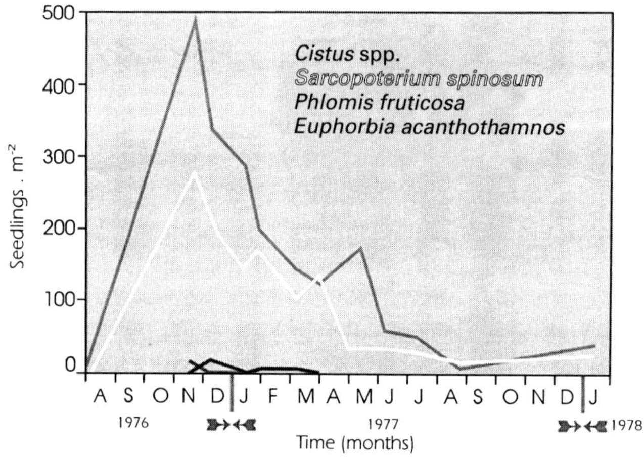

Fig. 16.4. Seedling density (indiv. m^{-2}) during early post-fire succession in a phryganic ecosystem in Greece. (Arianoutsou and Margaris 1981a)

year. Seed germination is not only enhanced by fire in many species but it results in significantly higher densities sometimes reaching 300 seedlings m^{-2}, as in the case of *Cistus* species. Seed germination for the phryganic species tested so far reveals a mechanical rupture of the hard seedcoat for members of the Cistaceae (Arianoutsou and Margaris 1981b; Vuillemin and Bullard 1981; Thanos and Georghiou 1988; Trabaud and Oustric 1989; Roy and Sonie 1992) and for members of the Leguminosae (Doussi and Thanos 1993). Light quality has also been found as cueing seed germination for the phryganic species *Sarcopoterium spinosum* (Roy and Arianoutsou 1985).

From the few demographic studies in post-fire communities in the Mediterranean Basin a decrease in seedling density of woody taxa during the first post-fire year has been shown (Naveh 1974; Papanastasis 1977; Arianoutsou and Margaris 1981; Trabaud and Oustric 1989). No evidence of remarkable seed germination exists so far for these woody species during the second post-fire year, while there is very strong evidence that this does occur massively for several herbaceous ones, such as in the case of legumes (Fig. 16.5).

The juvenile phase for most of the woody species lasts only two years, that is they are at reproductive age when they are in the second year of their lives (pers. observ.). Data for seedling emergence of the dominant woody phryganic species in the field reveal that seed germination normally occurs even without fire (Figs. 16.6–16.10). It seems that fire plays an important role not in seed germination per se, since it is happening anyway, but in massive seed germination as a result of the creation of a micro-

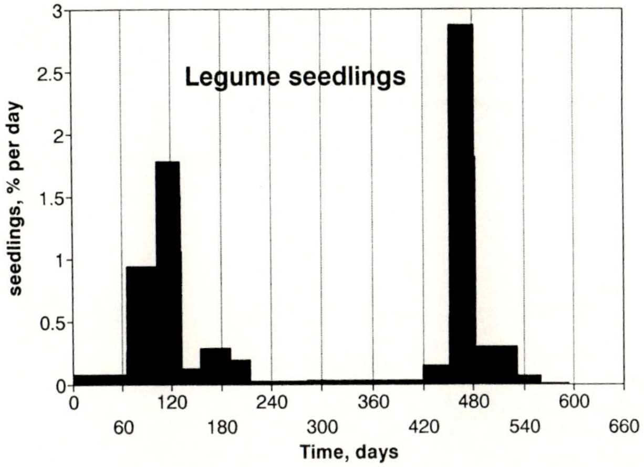

Fig. 16.5. Percentage increment of legume seedlings appearing on a burned Aleppo pine forest in Attica, Greece, during the first two post-fire years. (Papavassiliou and Arianoutsou, in prep)

environment making all of the conditions which are necessary for germination more available. This massive germination leads to the formation of a seedling reservoir from which community recruitment takes place much more easily. In other words, and as far as it concerns the dominant woody phryganic species, germination is regularly and continuously occurring, with a time lag of 1–2 years possibly, following the initial burst. The result of such a continuous seedling production is the existence of mixed-aged stands. The pyramid of age classes of *P. fruticosa, E. acanthothamnos* and *Cistus* spp. (Fig. 16.11) constructed with data drawn from a mature phryganic ecosystem, demonstrate this characteristic. Westman's data (1981) for coastal sage shrub formations are in full agreement with the above finding.

Roy and Sonie (1992) working with *Cistus monspeliensis* and *Cistus albidus* stands in southern France, have found that recruitment of both species occurred steadily during the five post-fire years, and then stopped. They claimed that seedling recruitment was related to canopy density. Data on similar perspectives are currently being gathered in Greece by our group.

It seems that the herbaceous taxa of these ecosystems generally follow the above pattern, being different in that they reach their reproductive maturity in the first post-fire year (Papavassiliou et al. 1994; Kazanis and Arianoutsou, unpubl. data).

Since almost all plants of the woody component in the evergreen sclerophyllous formations are resprouters we expect to have a mono-

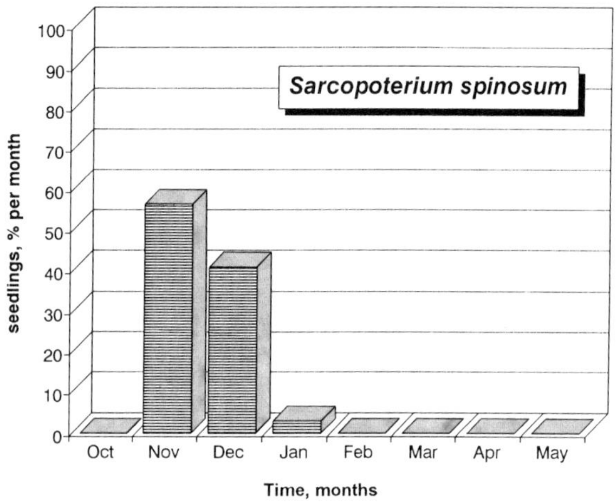

Fig. 16.6. Percentage increment of *Sarcopoterium spinosum* seedlings appearing in a mature phryganic ecosystem of Attica, Greece. (Data adapted from Argyris 1977)

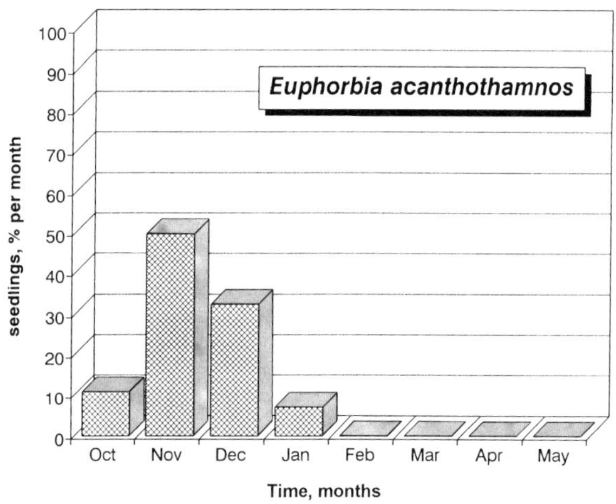

Fig. 16.7. Percentage increment of *Euphorbia acanthothamnos* seedlings appearing in a mature phryganic ecosystem of Attica, Greece. (Data adapted from Argyris 1977)

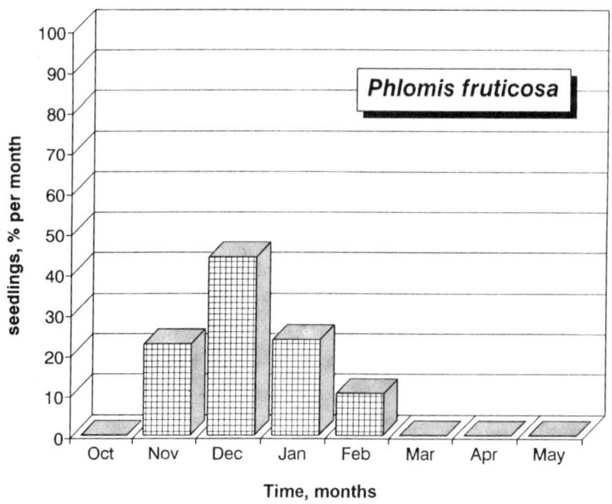

Fig. 16.8. Percentage increment of *Phlomis fruticosa* seedlings appearing in a mature phryganic ecosystem of Attica, Greece. (Data adapted from Argyris 1977)

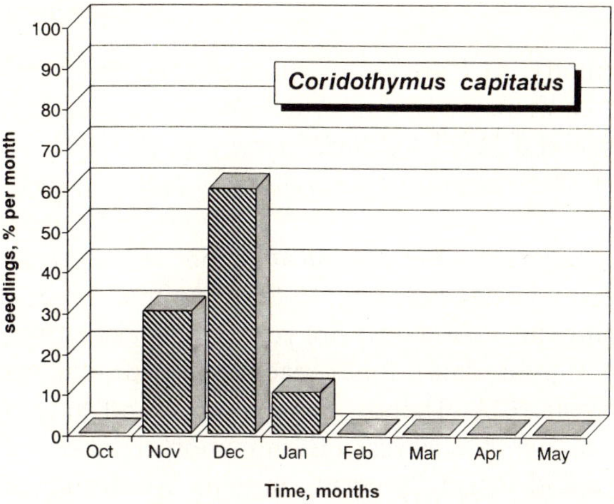

Fig. 16.9. Percentage increment of *Coridothymus capitatus* seedlings appearing in a mature phryganic ecosystem of Attica, Greece. (Data adapted from Argyris 1977)

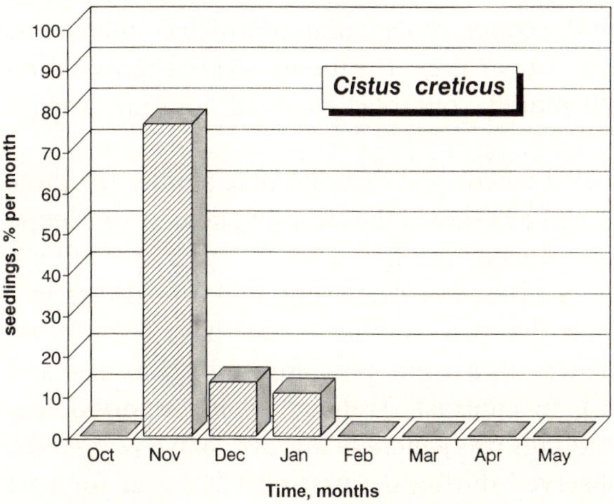

Fig. 16.10. Percentage increment of *Cistus creticus* seedlings appearing in a mature phryganic ecosystem of Attica, Greece. (Data adapted from Argyris 1977)

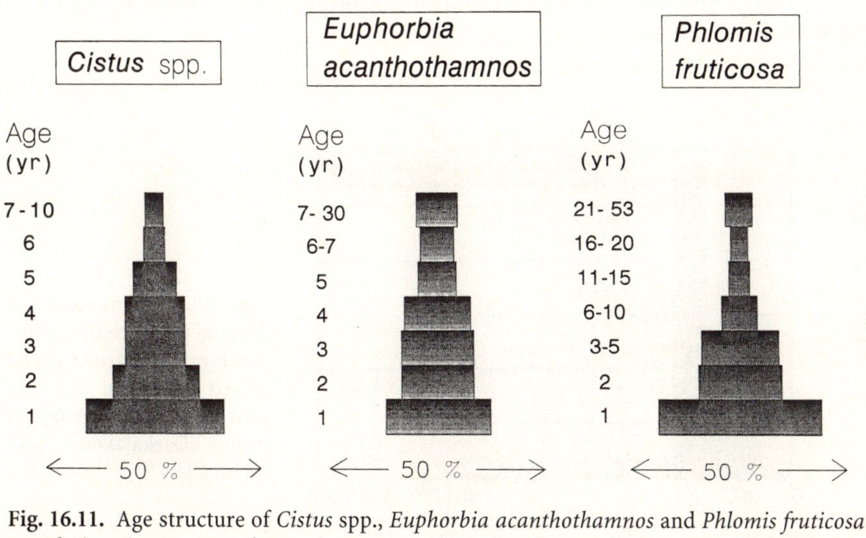

Fig. 16.11. Age structure of *Cistus* spp., *Euphorbia acanthothamnos* and *Phlomis fruticosa* populations in a mature phryganic ecosystem in Attica, Greece. (Arianoutsou and Margaris 1982)

layered single-aged community, at least as far as concerns the above-ground plant parts. Although there is some information on shoot growth in these plants (Arianoutsou and Mardiris 1985; Perreira et al. 1985), almost nothing is available on shoot recruitment versus age – and this has to be done.

The situation in the pine forests is more complicated. Pine seeds germinate in a wave during the first post-fire year (Fig. 16.12) and almost no new seedlings are observed during the second post-fire year (Daskalakou 1996; Kazanis and Arianoutsou, unpubl. data). Pines reach their reproductive maturity after their sixth year (C.A. Thanos, pers. comm.) and have a considerably long life span, longer than 50 years. Germination in unburned pine forest is rather scarce and it usually occurs in gaps. According to Skordilis (1992), germination of pine seeds is light-controlled. Fire causes a massive seed release from the cones, the opening of which is induced by high temperatures developing during fires. The increased seed germination thus may be attributed to the more favourable light conditions (an ameliorated open environment).

There are not many detailed diachronic demographic studies for these pine species. Daskalakou (1996) has shown that seedling density of Aleppo pine is decreased drastically during the first post-fire year, reaching a number of 2.5 seedlings m^{-2} and further decreasing to 1.5 seedlings m^{-2} at the end of the second post-fire year. Similar results have been found in a post-fire chronosequence of Aleppo pine forest by Kazanis and Arianoutsou (unpubl. data). In contrast, Trabaud (1988) reported that mortality of Aleppo pine seedlings was generally low over a five year study. High mortality was also observed during the first post-fire year for East Mediterranean pine (Thanos et al 1989; Thanos and Marcou 1991, 1993; Spanos 1992, 1994).

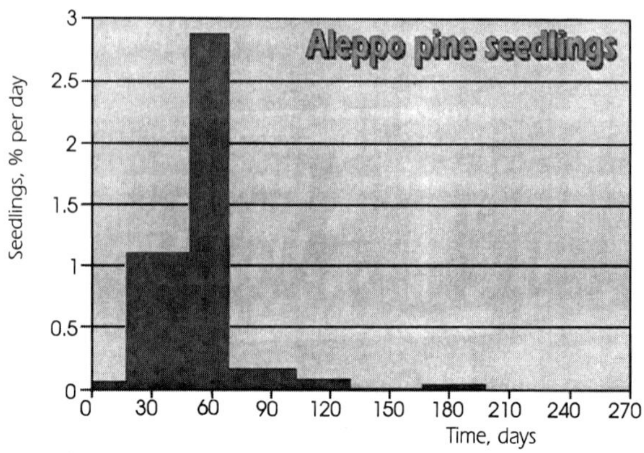

Fig. 16.12. Increment of Aleppo pine seedlings appearing in a burned forest of Attica during the first post-fire year. (Adapted from Daskalakou and Thanos 1997)

All evergreen sclerophyllous species of the understorey regenerate vegetatively, actively posing an interspecific competition upon the young pine seedlings. Apart from these, the few representatives of seasonal dimorphic phrygana are mainly species of the Cistaceae, reestablishing mainly by seeds. Herbaceous annuals also germinate, while most perennial herbs resprout.

Gradually, as the canopy of the forest closes, the understorey taxa are restricted but they become more abundant again at the more mature stages, when the tree floor starts to thin out due to senescence (in the absence of fire) (Kazanis and Arianoutsou 1996).

16.7 Seed Bank Dynamics

Westman (1981) in his work with coastal sage shrubs claimed that a large proportion of the colonising flora, at least the herbaceous species, arrive by immigration from the adjacent unburned stands. This is unlikely to happen for many reasons, mainly because of the distances involved, the weakness of wind dispersal (Harper 1977; Cook 1980) and the rapid appearance of the herbs in the burned sites (Keeley 1980; Keeley et al. 1981). It is likely that the fire annuals of the chaparral behave like fugitive species, but since any single chaparral site is inevitably subjected to fire there is little selection towards long-range dispersal. In South African fynbos and in Australian heathlands the dominant shrubs are adapted to a short-range dispersal rather than to long-range modes (Williams 1972; Bond 1980; Berg 1981). Evidence from our studies in Greece confirms the latter theory, although some species may have evolved long-range dispersal through zoochory, as for example some legumes whose dispersal units "stick" to animals which graze the soft young post-fire growth, or seeds of the fleshy fruits of some evergreen sclerophyllous plants dispersed by birds.

Formation of soil seed banks is an alternative solution for those plants having short distance dispersal mechanisms (Fenner 1987). It seems that soil seed banks are the norm for the phryganic species which are obligate seed regenerators, as the Cistaceae, while canopy seed banks are the norm for the other obligate seed regenerators of the Mediterranean plant communities of Greece, the pines (Table 16.4). Not much research has been done so far for either of these, with some exception for *Pinus halepensis* (Daskalakou 1996), for some Cistaceae (Skourou and Arianoutsou, unpubl. data) and for pioneer annual legumes (Papavassiliou et al. 1994).

Searching the literature on seed bank dynamics in Mediterranean-climate plant communities revealed only a few studies (Troumbis and

Trabaud 1987; Zammit and Zedler 1988; Parker and Kelly 1989; Pierce and Cowling 1991). The topic of the timing of the formation of these reservoirs remains an open field for most of the Mediterranean communities.

16.8 Life Span and Fecundity of Seeding Plants

Among the woody species which are obligate seed regenerators, members of the Cistaceae have a relatively short life span of about 15 years (Fig. 16.11), while pines have a long life cycle. None of the facultative seed regenerators of the phryganic species examined so far appear to be short-lived. On the contrary there were individuals of *Euphorbia acanthothamnos* more than 50 years old (Fig. 16.11).

Australian and fynbos heathland seeding species have relatively short life spans (Kruger 1983), while life span of Californian *Ceanothus* species – many of which are typical seeding taxa – is reported to be relatively limited, although it can reach to 40–60 years (Hanes 1977; Keeley 1975).

At this point, I think it is worthwhile discussing what is a short and what is a long life span, as well as the signficance of life span in relation to fire. If we were not referring to a fire prone environment, life span would probably have less importance. But in the Mediterranean climate it becomes crucial, especially if some plants rely upon recurrent fires for their recruitment and existence.

None of the Mediterranean plants of Greece seem to be strictly dependent upon fire for its recruitment. Even post-fire obligate seed regenerators such as rockroses and pines do germinate without fire, as Fig. 16.9 shows for the former and repeated observations in forest gaps and sideways have shown for the latter. It is beyond any doubt that massive population recruitment for these species is greatly facilitated by fire, but would they die if it does not happen? I believe that they would not. On the contrary, the problem lies to those long-lived species which have extended juvenile periods, such as pines, which are expected to become extinct not in the absence of fire but in very frequent fires at intervals shorter than the minimum time which is necessary for full regeneration maturity.

This brings us to the next very important issue of plant demography, which is the fecundity schedule. Very few data are available for either Mediterranean-type ecosystems. Specht et al. (1958) showed accelerating fruit production from about nine years to between 20 and 25, with some decline afterwards. Bond (1980), however, did not find a similar trend in fynbos. Keeley and Keeley (1977) reported that reproductive vigour is maintained with advancing age in the dominant seeding chaparral species.

For seeders in Greek Mediterranean formations, a broad survey has recently started within the frame of the two research projects mentioned earlier, namely MODMED and PROMETHEUS which are expected to produce such data.

Finally, although there is much evidence that initially high post-fire density of all seeding species gradually becomes lower (see above-cited references), almost no data are available on density-dependent or independent mortality of seedlings.

16.9 Conclusions

It is evident from the above that there is a considerable gap of knowledge in vegetation dynamics, species replacement patterns and demography in the fire prone Mediterranean communities of Greece and of the Mediterranean Basin. The need to fill such a gap does not only originate from a merely scientific interest. Rational landscape management must be strongly dependent on its carrying capacity, which is defined by community structure, diversity and plant demographic features, as they are all influenced by the seasonal and geographical variation of the abiotic environment. Increased attention has recently been given to the issues of degradation and desertification threats, especially under a changing environment, both in socio-economic terms as well as climatically. None of the proposed models prepared to describe these processes, regardless of how sophisticated they might be, will be reliable unless they accurately represent the complexity of the Mediterranean landscapes and the ecosystems they support. And this can only be achieved through the evaluation of real data. The challenge is there and I hope that we will meet it.

Acknowledgements. The author expresses her thanks to the European Union for funding under the projects MODMED (EV-CT94-0489, ENV4-CT95-0139) and PROMETHEUS (EV-CT94-0482).

References

Argyris J (1977) The ecology of seeds of some phryganic species. PhD Thesis, University of Athens

Arianoutsou M (1979) Biological activity after fire in a phryganic ecosystem. PhD Thesis, University of Thessaloniki

Arianoutsou M (1984) Post-fire successional recovery of a phryganic (East Mediterranean) ecosystem. Acta Oecol (Oecol Plant) 59:387–394

Arianoutsou M, Margaris NS (1981a) Producers and the fire cycle in a phryganic ecosystem. In: Margaris NS, Mooney HA (eds) Components of productivity of mediterranean-climate regions - basic and applied aspects. Dr W. Junk, The Hague, pp 181–190

Arianoutsou M, Margaris NS (1981b) Early stages of regeneration after fire in a phryganic ecosystem (East Mediterranean). I. Regeneration by seed germination. Ecol Medit 7:119–128

Arianoutsou M, Margaris NS (1982) Phryganic (East Mediterranean) ecosystems and fire. Ecol Medit 8:473–480

Arianoutsou M, Mardiris Th (1985) Observations on the phenology of the dominant plants of the Greek maquis. In: Catarino F, Lange OL, Oechel WL (eds) Plant response to stress. Functional analysis in mediterranean ecosystems. Springer, Berlin Heidelberg New York, pp 515–519

Barry J (1960) Contribution a l' étude de la vegetation de la region de Nimes. III. Les stades preforestiers. Ann Biol 36:311–540

Berg RY (1981) The role of ants in seed dispersal in Australian lowland heathland In: Specht RL (ed) Ecosystems of the world, vol 9B. Heathlands and related shrublands. Elsevier, Amsterdam, pp 51–59

Bond W (1980) Fire and senescent fynbos in Swartberg, southen Cape. S Afr For J 114:68–71

Christensen NL, Muller CH (1975) Effects of fire on factors controlling plant growth in *Adenostoma* chaparral. Ecol Monogr 45:29–55

Cook R (1980) The biology of seeds in the soil In: Solbrig OT (ed) Demography and evolution in plant populations. Blackwell, Oxford, pp 51–59

Daskalakou E (1996) Ecophysiology of post-fire regeneration of Aleppo pine. PhD Thesis, University of Athens

DeLillis M, Testi A (1990) Post-fire dynamics in a disturbed mediterranean community in central Italy. In: Goldamner JG, Jenkins MJ (eds). Fire in ecosystem dynamics. Mediterranean and northern perspectives. SPB Academic, The Hague, pp 53–62

Doussi M, Thanos CA (1993) The ecophysiology of fire-induced germination in hard-seeded plants. In: Come D, Corbineau F (eds) Basic and applied aspects of seed biology. ASFIS, Paris, pp 455–460

Dugrand R (1964) La garrigue montpellieraine. PVF, Paris.

Daskalakou E, Thanos CA (1997) Postfire establishment and survival of Aleppo pine seedlings. In: Balabanis P, Eftichidis G, Fantechi R (eds) Proc European School of Climatology and Natural Hazards Course, Porto Carras, Halkidiki, Greece. European Commission, pp 357–369

Espirito Santo MD, Rego F, Costa JC (1993) Vegetation dynamics in the Serra dos Candeeiros (central Portugal).In: Trabaud L, Prodon R (eds) Fire in Mediterranean ecosystems. Ecosystem Research Report no 5, Commission of the European Communities, Brussels, pp 29–46

Faraco AM, Fernandez F, Moreno JM (1993) Post-fire vegetation dynamics of pine woodlands and shrublands in the Sierra de Gredos. In: Trabaud L, Prodon R (eds) Fire in Mediterranean ecosystems. Ecosystem Research Report no 5. Commission of the European Communities, Brussels, pp 101–112

Fenner M (1987) Seed characteristics in relation to succession. In: Gray AJ, Crawley MJ, Edwards PJ (eds) Colonisation, succession and stability. Blackwell, Oxford, pp 103–114

Gill GM, Groves RH (1981) Fire regimes in heathlands and their plant ecological effects. In: Specht RL (ed). Ecosystems of the world, vol 9B. Heathlands and related shrublands. Elsevier, Amsterdam, pp 61–84

Hanes TL (1971) Succession after fire in the chaparral of southern California. Ecol Monogr 41:27–52

Hanes TL (1977) Chaparral. In: Barbour MG, Major J (eds). Terrestrial vegetation of California. Wiley, New York, pp 417–469

Harper JL (1977) Population biology of plants, Academic Press, London.

Higgs EJ, Vita-Finci C, Harris DR, Fagg AE (1967) The climate, environment and industries of Stone Age, Part III. Proc Prehist Soc 33:1–29

Horton JS, Kraebel CJ (1955) Development of vegetation after fire in the chamise chaparral of Southern California. Ecology 36:244–262

Kailidis D (1992) Forest fires in Greece. In: Ministry of Agriculture, Secretariat General for Forests and Natural Environment (tech coord). Proc Int Seminar on Forest Fire prevention, land use and people, Athens, Greece, pp 27–40

Kazanis D, Arianoutsou M (1994) Vegetation structure in a post-fire successional gradient of *Pinus halepensis* forests. In: Viegas DX (ed) Proc 2nd Int Conf on Forest Fire Research. Portugal, pp 979–993

Kazanis D, Arianoutsou M (1996) Vegetation composition in a post-fire successional gradient of *Pinus halepensis* forests in Attica, Greece, Int J Wildland Fire, 6:83–91

Keeley JE (1975) Longevity of non-resprouting *Ceanothus*. Am Midl Nat 93:504–507

Keeley JE (1977a) Fire-dependent reproductive strategies in *Arctostaphylos* and *Ceanothus*. In: Mooney HA, Conrad CE (eds). Proc Symp on the Environmental consequences of fire and fuel management in mediterranean ecosystems, Palo Alto, California. USDA Forest Service Gen Tech Rep WO-3, pp 391–396

Keeley JE (1977b) Seed production, seed populations in the soil and seedling production after fire for two congeneric parts of sprouting and non-sprouting chaparral shrubs. Ecology 58:820–829

Keeley JE, Keeley S (1977) Energy allocation patterns of sprouting and a non-sprouting species of *Arctostaphylos* in the California chaparral. Am Midl Nat 98:1–10

Keeley JE, Zedler PH (1978) Reproduction of chaparral scrubs after fire: a comparison of sprouting and seeding strategies. Am Midl Nat 99:142–161

Keeley S, Keeley JE Hutchinson SM, Johnson AW (1981) Postfire succession of the herbaceous flora in Southern California. Ecology 62:524–530

Keeley JE, Nitzberg ME (1984) The role of charred wood in the germination of the chaparral herbs *Emmenanthe penduliflora* and *Eriphyllum confertiflorum*. Madroño 31:208–218

Keeley JE, Keeley S (1989) Allelopathy and the fire-induced herb cycle. In: Keeley JE (ed). The California chaparral. Paradigms re-examined. Natural History Museum of Los Angeles County, Los Angeles, pp 65–72

Kruger FJ (1977) Ecology of Cape Fynbos in relation to fire. Proc Symp on the Environmental consequences of fire and fuel management in mediterranean ecosystems, Palo Alto, California. USDA Forest Service Gen Tech Rep WO-3, pp 65–72

Kruger FJ (1979) Fire. In: Day J, Siegfried NR, Jarman ML (eds) Fynbos ecology: a preliminary synthesis. S Afr Nat Sci Prog Rep no 40, CSIR, Pretoria, pp 43–57

Kruger FJ (1983) Plant community diversity and dynamics in relation to fire. In: Kruger FJ, Mitchell DT, Jarvis JUM (eds) Mediterranean-type ecosystems: the role of nutrients. Springer, Berlin Heidelberg New York, pp 446–472

Liacos LG (1973) Present studies and history of burning in Greece. In: Proc 13th Annu Tall Timbers fire ecological conference, no 13, Tallahassee, Florida, pp 65–96

LeHouerou HN (1973) Fire and vegetation in the Mediterranean basin. In: Proc 13th Annu Tall Timbers fire ecological conference, no 13, Tallahassee, Florida, pp 237–277

Lucchesi S, Giovannini G (1993) Plant community dynamics following a fire: a case study in Toscany. In: Trabaud L, Prodon R (eds) Fire in Mediterranean ecosystems. Ecosystem Research Report no 5, Commission of the European Communities, Brussels, pp 47–54

Mazzoleni S, Pizzolongo P (1990) Post-fire regeneration patterns of Mediterranean shrubs in the Campania region, southern Italy. In: Goldamner JG, Jenkins MJ (eds). Fire in

ecosystem dynamics: Mediterranean and northern perspectives. SPB Academic, The Hague, pp 43–52
Mariolopoulos I, Karapiperis L (1955) Rainfall in Greece. Ministry for the Co-ordination
Mavromatis G (1978) Vegetation map of Greece. Institute of Forest Research. Athens, Greece
Mooney HA, Conrad CE (tech coord) (1977) Proc Symp on the Environmental consequences of fire and fuel management in mediterranean ecosystems, Palo Alto, California, USDA Forest Service, Washington DC, General Tech Rep WO-3
Naveh Z (1974) Effects of fire in the Mediterranean region. In: Kozlowski TT, Ahlgren CE (eds). Fire and ecosystems. Academic Press, New York, pp 410–434
Naveh Z (1975) Evolutionary significance of fire in the Mediterranean region. Vegetatio 29:199–208
Ne'eman G, Lahav H, Izhaki I (1993) The resilience of vegetation to fire in an east-Mediterranean pine forest on Mount Carmel, Israel: the effect of post-fire management. In: Trabaud L, Prodon R (eds) Fire in Mediterranean ecosystems. Ecosystem Research Report no 5, Commission of the European Communities, Brussels, pp 127–141
Papanastasis VP (1977) Fire ecology and management of phrygana communities in Greece. In: Mooney HA, Conrad CE (tech coord). Proc Symp on the Environmental consequences of fire and fuel management in mediterranean ecosystems, Palo Alto, California, USDA Forest Service, Washington DC, General Tech Rep WO-3, pp 476–482
Papavassiliou S, Arianoutsou M, Thanos CA (1994) Aspects of the reproductive biology of fire following species of Leguminosae in a *Pinus halepensis* Mill. Forest. In: Viegas DX (ed) Proc 2nd Int Conf on Forest Fire Research, Portugal, suppl.
Parker VT, Kelly VP (1989) Seed banks in California chaparral and other Mediterranean climate shrublands. In: Leck MA, Parker VT, Simpson RL (eds) Ecology of seed banks. Academic Press, San Diego, pp 231–255
Perreira JS, Beyschlag G, Lange OT, Beyschlag W, Tenhunen JD (1985) Comparative phenology of four mediterranean shrub species growing in Portugal. In: Catarino F, Lange OL, Oechel WC (eds). Plant response to stress. Functional analysis in mediterranean ecosystems. Springer, Berlin Heidelberg New York, pp 503–514
Pierce SM, Cowling RM (1991) Dynamics of soil-stored seed banks of six shrubs in fire-prone dune fynbos. J Ecol 79:731–747
Purdie RW (1977) Early stages of regeneration after burning in dry sclerophyllous vegetation. I. Regeneration of the understory by vegetative means. II. Regeneration by seed germination. Aust J Bot 25:21–34, 35–46
Quick CR (1959) *Ceanothus* seeds and seedlings on burns. Madroño 3:135–140
Roy J, Arianoutsou M (1985) Light quality as the environmental trigger for the germination of the fire-promoted species *Sarcopoterium spinosum* L. Flora 177:345–349
Roy J, Sonnie L (1992) Germination and population dynamics of *Cistus* species in relation to fire. J Appl Ecol 29:647–655
Rundel PW (1981) Fire as an ecological factor. In: Lange OL, Nobel PS, Osmond CB. Ziegler H (eds) Encyclopedia of plant physiology. vol 12A. Physiological plant ecology. Springer, Berlin Heidelberg New York, pp 501–538
Russell RP, Parsons RF (1978) The effect of time since fire on heath floristics at Wilson's Promontory, southern Australia. Aust J Bot 26:53–61
Schlesinger WH, Gill DS (1978) Demographic studies of the chaparral shrub, *Ceanothus megacarpus,* in the Santa Ynez Mountains, California. Ecology 59:1256–1263
Skordilis A (1992) Seed germination and seedling development in *Pinus halepensis* and *P. brutia:* physiological and ecological approach. PhD Thesis, University of Athens
Spanos YA (1992) Structure analysis and regeneration of Brutia pine in Thasos island. Ph.D. Thesis, University of Thessaloniki

Spanos YA (1994) Natural regeneration of *Pinus brutia* in northwest areas of Thasos island, burned in 1989. Geot Epistim Themata (Greece) 4:33–39

Specht RL (1981) Responses to fires of heathlands and related shrublands. In: Gill AM, Groves RH, Noble IR (eds) Fire and the Australian biota. Australian Academy of Sciences, Canberra, pp 395–414

Specht RL, Rayson P, Jackman ME (1958) Dark island heath (Ninety-Mile Plain, South Australia). VI. Pyric succession: changes in composition, coverage, dry weight and mineral nutrient status. Aust J Bot 6:193–202

Thanos CA, Georghiou K (1988) Ecophysiology of fire-stimulated seed germination in *Cistus incanus* ssp. *creticus* (L.) Heywood and *C. salvifolius* L. Plant Cell Environ 11:841–849

Thanos CA, Marcou S (1991) Post-fire regeneration in *Pinus brutia* forest ecosystems of Samos island (Greece): 6 years after. Acta Oecol (Oecol Plant) 12:633–642

Thanos CA, Marcou S (1993) Natural regeneration of *Pinus brutia* forests of Samos island. Key events during the first 10 years of the postfire period. Int Symp on *Pinus brutia*. Ten, Turkey, pp 176–183

Thanos CA, Marcou S, Christodoulakis D, Yannitsaros A (1989) Early post-fire regeneration in *Pinus brutia* forest ecosystems of Samos island (Greece). Acta Oecol (Oecol Plant) 10:70–94

Trabaud L (1980) Influence du feu sur les semences enfouies dans les couches superficielles du sol d'une garrigue de Chene kermes. Naturalia monspeliensia. Ser Bot 39:1–12

Trabaud L (1988) Survie de jeunes plantules de pin d' Alep apparus apres incendie. Stud Oecol 5:151–170

Trabaud L, Oustric J (1989) Heat requirements for seed germination of three *Cistus* species in the garrigue of southern France. Flora 183:321–325

Troumbis A, Trabaud L (1987) Dynamique de la banque de graines de deux especes de Cistes dans le maquis grec. Acta Oecol (Oecol Plant) 8:167–179

Vuillemin J, Bullard C (1981) Ecophysiologie de la germination de *Cistus albidus* L. et *Cistus monspeliensis* L. Naturalia monspeliensia. Ser Bot 46:1–11

Westman WE (1981) Coastal sage scrub succession. In: Mooney HA, Conrad CE (tech coord). Proc Symp on the Environmental consequences of fire and fuel management in mediterranean ecosystems, Palo Alto, California, USDA Forest Service, Washington DC, General Tech Rep WO-3, pp 91–96

Williams IJM (1972) A revision of the genus *Leucadendron* (Proteaceae). Contrib Bolus Herb 3:1–425

Zammit CA, Zedler PH (1988) The influence of dormant shrubs, fire and time since fire on soil seed banks in mixed chaparral. Vegetatio 75:175–187

17 Ecophysiological Processes and Demographic Patterns in the Structuring of California Chaparral

S.D. Davis, K.J. Kolb and K.P. Barton

It is now becoming clear that xylem dysfunction induced by drought is a serious problem for plants. Therefore, resistance of xylem to cavitation events is an important (perhaps the most important) parameter that determines drought resistance.

(Tyree and Ewers 1991)

17.1 Introduction

Demographic patterns in California chaparral are greatly influenced by two environmental stresses that interact to regulate community structure. These two factors are recurrent wildfire and persistent summer drought (Keeley and Keeley 1988; Axelrod 1989; Moreno and Oechel 1991).

In regions of high fire frequency such as the Santa Monica Mountains of southern California, chaparral is dominated by shrub species that are disturbance dependent (see Keeley, chapter 14, this volume). Such species are either non-sprouters after fire, relying solely on seed germination for reestablishment, or they are facultative sprouters after fire, utilizing two modes of post-fire recovery: seed germination and vegetative resprouting (Keeley 1986). Both facultative sprouters and non-sprouters recruit new individuals into their population almost exclusively during the first year following a fire event (Keeley 1992). Their seeds are not only refractory, but require a fire cue to promote germination, such as high temperatures to scarify their thick seed coats, or chemical signals released from charred wood to break dormancy (Keeley 1991).

Summer drought interacts with wildfire as a major determinant of establishment patterns because newly germinated seedlings, possessing only a diminutive root system, must survive a 4 to 6 month rainless period during their first critical year of establishment (Frazer and Davis 1988; Thomas and Davis 1989; Mills and Kummerow 1989; Moreno and Oechel 1991). High seedling mortality is generally the rule but available data suggest that seedlings from non-sprouting species have consistently lower mortality than seedlings of sprouting species (Sampson 1944; Musick 1972; Keeley and Zedler 1978; Jacks 1984, Frazer and Davis 1988; Thomas and Davis 1989; Davis 1989; Mills and Kummerow 1989).

Ecological Studies, Vol. 136
Rundel et al. (eds.) Landscape Degradation and Biodiversity in Mediterranean-Type Ecosystems

At first glance, it appears that chaparral sprouters would have a distinct advantage over non-sprouters because they have two mechanisms of post-fire recovery, one which holds previously acquired position (resprouts), the other allowing population expansion (seedlings). However, it appears that non-sprouters persist because they have a genetic advantage (Wells 1969). That is, with each fire cycle they produce an entirely new pool of genetically recombined individuals upon which natural selection can operate. Thus, non-sprouters have greater genetic diversity (greater speciation), and more quickly paced evolution, that closely tracks the recent advent of a drier, more fire prone climate to southern California (Wells 1969; Raven 1977; Axelrod 1989).

It follows that non-sprouting chaparral species may be more adapted to water stress than sprouting species. Saruwatari and Davis (1989) have tested this hypothesis by comparing the tissue water relations of post fire seedlings of *Ceanothus megacarpus* (non-sprouter), *C. spinosus* (facultative sprouter), and *Malosma laurina* (facultative sprouter) growing together in a mixed stand in the Santa Monica Mountains. However they found no evidence using standard pressure volume curves and dew point hygrometry to support the notion that non-sprouters were more drought tolerant. All seedling species examined experienced similar water stress (minimum seasonal water potential and osmotic potential) and all species lost turgor during summer drought. Furthermore, these negative results were consistent with earlier work on four additional species of chaparral shrubs (Davis and Mooney 1985, 1986a,b). It appears that non-sprouting species like *C. megacarpus* have an ability to dehydrate beyond the turgor loss point and survive whereas this is not true for sprouters such as *C. spinosus* and *M. laurina*, even though they often grow together in mixed stands.

One way to further explore this possibility that non-sprouters are more drought tolerant is to test the hypothesis that they are more resistant to water stress-induced embolism of their xylem tissues. It may be that the apoplast (xylem transport system) is the weak link in resistance to water stress, much more so than the symplast system (cellular water potential, osmotic potential, and turgor potential) measured by Saruwatari and Davis (1989). Under conditions of moderate water stress, some of the normally water-filled xylem conduits of chaparral shrubs may cavitate, fill with air (embolize), and become inactive in water transport. In cases of severe water stress, embolism may become extensive enough to seriously hamper hydraulic supply to shoot tissues, affecting growth, productivity, and even causing branch dieback or whole shoot mortality (Tyree and Sperry 1989; Tyree and Ewers 1991). Resistance to embolism may be especially important for successful seedling establishment in arid environments (Williams et al. 1997).

From studies of descriptive wood anatomy, Carlquist argued that the xylem of some chaparral species is more resistant to dysfunction (cavitation) caused by water stress than others (Carlquist 1985, 1989; Carlquist and Hoekman 1985). Tyree and Ewers (1991) have suggested that the vulnerability of stem xylem to water stress-induced embolism may be a useful index of drought tolerance. We have recently pursued this possibility among chaparral species with good success (Kolb and Davis 1994; Hargrave et al. 1994; Tyree et al. 1994; Jarbeau et al. 1995; Redtfeldt and Davis 1996). The vulnerability of the hydraulic transport system (xylem) to water stress now appears to be at least a partial explanation as to why some chaparral species are more susceptible to water stress than others.

Here we will first focus on four chaparral species that can be found growing together in mixed stands in the Santa Monica Mountains: *M. laurina*, *C. megacarpus*, *Adenostoma fasciculatum*, and *A. sparsifolium* (Fig. 17.1). We chose these species because they are known to have different life history characteristics (sprouters and non-sprouters), seasonal water status, rooting depth, and post-fire establishment patterns (Thomas and Davis 1989; Redtfeldt and Davis 1996). Secondly, we briefly examine three pairs of sprouting and non-sprouting *Ceanothus* that occur within different microsites in the Santa Monica Mountains (Nicholson 1993; Davis, et al., 1999).

17.2 Field Study Area

Figure 17.1A depicts a typical mixed chaparral stand in the Santa Monica Mountains which is similar to that found in other parts of southern California. This particular stand occurs at the western end of the mountain range at an elevation of 600 m. Moving from right to left in Figure 17.1A, the stand is comprised of *Malosma laurina* (large leaved shrub in the bottom right), *Adenostoma sparsifolium* (yellow-green leaves), *Adenostoma fasciculatum* (shrubs with white blossoms) and *Ceanothus megacarpus* (interspersed between *A. fasciculatum* and *A. sparsifolium*). *Ceanothus megacarpus* is a non-sprouter after fire whereas the other three species are facultative sprouters (Table 17.1). All four species recruit new individuals into their population during the first year after wildfire. *Ceanothus megacarpus* flowers in February during the wettest month of the year, *A. fasciculatum* flowers in springtime during the onset of summer drought, but *A. sparsifolium* and *M. laurina* delay flowering until the peak of the summer drought in August (Table 17.1). This flowering pattern possibly reflects the

Fig. 17.1A. A typical mixed stand of chaparral shrubs in a coastal exposure of the Santa Monica Mountains of southern California. The stand is dominated by *Ceanothus megacarpus* (*C*), *Adenostoma fasciculatum* (*F*), *Adenostoma sparsifolium* (*S*), and *Malosma laurina* (*M*). **B** The same stand immediately after a 14000 ha wildfire of October, 1993. **C** The same stand one and one-half years after the wildfire, which is now dominated by a "fire-following", herbaceous perennial (*Dicentra ochroleuca*) (*D*)

shallower rooting depths of *C. megacarpus* and *A. fasciculatum* in comparison to *A. sparsifolium* and *M. laurina* and is consistent with the lower seasonal water potentials reported for *C. megacarpus* and *A. fasciculatum* (Table 17.1, Fig. 17.2). Post-fire seedling survivorship is much greater for *C. megacarpus* than for *M. laurina* (Table 17.1, Fig. 17.3A) and seedling survivorship for *A. fasciculatum* is greater than for *A. sparsifolium* (Table

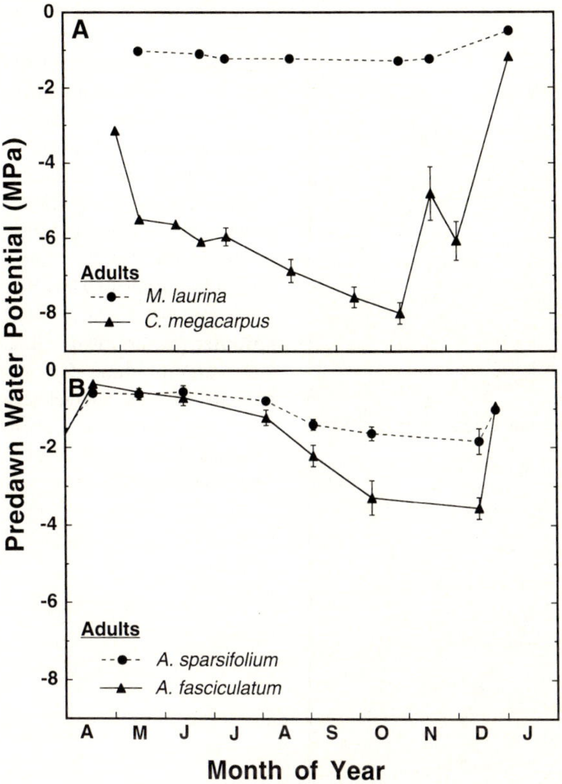

Fig. 17.2. Seasonal changes in predawn water potentials for co-occurring mature shrubs of: ***A*** *Malosma laurina/Ceanothus megacarpus*; and ***B*** *Adenostoma sparsifolium/Adenostoma fasciculatum* growing in the Santa Monica Mountains of southern California. Data taken from Thomas and Davis (1989) and Redtfeldt and Davis (1996)

Table 17.1. Four species of chaparral shrubs that dominate much of the coastal exposures of the Santa Monica Mountains of southern California are *Ceanothus megacarpus*, *Adenostoma fasciculatum*, *A. sparsifolium*, and *Malosma laurina*. Although growing together in a mixed stand (cf. Fig. 17.1), phenologies are very different and species display differential susceptibility to disturbance by drought and wildfires

	C. megacarpus	*A. fasciculatum*	*A. sparsifolium*	*M. laurina*	References[a]
Post-fire sprouter	No	Yes	Yes	Yes	a,b
Post-fire seed germination	Yes	Yes	Yes	Yes	a,c
Month of peak flowering	February	May	August	August	d,e
Maximum root depth (m)	2.4	>2.4	>6.0	>13.2	a,b
Minimum seasonal Ψ (MPa)	−9	−7	−4	−2	a,b,f
Post-fire seedling survival (%)	63.0	20.1	8.2	1.1	a,c
Water potential at 50% embolism	−11.0	−8.3	−5.9	−1.9	b,f,g
Specific conductivity ($m^2\ MPa^{-1}\ s^{-1}$)	1.1×10^{-3}	0.7×10^{-3}	1.8×10^{-3}	5.7×10^{-3}	b,f,g

a, Thomas and Davis (1989); b, Redtfeldt and Davis (1996); c, Barton (1995); d, Bauer (1936); e, Hanes (1965); f, Kolb and Davis (1994); g, Jarbeau et al. (1995).

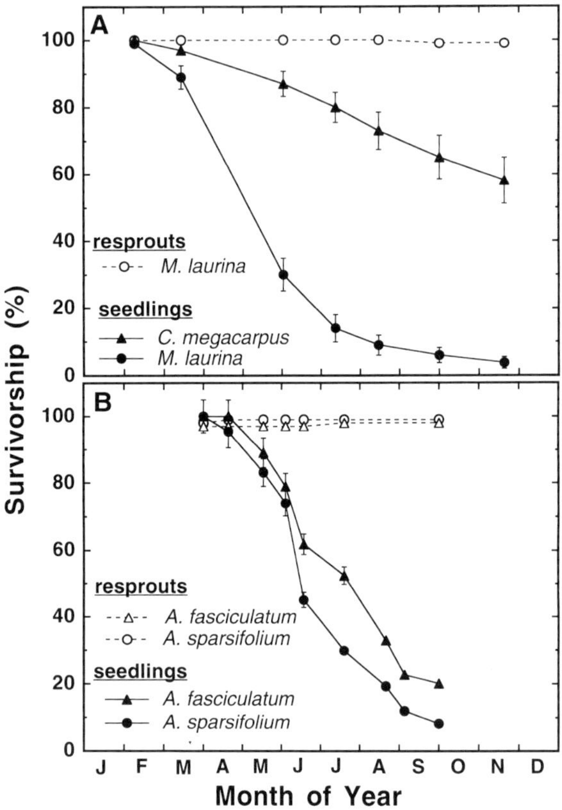

Fig. 17.3. Post-fire survivorship of co-occurring seedlings and resprouts of: *A* *Malosma laurina*/*Ceanothus megacarpus*; and *B* *Adenostoma fasciculatum*/*Adenostoma sparsifolium*, growing in the Santa Monica Mountains of southern California. *C. megacarpus* is a non-sprouter after fire, thus survivorship data for resprouts of this species are not given. Data taken from Thomas and Davis (1989) and Barton (1995)

17.1, Fig. 17.3B). Seedling survivorship is much lower than that of resprouts at the same study site (Fig. 17.3A, 17.3B).

The study site shown in Figure 17.1B burned in late October of 1993 in a 14000ha wildfire that consumed all above-ground vegetation. About 1.5 years after the fire, the site is dominated by post-fire herbaceous species, primarily *Dicentra ochroleuca*, which have an average height of about 2m (Fig. 17.1C). Chaparral seedlings are much shorter than this and thus must not only compete for available moisture during summer drought but also light and space (cf. Thomas and Davis 1989).

17.3 Water Stress and Embolism

Available data indicate that predawn water potentials can be very low for seedlings of all four species during their first summer drought after fire,

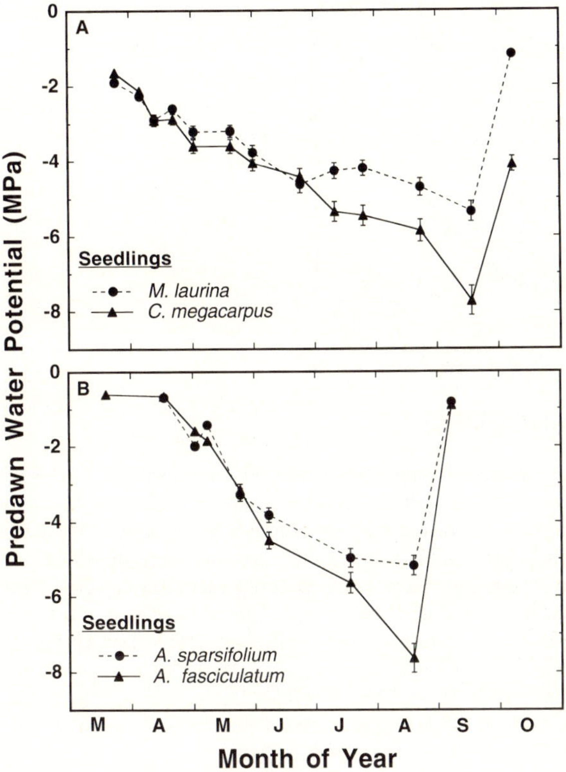

Fig. 17.4. Seasonal change in predawn water potentials for first year, post-fire seedlings of co-occurring: *A* *Malosma laurina/Ceanothus megacarpus*; and *B* *Adenostoma sparsifolium/Adenostoma fasciculatum* growing in the Santa Monica Mountains of southern California. Data taken from Frazer and Davis (1988) and Barton (1995)

presumably due to limited root development and limited access to deep moisture resources (Frazer and Davis 1988; Thomas and Davis 1989; Barton 1995; Fig. 17.4). However, water potentials of *M. laurina* seedlings do not go as low as that of co-occurring *C. megacarpus* (minimum water potential of −4 versus −8 MPa) nor do seedlings of *A. sparsifolium* have water potentials as low as that of *A. fasciculatum* (minimum water potential of −5 versus −8 MPa). We suspect that the reason for this difference was that *M. laurina* and *A. sparsifolium* seedlings are more susceptible to water stress-induced embolism of their xylem, thus a greater percent of the seedlings die at a higher water potential (cf. Fig. 17.3). This is supported by the observation that the seasonal decrease in water potential levels off in June for both *M. laurina* and *A. sparsifolium*, whereas the water potentials continue to decline for *C. megacarpus* and *A. fasciculatum* (Fig. 17.4). Furthermore, this is consistent with the observed differences in susceptibility to water stress-induced embolism measured for adult shrubs (Table 17.1, Fig. 17.5). With reference to Fig. 17.5, if seedlings of *M. laurina* and *A. sparsifolium* were to achieve water potentials of −8 MPa, the xylem of both

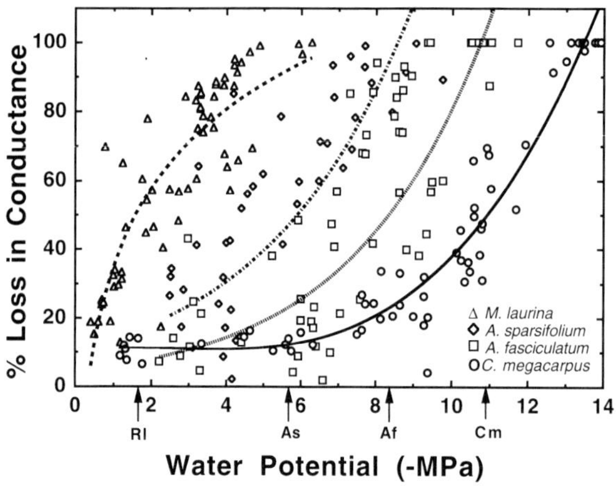

Fig. 17.5. Vulnerability of stem xylem to water stress-induced embolism for *Malosma laurina*, *Adenostoma sparsifolium*, *Adenostoma fasciculatum* and *Ceanothus megacarpus* that commonly grow together in mixed stands in the Santa Monica Mountains of southern California (cf. Fig. 17.1). **Arrows** along the x-axis indicate the water potential where 50% loss in hydraulic conductance due to air embolism occurs for each representative species. Data for each species were taken from Jarbeau et al. (1995) for *M. laurina*, Redfeldt and Davis (1995) for *A. sparsifolium* and *A. fasciculatum*, and Kolb and Davis (1994) for *C. megacarpus*. Curves were fit exponentially for *A. sparsifolium* ($y = 10.85 \times 10 - 0.113X$, $r^2 = 0.42$) and *A. fasciculatum* ($y = 4.55 \times 10 - 0.125x$, $r^2 = 0.53$) but logarithmically for *M. laurina* ($y = 75.56 \log + 34.2$, $r^2 = 0.74$) and with a third degree polynomial for *C. megacarpus* ($y = 0.072X^3 - 0.575X^2 + 1.25X + 10.49$, $r^2 = 0.924$)

species would be 100% embolized, thus blocking all water transport to leaves.

For adult shrubs growing in the Santa Monica Mountains, we found susceptibility to water stress-induced embolism (water potential causing 50% embolism) to vary between −11 MPa for *C. megacarpus* to −1.9 MPa for *M. laurina* (Fig. 17.5). *A. fasciculatum* (−8.3 MPa) and *A. sparsifolium* (−5.9 MPa) were intermediate. The increasing order of susceptibility to embolism (*C. megacarpus* < *A. fasciculatum* < *A. sparsifolium* < *M. laurina*) corresponds to the order in which post-fire seedlings were susceptible to summer drought: *C. megacarpus* (63% survival), *A. fasciculatum* (20% survival), *A. sparsifolium* (8% survival), and *M. laurina* (1% survival) (Table 17.1). Furthermore, the order corresponded to observed differences in minimum seasonal water potential and presumed rooting depth of adults (Table 17.1). These results suggest that differential tolerance to water stress and post-fire drought may be linked to life history characteristics, rooting depth, and niche segregation among chaparral species. The greater efficiency in xylem water transport for *M. laurina* and *A. sparsifolium* (greater specific conductivity, Table 17.1) also suggests a pos-

Table 17.2. Leaf characteristics representative of two taxonomic sections, *Cerastes* and *Euceanthus*, of the genus *Ceanothus*. (Barnes 1979)

Genus	*Ceanothus*			
Section	*Cerastes*		*Euceanothus*	
Species	*C. ferrisae*	*C. ramulosus*	*C. thrysiflorus*	*C. incanus*
Stomatal crypts	yes	yes	no	no
Ψ at stomatal closure (MPa)	−5.5	−4.5	−3.0	−2.0
Leaf-specific mass ($g\,m^{-2}$)	280	317	163	161
Leaf thickness (μm)	558	605	222	352
Cuticle thickness (μm)	4.5	8.3	1.6	3.4

Ψ, Xylem pressure potential.

sible tradeoff between xylem safety against embolism and xylem efficiency in water transport (cf. Jarbeau et al. 1995).

In our second example we examine species pairs from two subgenera of sprouting (*Euceanothus*) and non-sprouting (*Cerastes*) *Ceanothus*. Nobs (1963) and more recently Barnes (1979) have shown that the non-sprouting subgenus *Cerastes* has representative species with morphological features suggestive of greater drought tolerance than the subgenus *Euceanothus* (Table 17.2). For example, species in *Cerastes* have leaves with stomata sunken into epidermal crypts, their stomata close at significantly lower water potentials, leaf specific weight and thickness are higher, and cuticle thickness is greater (Table 17.2). Here, we focus on three pairs of species from each subgenus, *Cerastes* and *Euceanothus* respectively: *C. megacarpus*/*C. spinosus*; *C. cuneatus*/*C. oliganthus*; and *C. crassifolius*/*C. leucodermis*. Each of these pairs commonly grow together in the transverse mountain ranges of southern California (Santa Monica, Santa Ynez, Topatopa, and San Gabriel Mountains). Phyllis Nicholson (1993) has recently shown that these pairs of *Ceanothus* are sequentially replaced along an elevational gradient, that is, *C. megacarpus*/*C. spinosus* occur at lower elevation, coastal sites; *C. cuneatus*/*C. oliganthus* at sites intermediate in elevation; and *C. crassifolius*/*C. leucodermis* at high elevations. In general, the length of summer drought decreases and the amount of rainfall increases with increasing elevation in these mountain ranges (Nicholson 1993).

When these species pairs of *Ceanothus* were compared within the same microsites by Davis et al. (1999), they were found to have a consistent pattern in vulnerability to water stress-induced xylem dysfunction (Fig. 17.6). In all cases, the non-sprouters in the subgenus *Cerastes* were found to be much more resistant to xylem embolism caused by water stress than

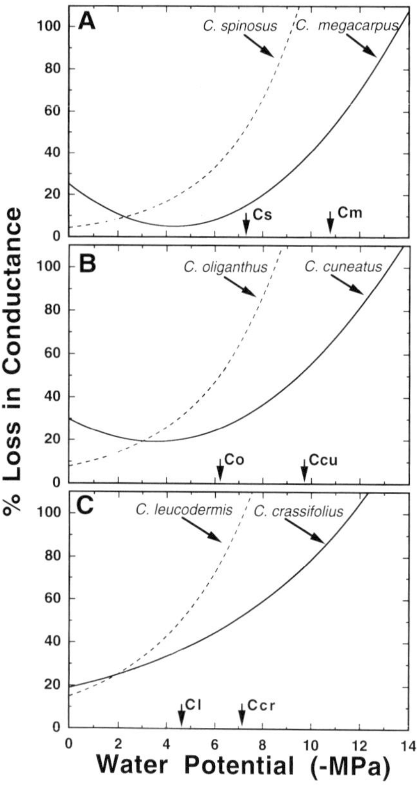

Fig. 17.6. Vulnerability of stem xylem to water stress-induced embolism for three pairs of sprouting (subgenus *Euceanothus*) and non-sprouting (subgenus *Cerastes*) *Ceanothus*. Species pairs commonly grow along an elevational gradient starting with ***A*** *Ceanothus spinosus*/*C. megacarpus* at low elevations, ***B*** *C. oliganthus*/*C. cuneatus* at intermediate elevations and **C** *C. leucodermis*/*C. crassifolius* at high elevations in the transverse mountain ranges of southern California (Santa Monica, Santa Ynez, Topatopa, and San Gabriel Mountains). **Arrows** along the x-axis indicate the water potential where 50% loss in hydraulic conductance due to air embolism occurs for each representative species. Curves were fit exponentially for *C. spinosus* ($y = 4.27 \times 10\ 0.148x$, $r^2 = 0.81$), *C. megacarpus* ($y = 5.46 \times 10\ 0.085x$, $r^2 = 0.75$), *C. oliganthus* ($y = 7.79 \times 10\ 0.131x$, $r^2 = 0.72$), *C. cuneatus* ($y = 13.32 \times 100.055x$, $r^2 = 0.45$), *C. leucodermis* ($y = 15.0 \times 100.114x$, $r^2 = 0.362$), and *C. crassifolius* ($y = 18.97 \times 100.062x$, $r^2 = 0.36$). Unpublished data from Davis et al

species in the subgenus *Euceanothus* (cf. Nobs 1963; Barnes 1979). The level of water stress causing 50% loss in hydraulic conductance varied between −7.1 to −10.8 MPa for *Cerastes* but −4.6 to −7.3 MPa for *Euceanothus* (Fig. 17.6). It appears that as the pairs move from low elevation (low precipitation) to higher elevation (higher precipitation) that susceptibility to embolism also increases as indicated by the shift in the water potential where there is a predicted 50% loss in hydraulic conductance (Fig. 17.6).

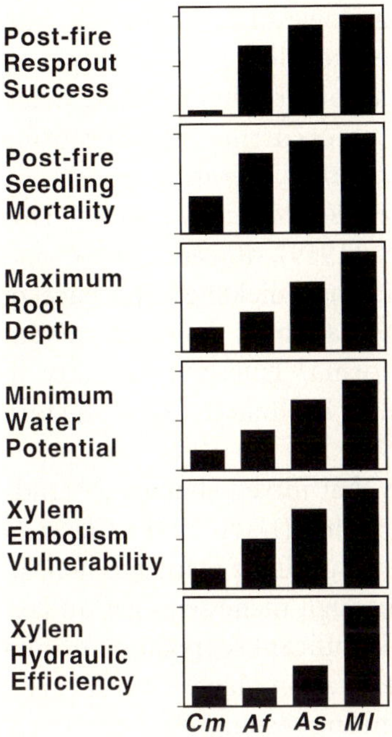

Fig. 17.7. Summary graph showing the relationships between water stress factors (xylem embolism, minimum seasonal water potential, maximum rooting depth, hydraulic transport efficiency of stem xylem) and wildfire factors (post-fire resprout success, post-fire seedling mortality) that contribute to demographic patterns in mixed chaparral communities of the Santa Monica Mountains of southern California (cf. Fig. 17.1). The species compared are **Cm**, *Ceanothus megacarpus*, **Af**, *Adenostoma fasciculatum*, **As**, *Adenostoma sparsifolium* and **Ml**, *Malosma laurina*

17.4 Conclusions

From the results presented in this chapter, it appears that vulnerability to water stress-induced embolism of stem xylem may be a useful index of drought tolerance among species of chaparral shrubs of southern California. Within a mixed stand of chaparral composed of four dominant species, with known differences in minimum seasonal water potential, rooting depth, and post-fire seedling survivorship, vulnerability to embolism was a good predictor of each pattern. Shrub species with lower seasonal water potential were the least susceptible to embolism. Species with deep roots were the most vulnerable to embolism but the most efficient at water transport (high specific conductivity). Species with high seedling survivorship during summer drought were more resistant to water stress-induced embolism of their stem xylem (Fig. 17.7).

It is probably not a coincidence that the shallowest rooted species, with the lowest seasonal water potential, the highest seedling survivorship, and the highest resistance to embolism were non-sprouters after wildfire. When a comparison is made between pairs of co-occurring non-sprouting (*Cerastes*) and sprouting (*Euceanothus*) *Ceanothus*, *Cerastes* species are consistently more resistant to embolism than *Euceanothus* species. This is consistent with the arguments of Wells (1969), Raven (1977), and Axelrod (1989) that the non-sprouting trait has quickened the pace of evolution, allowing such species to more closely adapt to the recent advent of a summer-dry climate to southern California. Thus vulnerability to embolism among chaparral species may also be linked to life history characteristics.

These patterns further support the notion that mixed chaparral stands are comprised of species occupying different niches (Davis 1991). Classical competition theory suggests that coexistence of species is made possible by reduced competition for limited resources through niche segregation (resource partitioning). It may be that the most significant segregation occurs at the post-fire reestablishment phase or regeneration niche (Grubb 1977). Most agree that one of the most important limiting resources for chaparral shrubs is water, especially during summer drought and most significantly for post-fire seedlings that have not yet established a root system to tap deep moisture reserves. It is obvious that the greatest below ground niche segregation among chaparral species occurs during the first summer drought after wildfire when resprouts of sprouting species retain their pre-established root system (>13.4 m deep) but grow adjacent to seedlings of non-sprouting species with a maximum root depth of 0.6 m (Frazer and Davis 1988, Thomas and Davis 1989). Kummerow et al. (1985) have shown that there is no evidence of competition for soil moisture between chaparral resprouts and seedlings. Furthermore, it appears that such dramatic differences during the regeneration phase after fire are retained to a lesser degree in mature stands in such a way that flowering, minimum seasonal water potential, and water potentials where stomata close are also different (Tables 17.1 and 17.2). Again, vulnerability to embolism appears to be a good predictor of such differences (Fig. 17.7).

References

Axelrod DI (1989) Age and origin of chaparral. In: Keeley SC (ed) The California chaparral: paradigms reexamined. Natural History Museum of Los Angeles County, Los Angeles, pp 7–19

Barnes FT (1979) Water relations of four species of *Ceanothus*. Masters Thesis, California State University, San Jose, California

Barton KP (1995) Reestablishment of *Adenostoma fasciculatum* and *A. sparsifolium* chaparral after the Malibu wildfire of 1993. Honors Thesis, Pepperdine University, Malibu, California

Bauer HL (1936) Moisture relations in chaparral of the Santa Monica Mountains, California. Ecol Monogr 6:409–454

Carlquist S (1985) Vasicentric tracheids as a drought survival mechanism in the woody flora of Southern California and similar regions; review of vasicentric tracheids. IAWA Bull 11:37–68

Carlquist S (1989) Adaptive wood anatomy of chaparral shrubs. In: Keeley SC (ed) The California chaparral: paradigms reexamined. Natural History Museum of Los Angeles County, Los Angeles, pp 25–36

Carlquist S, Hoekman DA (1985) Ecological wood anatomy of the woody southern California flora. IAWA Bull 6:319–347

Davis SD (1989) Patterns in mixed chaparral stands: differential water status and seedling survival during summer drought. In: Keeley SC (ed) The California chaparral: paradigms reexamined. Natural History Museum of Los Angeles County, Los Angeles, pp 97–105

Davis SD (1991) Lack of niche differentiation in adult shrubs implicates the importance of the regeneration niche. Trends Ecol Evol 9:272–274

Davis SD, Ewers FW, Wood J, Reeves JJ (1999) Differential susceptibility to xylem cavitation among three pairs of *Ceanothus* species in the Transverse Mountain Ranges of southern California. Ecoscience. In Press

Davis SD, Mooney HA (1985) Comparative water relations of adjacent California shrub and grassland communities. Oecologia 66:530–535

Davis SD, Mooney HA (1986a) Water use patterns of four co-occurring chaparral shrubs. Oecologia 70:172–177

Davis SD, Mooney HA (1986b) Tissue water relations of four co-occurring chaparral shrubs. Oecologia 70:527–535

Frazer JM, Davis SD (1988) Differential survival of chaparral seedlings during the first summer drought after wildfire. Oecologia 76:215–221

Grubb PJ (1977) The maintenance of species-richness in plant communities: The importance of the regeneration niche. Biol Rev 52:107–145

Hanes TL (1965) Ecological studies on two closely related chaparral shrubs in southern California. Ecol Monogr 35:213–235

Hargrave KR, Kolb KJ, Ewers FW, Davis SD (1994) Conduit diameter and drought induced embolism in *Salvia mellifera* Greene (Labiatae). New Phytol 126:695–705

Jacks PM (1984) The drought tolerance of *Adenostoma fasciculatum* and *Ceanothus crassifolius* seedlings and vegetation change in the San Gabriel chaparral, San Diego State University, San Diego

Jarbeau JA, Ewers FW, Davis SD (1995) The mechanism of water stress-induced embolism in two species of chaparral shrubs. Plant Cell Environ 18:189–196

Keeley JE (1986) Resilience of Mediterranean shrub communities to fires. In: Dell B, Hopkins AJM, Lamont BB (eds) Resilience in mediterranean-type ecosystems, W. Junk Publishers, The Hague, pp 95–112

Keeley JE (1991) Seed germination and life history syndromes in the California chaparral. Bot Rev 57:81–116

Keeley JE (1992) Recruitment of seedlings and vegetative sprouts in unburned chaparral. Ecology 73:1194–1208

Keeley JE, Keeley SC (1988) Chaparral. North American terrestrial vegetation, Cambridge University Press, Cambridge, pp 165–207

Keeley JE, Zedler PH (1978) Reproduction of chaparral shrubs after fire: a comparison of sprouting and seeding strategies. Amer Midl Nat 99:142–161

Kolb KJ, Davis SD (1994) Drought-induced xylem embolism in co-occurring species of coastal sage and chaparral of California. Ecology 75:648–659

Kummerow J, Ellis BA, Mills JN (1985) Post-fire seedling establishment of *Adenostoma fasciculatum* and *Ceanothus greggii* in southern California chaparral. Madroño 32:148–157

Mills JN, Kummerow J (1989) Herbivores, seed predators, and chaparral succession. In: Keeley SC (ed) The California chaparral: paradigms reexamined. Natural History Museum of Los Angeles County, Los Angeles, pp 97–105

Moreno JM, Oechel WC (1991) Fire intensity effects on germination of shrubs and herbs in southern California chaparral. Ecology 72:1993–2004

Musick HB (1972) Post-fire seedling ecology of two *Ceanothus* species in relation to slope exposure. Masters Thesis, University of California, Santa Barbara

Nicholson P (1993) Ecological and historical biogeography of *Ceanothus* (Rhamnaceae) in the transverse ranges of southern California, PhD Thesis, University of California, Los Angeles

Nobs MA (1963) Experimental studies on species relationships in *Ceanothus*. Carnegie Inst Wash Publ 623

Raven PH (1977) The California flora. In: Barbour, MB, Major J (eds) Terrestrial vegetation of California. Wiley, New York, pp 109–138

Redtfeldt RA, Davis SD (1996) Physiological evidence of niche segregation between two co-occurring species of *Adenostoma* in California chaparral. Ecoscience 3:290–296

Sampson AW (1944) Plant succession on burned chaparral lands in Northern California. Univ Calif Agric Sta Bull No 685, pp 1–144

Saruwatari MW, Davis SD (1989) Tissue and water relations of three chaparral shrub species after wildfire. Oecologia 80:303–308

Thomas, CM, Davis SD (1989) Recovery patterns of three chaparral shrub species after wildfire. Oecologia 80:309–320

Tyree MT, Sperry JS (1989) Vulnerability of xylem to cavitation and embolism. Ann Rev Plant Physiol Mol Bio 40:19–38

Tyree MT, Ewers FW (1991) Hydraulic architecture of trees and other woody plants. New Phytol 119:345–360

Tyree MT, Davis SD, Cochard H (1994) Biophysical perspectives on xylem evolution: is there a tradeoff of hydraulic efficiency for vulnerability to dysfunction? IAWA Bull 15:335–360

Wells PV (1969) The relation between mode of reproduction and extent of speciation in woody genera of California chaparral. Evolution 23:264–267

Williams JE, Davis SD, Portwood KA (1997) Xylem embolism in seedlings and resprouts of *Adenostoma fasciculatum* after fire. Austr J Bot 45:291–300

Part V

Landscape Disturbance and Animal Diversity

18 Insect Population Changes and Conservation in the Disturbed Landscapes of Mediterranean-Type Ecosystems

M.J. Samways

18.1 Introduction

The Insecta is the most speciose taxonomic class of animals. There are great differences in estimates of the total number of species, but a figure of just under 10 million is probably fairly close (Hammond 1992; Stork 1993). Many of these species are narrow endemics, especially in the tropics (Erwin 1988) and the Southern Hemisphere (Darlington 1965; Platnick 1991; Samways 1992). The southwestern Cape Mediterranean-type ecosystem (MTE) is among the most well-researched of these southern areas and is an epicentre for many endemic insects (Endrödy-Younga 1988; Gess 1992). It is also an area, especially around Cape Town, where many insects, especially butterflies (Rebelo 1992) are highly threatened.

Loss of global insect species, principally from landscape fragmentation and habitat loss, is hard to estimate, but Mawdsley and Stork (1995) suggest it might be between 7 and 30 species every week.

Disturbed or fragmented landscapes induce certain insect species to increase in abundance while others decrease or locally disappear (Samways et al. 1996). In addition, some insects are new invaders, either through natural range expansion (e.g. *Danaus plexippus* reaching the Canaries in the 1860s and Spain in recent years; Vane-Wright 1986), or carried by humans, both accidentally (exotic invaders) or deliberately (biological control agents).

Against this background, six points will be discussed here on how insects respond to landscape disturbance and fragmentation in MTEs and how these responses relate to conservation:

1. Insects as indicators of anthropogenic change,
2. Threatened insects,
3. Threats to insect mutualisms,
4. Threatened plants as a threat to insects,

Ecological Studies, Vol. 136
Rundel et al. (eds.) Landscape Degradation and Biodiversity in Mediterranean-Type Ecosystems

5. Insects as threatening or invasive agents,
6. Beneficial insects,

These points are then used to make research, policy and management recommendations.

As the genetics of insects are so variable, as illustrated by different morphs and subspecies, the concept of "evolutionary significant units" (ESUs) is recognised here (Moritz 1994; Vogler and DeSalle 1994). Before discussing these six points, there are three provisos.

18.2 Three Provisos

18.2.1 Landscape Disturbance: Fragmentation or Variegation

The landscape is frequently said to be "fragmented", with clear visible boundaries to the human eye (e.g. Forman and Godron 1986). McIntyre and Barrett (1992) have questioned this model, suggesting that the form and extent of the modified landscape vary. They suggest that landscape elements create a fuzzy-edged mosaic that acts as a differential filter with

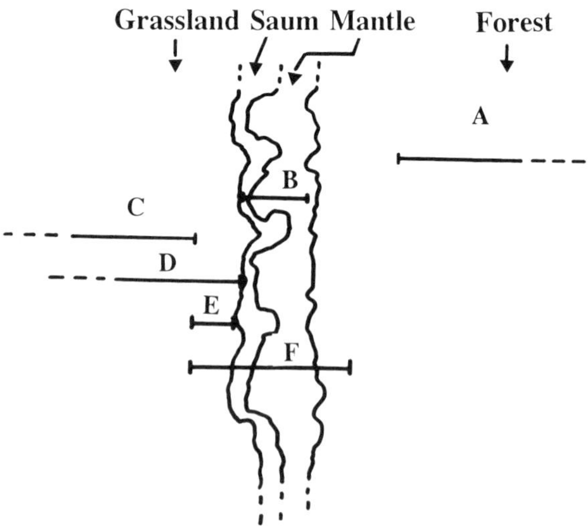

Fig. 18.1. The more realistic variegation model (i.e. fuzzy-edges) rather than the fragmentation model of the landscape, as determined by the local distribution of animal species. Illustrated here is the edge of a forest adjacent to a grassland, with a herbaceous saum and a bushy mantle. Species *A* interior forest species; *B* stenotopic edge species; *C* interior grassland species; *D* eurytopic grassland species; *E* stenotopic grassland species that prefers the grass edges; *F* eurytopic edge species. Based on findings in Ingham and Samways (1996) and Samways and Moore (1991)

variegation of species' local distributions. Indeed, each butterfly species sees and responds to landscape elements differently (Wood and Samways 1991), while other insects (and arachnids) inhabit the landscape as if it was fuzzy-edged (Ingham and Samways 1996; see Fig. 18.1).

18.2.2 Are Native Insects in MTEs Under More Threat than Those in Other Ecosystems?

Hannah et al. (1995) emphasize that MTEs are under enormous anthropogenic pressure. As regards insects, there are two important considerations. The first is that for highly localized endemic species, threats are an all-or-none affair. If an endemic occurs at the proposed site for an urban development it may be doomed as a species. On the other hand, being "pre-adapted" to living in a small area, it may well survive in a remnant patch surrounded by anthropogenic disturbance. Also, MTEs are relatively small in size, and mostly highly desirable for agriculture and human habitation. This means that although the total current loss of populations and species may not equal those of the tropics (Mawdsley and Stork 1995), the number of current, known extinctions is apparently high (e.g. Hafernik 1992).

18.2.3 Do the Differences in Geographical Position, Pre-Historical Climate and Historical Land Use Patterns Affect the View of Insect Conservation in the Different MTEs?

The world's biodiversity is pear-shaped, with a concentration in the tropics and in the south (Darlington 1965; Platnick 1991; Samways 1992). Insects follow this trend with particularly high levels of narrow endemism in MTEs of Chile (Platnick 1992), South Africa (Samways 1994) and Australia (CSIRO 1991). However, there are also high levels of endemism, at least at the species and subspecies levels, in the MTEs of North America (Hafernik 1992). In the Mediterranean, the proportion of narrow endemics tends to be relatively small, but high for some taxa on islands. Even on the mainland, where topography permits, some taxa have radiated into several ESUs. *Decticus verrucivorus* has a distinct French Mediterranean subspecies *D. v. monspeliensis* and there is another from Sardinia, *D. v. insularis.* There is also the South Italian endemic *D. soudoni* and the Apennine endemic, *D. aprutianus* (Harz 1969). Yet, on the other hand, surrounding the whole of the Mediterranean basin is the widespread, common and disturbance-tolerant *D. albifrons* (Fabricius) (Samways and Harz 1982).

Although making comparisons of levels of endemism are difficult (because of differing variables such as area of endemism, topography, taxa being considered, level of the endemic taxon, lack of taxonomic knowledge etc.) there appears to be a general trend from low to high proportions of endemic ESUs from Mediterranean and North America to Chile, South Africa and Australia, with the last three not necessarily in that order. Comparative endemism among the insects of the various MTEs requires much more research.

18.3 Six Points on Insects and Conservation in MTEs

18.3.1 Insects as Indicators of Anthropogenic Change

Indicator taxa are those that reflect the quality and changes in particular environmental conditions, as well as aspects of community composition (Soulé and Kohm 1989; Noss 1990; Kremen et al. 1993; Samways 1994). Insects are particularly useful for fragmentation studies as they are: (1) numerous, i.e. good working numbers; (2) consist of a range of species and developmental morphs sensitive to different aspects of environmental change; (3) often have small home ranges and definite microhabitat requirements and so can indicate changes in point conditions; and, (4) often strongly associated with other organisms (e.g. as parasites, parasitoids, herbivores, pollinators etc.), and therefore can reflect changes in ecosystem processes. The important point is that the right indicator must be chosen for the specific question being asked or for monitoring the appropriateness of management practices.

Various taxa have been used for monitoring fragmentation in MTEs. Ants (Formicidae) are particularly useful and have been used for monitoring Western Australian roadside corridors (Keals and Majer 1991) and for making comparisons of assemblages in different types of Australian forest (Majer et al. 1994). They have also been used as indicators of the changes in natural South African fynbos when it is planted to pines (Donnelly and Giliomee 1985; see Fig. 18.2). Odonata have been used in MTEs for indicating the impact of introduced rainbow trout and that of exotic invasive plants (Samways 1995).

Many other taxa are also plausible candidates, depending on the questions being asked. In southern France, for example, the relative Orthoptera species compositions of different landscape fragments indicate the extent to which these modified fragments are still able to support species (Samways 1989). In other words, the vegetation composition, management and size of fragment has a clear impact on the insect fauna. Also, the Australian Collembola, as well as other taxa, have been used to illustrate

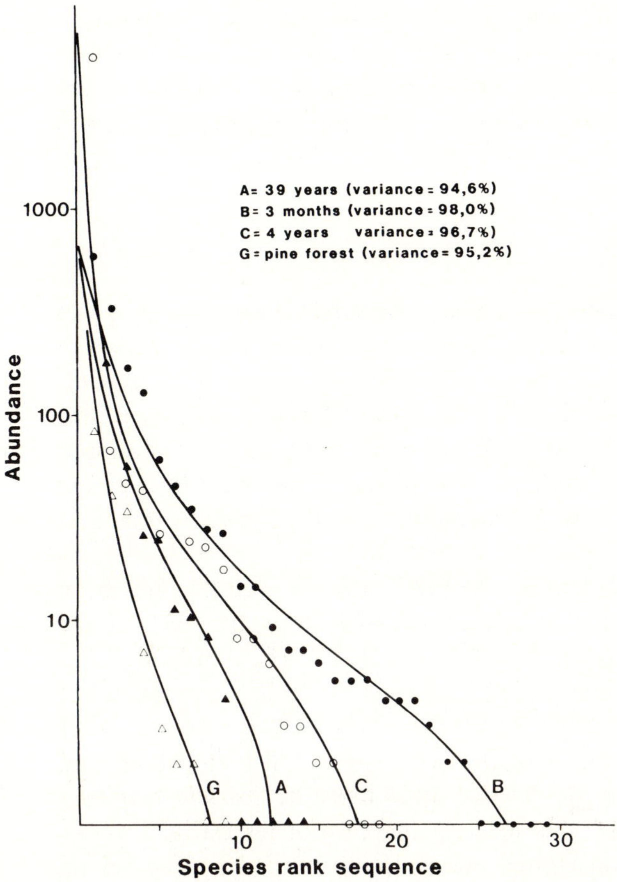

Fig. 18.2. Rank-abundance plots, with best fit log-normal curves for four assemblages of ants, three in fynbos unburnt for 3 months (*B*), 4 years (*C*) and 39 years (*A*), and one in pine plantation planted in fynbos (*G*). Note the poor species richness in pine plantation (*G*) and long-unburnt fynbos (*A*) (Donnelly and Giliomee 1985)

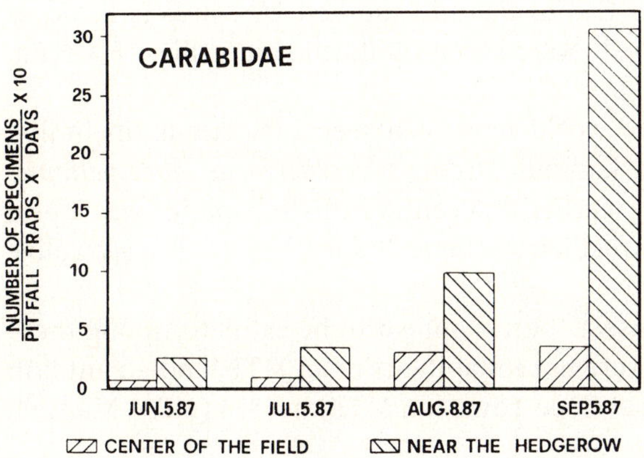

Fig. 18.3. Numbers of Carabidae specimens caught in the centre of an experimental soybean field and at the edge of the field adjacent to a hedgerow in Friuli, Italy (Nazzi et al. 1989)

the effectiveness of re-establishing partially native ecosystems (Greenslade 1992). Carabidae, as well as other taxa caught in pitfall traps, have also been used to illustrate the importance of maintaining relict hedgerows surrounding crop fields in Italy towards promoting a diverse biota (Nazzi et al. 1989; see Fig. 18.3).

18.3.2 Threatened Insects

Hafernik (1992) illustrates how the expanding urban development of San Francisco Bay and Los Angeles has been responsible for the extinction of butterfly species. Early extinctions were the satyr butterfly *Cercyonis sthenele sthenele* and the xerces blue butterfly *Glaucopsyche xerces*, followed in the 1970s by local (not total) extinction of the callippe silverspot butterfly *Speyeria callippe callippe*, the pheres blue butterfly *Icaricia icariodes pheres*, and the bay checkerspot butterfly *Euphydryas editha bayensis* (Hafernik 1992).

It is sand-dune development in the San Francisco area that has been particularly devastating, having caused the extinction, by 1937, of the Antioch Dune katydid *Neduba extincta* (Rentz 1993). The remaining 21 ha of highly altered dune habitat support the last remaining population of Lange's metal-mark butterfly (*Apodemia mormo langei*). But other biotopes are also threatened, with development of hill biotopes around San Francisco threatening the mission blue butterfly *Icaricia icariodes missionensis* (Hafernik 1992). Indeed, in the case of terrestrial invertebrates from the continental United States, California is the area having had the most extinctions (20) as well as those listed as "endangered" (10) and "candidate" (122) species. In fact, the first recorded insect extinctions in the US were from California, and the first U.S. insects listed as "endangered" or "threatened" were seven California butterflies (Hafernik 1992).

Other MTEs around the world have also seen extinctions, including the magnificent flightless katydid *Decticus verrucivorus monspeliensis* (Samways 1989). Ironically, earlier this century this subspecies was a pest of cereal fields around Montpellier, southern France, but was last seen alive in 1972.

In South Africa, three insect species appear to be extinct. Among these are two butterflies, which were not inhabitants of an MTE, and an ant lion *Halterina purcelli* last seen at Cape Town (an MTE) in 1884 (M.W. Mansell, pers. comm.).

In the MTEs of Australia, no insect species appears to have become extinct (New 1984; Hill and Michaelis 1988; Yen and Butcher 1994). Also there does not seem to be any information supporting or refuting insect extinctions in Chilean MTEs.

The point is that a species extinction occurs as the last population disappears. In the meantime, fragmentation with concurrent habitat destruction is the most frequent threat to insects (New et al. 1995). And so, many species are surviving in isolated geographical pockets. In California, species such as the San Francisco forktail damsel fly, *Ischnura gemina* (Hafernik 1989), the curve-foot hygrotus diving beetle, *Hygrotus curvipes* (Hafernik 1989), and the bay checkerspot butterfly, *Euphydryas editha bayensis* (Murphy and Weiss 1988) are all characterized by a metapopulation structure, the gene flow of which is being threatened by the fragmented and modified landscape.

Elsewhere, the picture appears to be the same. The local extinction of *Platycleis fedtshenkoi azami* (as well as *Decticus verrucivorus monspeliensis*) around Montpellier in southern France appears to have been caused by the compounding effect of fragmentation (particularly with the development of Montpellier airport) and other disturbance factors such as dumping of dredged silt (in the case of *P.f. azami*) and the planting of vines up to the roadside edge (in the case of *D. v. monspeliensis*).

Threats may also be to overwintering sites. In Europe, the butterfly *Colotis evagore*, which overwinters in a small area of the Mediterranean coast in southern Spain, is locally threatened (Jordano et al. 1991). Similarly, the Mexican and California roosting sites of *Danaus plexippus* are under pressure because of human interference at these localized sites (Brower and Malcolm 1989).

An important Mediterranean insect under threat from habitat and food-plant destruction, as well as specialist collecting, is the Corsican swallowtail butterfly, *Papilio hospiton*, of Corsica and Sardinia (New and Collins 1991). This species may well be an umbrella signalling similar threats to other less conspicuous, sympatric insects. A similar situation occurs among butterflies in Mexico, where not only habitat destruction, but also collecting is proving a threat, with four species selling at over $1500 per specimen (Llorente-Bousquets and Luis-Martínez 1993).

In South Africa, threats to MTE damselflies include removal of oviposition vegetation, excessive water extraction for agriculture, as well as predation from the exotic rainbow trout *Oncorhyncus mykiss* and shading out of habitats by exotic *Eucalyptus* and *Pinus* (Samways 1995; see Fig. 18.4).

Fig. 18.4. Destruction of endemic damselfly habitat on the Breede River, near Worcester, southwestern Cape. Excessive water extraction and invasion by exotic plants, especially *Acacia* spp., has made this locality now unsuitable for *Chlorolestes* spp. and *Ecchlorolestes peringueyi*

18.3.3 Threats to Insect Mutualisms

Disturbances to biotic interactions, such as that between aculeate wasps and plants, other insects and spiders, can have a cascade effect which inevitably can affect the whole ecosystem (Gess and Gess 1991). Mutualisms, particularly between plant and insect pollinators, are particularly threatened, especially where levels of endemism are high as in the southwestern Cape (Gess 1992) and parts of Mexico (Ayala et al. 1993).

The risk of plant extinctions can be assessed by considering three criteria: (1) probability of dispersal or pollinator failure; (2) reproductive dependence on the mutualism; and, (3) demographic dependence on seeds (Bond 1994). It appears that plants often compensate for high risk in one of the three categories by low risk in another. For example, self-incompatible plants with rare, specialist pollinators often propagate vegetatively. Of particular concern, is that at least for the South African MTEs, some plants lack these compensatory traits making extinction from failed mutualisms including pollination, as high as nearly 50% (Bond 1994). Not all the fynbos plants require specific pollinators with many pollinated by wind and a variety of insects, especially beetles (Hattingh and Giliomee 1989). At the other end of the spectrum, the butterfly *Aeropetes tulbaghia* is an important pollinator of a whole host of high elevation geophytes which have red flowers in the autumn (Johnson 1992). The point is that conservation of mutualistic interactions, whether pollination or seed dispersal (Slingsby and Bond 1985; Leigh 1994), is critical for the survival of the ecosystem, and this is critical for MTEs as any other type of ecosystem.

18.3.4 Threatened Plants as Threats to Insects

In the fynbos, it does not seem to be the plant taxonomic variety per se that has generated most of the insect diversity, but rather it is that the insects have speciated through allopatric isolation as did the plants themselves. Also, at least in the butterflies, there has not been a co-evolutionary radiation (Cottrell 1985). The Cicadellidae, which are much more intimately associated with their plant hosts, show much more specificity (Davies 1988 a,b). Endophages, which, by definition, are even more intimately associated with their hosts, show some highly endemic and tight associations (Wright 1993). This has important conservation implications because fire can make the plants non-apparent to the insects and may in some cases be an anti-herbivory strategy. This means that simply preserving a patch of vegetation does not guarantee the survival of the insects as the plants effectively vanish for a while after fire (Wright 1993).

There is of course always debate on the closeness and the exact nature of a herbivore–plant interaction (Gaston 1992; Moran et al. 1994), but, turning the situation around, some seed dispersers at least, keep some plants from becoming extinct (Leigh et al. 1993).

18.3.5 Insects as Threatening or Invasive Agents

It is well known that in South Africa (Richardson et al. 1992) and in Australia (New 1994) some invasive insect species are common to both continents e.g. *Vespula germanica* and *Iridomyrmex humilis* (now *Linepithema humile*) (New 1993; Bond and Slingsby 1984; Tribe and Richardson 1994). Of particular concern, is that such highly competitive and/or predacious insects will have a major impact on endemic species. Bond (1994) suggests that if *I. humilis* continues to spread, nearly half the Proteaceae may be lost as this invasive continues to displace native seed-burying ants. Similarly, *V. germanica* although apparently localized to 470 km^2 of the Cape Peninsula (Tribe and Richardson 1994) nevertheless is living in precisely one of the most faunistically interesting areas of southern Africa (especially the forests of Table Mountain) where at least 112 endemic invertebrate species occur (Picker and Samways 1996). Not that the wasp is necessarily a direct predator of these species but may eventually have a keystone effect on the local ecosystems, and may, if it does spread around the eastern coastal belt as Tribe and Richardson (1994) predict, could have major impacts elsewhere.

In Mexico, the Africanized bee (*Apis mellifera scutellata*) is not only threatening the local honeybee industry, but also native bees (Ayala et al. 1993).

Davis (1994) has looked at landscape fragmentation from the perspective of how insects might actually invade and increase in abundance. His conclusions are that the least biotope sensitive and least spatially discriminating taxa of dung beetles are the ones that have recently increased in range and moved into the Cape MTEs in response to landscape fragmentation as indigenous shrubland is replaced with pasture. These beetles are not threatening but simply able to take advantage of the disturbed landscape.

Not all threatening insects are exotic. Hegazy and Eesa (1991) record that the native seed predator, the bruchid beetle *Callosobruchus maculatus*, threatens the existence of *Ebenus armitagei*, a perennial woody species endemic to the western Mediterranean coast of Egypt and Libya.

18.3.6 Beneficial Insects

There have been several successful insect and weed biocontrol programmes in MTEs, especially in South Africa. The pertinent question here is to ask whether these biological control programmes have in any way misfired and have caused damage to non-target native faunas as they have in some other ecosystems (Samways 1988; Howarth 1991). To date there seems not to have been any noticed or recorded damage to non-target organisms in MTEs, despite some intensive programmes especially against weeds. However, this is not to say that unoticed interactions are not taking place.

The introduction of insect biocontrol agents against weeds into the MTE of South Africa has been fairly successful against *Hypericum perforatum* (Gordon and Kluge 1991), and partially successful against *Acacia longifolia* (Dennill and Donnelly 1991) and *Sesbania punicea* (Hoffman and Moran 1991).

There has been a positive effect of the introduction of the Australian galll-forming rust fungus *Uromycladium tepperianum* against *Acacia saligna* in the Cape of South Africa (Morris 1991; Fig. 18.5). *A. saligna* is highly invasive with dense stands outcompeting native fynbos and fragmenting the landscape. By 1995, at least at Cape Agulhas, the fungus was showing an effective impact, killing off seedlings as well as larger trees, with the beginnings of re-invasion by native vegetation and insects (Fig. 18.5).

Fig. 18.5. Dying *Acacia saligna* trees at Cape Agulhas, South Africa, after introduction of the successful fungal biocontrol agent *Uromycladium tepperianum*. The choked, local fynbos is being locally freed to recover

18.4 Synergism Between Landscape Disturbance and Other Factors Causing Changes in Insect Populations

It is not only narrow endemic insects that become extinct. This is a lesson well learnt from the Rocky Mountain grasshopper, *Melanoplus spretus* (Lockwood and De Brey 1990). This species was once widespread and so abundant that it was a serious agricultural pest in the western United States and Canada prior to the year 1880. Yet it was extinct by 1902 due to a combination of natural factors including predation and disease. After the population crash, the insect's range was reduced and fragmented, exacerbated by loss of rangeland to agriculture. In the late 1880s, oviposition and early nymphal development occurred almost exclusively in association with riparian vegetation. Then, intensive modification of these remnant patches by tillage, irrigation, loss of the beaver and introduction of cattle, plants and birds, caused the final extinction of the species.

The point is that it is not just habitat loss per se that is the problem. In the case of the bay checkerspot butterfly, landscape fragmentation that affects metapopulation dynamics plays an important role. Yet for other narrow endemics, the situation is rather more hit and miss. In other words, should a disturbance coincide with the small range of a species (or subspecies) then that species may become extinct. Should, for example, the

Fig. 18.6. An oasis in a matrix of suburbia. These monastery grounds of Maguelonne, southern France and the adjacent naturist beach are refugia for bush cricket species such as *Platycleis sabulosa* and *P. affinis*, which are otherwise being displaced by the expanding forms of Palavas and Carnon, seen in the background

Wynberg Caves of Table Mountain be disturbed (say, by encouraging the general public to enter the caves), then about twelve invertebrate species could become extinct.

From the evidence so far, loss of species and populations of insects from MTEs is coming about by the compound effects of direct fragmentation (through building and road construction, forestry and agriculture), and other related disturbance factors such as cattle grazing, mining, invasive plants and animals, and even the dumping of dredged sludge, etc.

But not all aspects of human disturbance are harmful. In southern France, for example, the grounds of the ancient cathedral of Maguelonne and the nearby naturist beach are islands of comparatively low ecosystem disturbance that provide refuge for rare species such as the katydid *Platycleis sabulosa* (Samways 1988; Fig. 18.6).

18.5 Discussion

Unquestionably all MTEs are among the most disturbed ecosystems in the world, as elegantly quantified by Hannah et al. (1995). As insects are sensitive responders to landscape disturbance and fragmentation (see Samways 1994), and yet they are also major components of ecosystems processes (see Price 1984), it is not surprising that their abundance patterns or their presence/absence status have changed dramatically in MTEs. This does, with the scant evidence available, appear to be the case with, for example, a decrease in native pollinating insects in Western Australia (New 1994).

Some insect species have become major exotic invaders threatening entire ecosystems (e.g. *Iridomyrmex humilis* (*Linepithema humile*) in the fynbos), yet, on the other hand, a few have been deliberately introduced as biocontrol agents and have benefited conservation (e.g. on *Hepericum perforatum* in South Africa). Reduction of exotic invasive vegetation is also being achieved with fungal pathogens. The outcome of these activities is the restoration of natural ecosystems. Such restoration may be through natural invasion and succession or by planting to mimic the natural plant community. In California, the restoration of riparian sites showed that arthropod communities can be encouraged to re-establish themselves, with detritivores appearing quickly, but predators and parasitoids still not fully established after three years (Williams 1993).

Restoration is a laudable reversal of landscape degradation. But restoration is expensive and does not guarantee communities will be restored to their full, original viability. A more efficient and parsimonious course is to prevent deterioration in the first place. This requires several steps:

1. That there be an awareness that diversity is important for increasing plant productivity (Naeem et al. 1994), which, in turn, supports all the rich, subtle mutualistic interactions (Bond 1994; Sapp 1994). Furthermore, increased primary productivity in more diverse plant communities is more resistant to, and recovers more fully from, a major disturbance such as drought (Tilman and Downing 1994).
2. That there be an awareness that increased size of spatial fragments is important. Where habitat destruction brings a community below a critical spatial size, the system can collapse (Hassell et al. 1994). Certainly, information from insects in another type of ecosystem in Sweden has shown that the katydid *Metrioptera bicolor* becomes locally extinct on a regular basis when patches are small ($<$0.5 ha) and relatively isolated ($>$100 m) from source patches (Kindvall and Ahlén 1992). Furthermore, the models of Tilman et al. (1994) suggest that even moderate habitat destruction will cause time-delayed but deterministic extinction of the dominant competitor in remnant patches. This means that the extinction debt (i.e. loss of species from the patches) associated with habitat destruction may have an insidious effect on ecosystems, even many generations after the fragmentation event.
3. That there be an inventory of the species particularly in the large remaining remnant patches of natural biota. These inventories should be

compared with those of smaller patches to ascertain the type and rate of local species loss. It may not however, be possible to do an all-taxa biodiversity inventory because of limiting resources. Certain taxa should be selected along the lines suggested by Kremen et al. (1993) and Samways (1994).

4. Disney (1986) has pointed out that inventories of just one insect order (e.g. Diptera) at one site in Britain is a massive task. For the much more species-rich MTEs, this means it is expedient to use flagship taxa that are well-known, high in numbers of endemics, and have high spatial and temporal apparency. Some taxa that fill this role in MTEs are Neuroptera (Mansell 1986), butterflies (New et al. 1995), and Odonata (Samways 1993) as well as other taxa appropriate to the precise conservation policy and the resources available.
5. Although some taxa are not particularly well-represented in MTEs, e.g. Cicindelid beetles (Pearson and Cassola 1992) and others are low in numbers of endemics, e.g. Carabid beetles (Pizzolotto and Brandmayr 1990), such groups can nevertheless be very useful in determining historical aspects (Darlington 1965; Noonan 1985) as well as landscape change in MTEs (Comandini and Vigna Taglianti 1990).
6. To establish whether these flagship/indicator groups also act as umbrella species for more cryptic species.
7. Such flagships can also be used along with other flagships (e.g. vertebrates, angiosperms) for undertaking gap analysis (Rebelo 1992). However, as some of these MTE insects have metapopulation dynamics (Hafernik 1992), maintaining them in a remnant patch will not necessarily guarantee survival in the long term. Clearly, increasing the size of fragments, improving linkages between patches and restoring areas would facilitate metapopulation activity and hence species survival.
8. As MTEs have been so disturbed, it means that natural environmental processes (e.g. water flow, fire frequency, soil formation etc.) have been interrupted. Also, as the management of entire landscapes is the most practicable umbrella option for conserving insects, it means that it is essential to restore or mimic these processes. This may involve regular burns, removal of exotic weeds, or containment of invasive insects. The important point then is to monitor the effectiveness of these management measures.

18.6 Corollary

The knowledge of insect inventories, ecology and conservation in MTEs is in its infancy, yet this is one of the most threatened types of ecosystems anywhere in the world. Many policy makers will not give high priority to these small animals, so the most practicable way to ensure their conservation is by conserving landscapes for their more glamorous fauna, flora and scenic wonders. The insects however, can then be monitored and determination made as to the effectiveness of this indirect conservation approach.

Acknowledgements. Ms Pamela Sweet kindly processed the manuscript. Financial support was from the Department of Environmental Affairs and Tourism, Foundation for Research Development, and the University of Natal Research Fund.

References

Ayala R, Griswold TL, Bullock SH (1993) The native bees of Mexico. In: Ramamoorthy TP, Bye R, Lot A and Fa J (eds) Biological diversity of Mexico: origins and distribution. Oxford University Press, New York, pp 179–227

Bond WJ (1994) Do mutualisms matter? Assessing the impact of pollinator and disperser disruption on plant extinction. Phil Trans R Soc Lond B 344:83–90

Bond WJ, Slingsby P (1984) Collapse of an ant-plant mutualism: the Argentine ant (*Iridomyrmex humilis*) and myrmecochorous Proteaceae. Ecology 65:1031–1037

Brower LP, Malcolm SB (1989) Endangered phenomena. Wings 14:3–10

Comandini F, Vigna Taglianti A (1990) Ground beetle communities in a Mediterranean area (Tolfa Mountains, Central Italy). In: Stork NE (ed) The role of ground beetles in ecological and environmental studies. Intercept, Andover, pp 171–179

Cottrell CB (1985) The absence of co-evolutionary associations with Capensis Floral Element plants in the larval/plant relationships of Southwestern Cape butterflies. In: Vrba ES (ed). Species and speciation. Transvaal Museum, Pretoria, pp 115–124

CSIRO (1991) The insects of Australia, 2nd edn. Melbourne University Press, Melbourne

Darlington PJ Jr (1965) Biogeography of the southern end of the world. Harvard University Press, Cambridge, Massachusetts

Davies D (1988a) Leafhoppers (Homoptera: Cicadellidae) associated with the Restionaceae. I. The tribes Cephalelini (Uloinae). J Entomol Soc S Afr 51:31–64

Davies D (1988b) Leafhoppers (Homoptera: Ciccidellidae) associated with the Restionaceae. II. The tribes Ashysanini and Paralimini (Euscelinae). J Entomol Soc S Afr 51:65–80

Davis ALV (1994) Habitat fragmentation in southern Africa and distributional response patterns in five specialist or generalist dung beetle families (Coleoptera). Afr J Ecol 32:192–207

Dennill GB, Donnelly D (1991) Biological control of *Acacia longifolia* and related weed species (Fabaceae) in South Africa. Agric Ecosyst Environ 37:115–135

Disney RHL (1986) Inventory surveys of insect faunas: discussion of a particular attempt. Antenna 10:112–116

Donnelly D, Giliomee JH (1985) Community structure of epigaeic ants in a pine plantation and in newly burnt fynbos. J Entomol Soc S Afr 48:259–265

Endrödy-Younga S (1988) Evidence for the low-altitude origin of the Cape Mountain Biome derived from the systematic revision of the genus *Colophon* Gray (Coleoptera, Lycanidae). Ann S Afr Mus 96:359–424

Erwin TL (1988) The tropical forest canopy: the heart of biotic diversity. In: Wilson EO (ed). Biodiversity. National Academy Press, Washington DC, pp 123–129

Forman RTT, Godron M (1986) Landscape ecology. Wiley, New York

Gaston KJ (1992) Regional numbers of insect and plant species. Funct Ecol 6:243–247
Gess FW, Gess SK (1991) Effects on increasing land utilization on species representation and diversity of aculeate wasps and bees in the semi-arid areas of southern Africa. In: LaSalle J, Gould ID (eds) Hymenoptera and biodiversity. CAB International, Wallingford, pp 83–113
Gess SK (1992) Biogeography of the masarine wasps (Hymenoptera: Vespidae: Masarinae), with particular emphasis on the southern African taxa and on correlations between masarine and forage plant distribution, J Biogeogr 19:491–503
Gordon AJ, Kluge RL (1991) Biological control of St John's Wort, *Hypericum perforatum* (Clusiaceae), in South Africa. Agric Ecosyst Environ 37:77–90
Greenslade P (1992) Conserving invertebrate diversity in agricultural, forestry and natural ecosystems in Australia. Agric Ecosyst Environ 40:297–312
Hafernik JE (1989) Surveys of potentially threatened Bay Area water beetles and the San Francisco Forktail damselfly. Technical Report of the United States Fish and Wildlife Agency, Washington DC
Hafernik JE Jr. (1992) Threats to invertebrate biodiversity: implications for conservation strategies. In: Fiedler PL, Jain SK (eds) Conservation biology: the theory and practice of nature conservation preservation and management. Chapman and Hall, London, pp 171–195
Hammond PM (1992) Species inventory. In: Groombridge B (ed) Global biodiversity: status of the earth's living resources. Chapman and Hall, London, pp 17–39
Hannah L, Carr JL, Lankerani A (1995) Human disturbance and natural habitat: a biome level analysis of a global data set. Biodiv Conserv 4:128–155
Harz K (1969) Die Orthopteren Europas I. Junk, The Hague
Hassell MP, Comins HN, May RM (1994) Species co-existence and self-organizing spatial dynamics. Nature 370:290–292
Hattingh V, Giliomee JH (1989) Pollination of certain *Leucadendron* species (Proteaceae). S Afr J Bot 55:387–393
Hegazy AK, Eesa NM (1991) On the ecology, insect seed-predation, and conservation of a rare and endemic plant species: *Ebenus armitagei* (Leguminosae). Conserv Biol 5:317–324
Hill L, Michaelis FB (1988) Conservation of Insects and Related Wildlife. Occasional Paper No 13. Australian National Parks and Wildlife Service, Canberra
Hoffman JH, Moran VC (1991) Biological control of *Sesbania punicea* (Fabaceae) in South Africa. Agric Ecosyst Environ 37:157–173
Howarth FG (1991) Environmental impacts of classical biological control. Ann Rev Entomol 36:485–509
Ingham DS, Samways MJ (1996) Application of fragmentation and variegation models to epigaeic invertebrates in South Africa. Conserv Biol 10:1353–1358
Johnson SD (1992) Plant-animal relationships. In: Cowling RM (ed) The ecology of fynbos: nutrients, fire and diversity. Oxford University Press, Cape Town, pp 175–205
Jordano D, Retamosa EC, Fernández Haeger J (1991) Factors facilitating the continued presence of *Colotis evagore* (Klug 1829) in southern Spain. J Biogeogr 18:637–646
Keals N, Majer JD (1991) The conservation status of ant communities along the Wubin-Perenjori corridor. In: Saunders DA, Hobbs RJ (eds). The role of corridors. Surrey Beatty, Chipping Norton, new South Wales, Australia, pp 387–393
Kindvall O, Ahlén I (1992) Geometrical factors and metapopulation dynamics of the bush cricket, *Metrioptera bicolor* Philippi (Orthoptera: Tettigoniidae). Conserv Biol 6:520–529
Kremen C, Colwell RK, Wrwin TL, Murphy DD, Noss RF, Sanjayan MA (1993) Terrestrial arthropod assemblages: their use in conservation planning. Conserv Biol 7:796–808
Leigh EG Jr (1994) Do insect pests promote mutualism among tropical trees? J Ecol 82:677–680

Leigh EG Jr, Wright SJ, Putz FE, Herre EA (1993). The decline of tree diversity on newly isolated tropical islands: a test of a null hypothesis and some implications. Evol Ecol 7:76–102

Llorente-Bousqets J, Luis-Martínez A (1993) Conservation-oriented analysis of Mexican butterflies: Papilionideae (Lepidoptera, Papilionoidea). In: Ramamoorthy T, Bye R, Lot A and Fa J (eds). Biological diversity of Mexico: origins and distribution. Oxford University Press, New York, pp 147–177

Lockwood JA, De Brey LD (1990) A solution for the sudden and unexplained extinction of the Rocky Mountain grasshopper (Orthoptera: Acrididae). Environ Entom 19:1194–1205

Majer JD, Recher HF, Postle AC (1994) Comparison of arthropod species richness in eastern and western Australian canopies: a contribution to the species number debate. Mem Queensl Mus 36:121–131

Mansell M (1986) Southern African ant-lions. Antennae 10:121–124

Mawdsley NA, Stork NE (1995) Species extinctions in insects: ecological and biogeographical considerations. In: Harrington R, Stork NE (eds) Insects in changing environment. Academic Press, London, pp 321–369

McIntyre S, Barrett GW (1992) Habitat variegation, an alternative to fragmentation. Conserv Biol 6:146–147

Moran VC, Hoffmann JH, Impson FAC, Jenkins JFG (1994) Herbivorous insect species in the tree canopy of a relict South African forest. Ecol Entom 19:147–154

Moritz C (1994) Defining "Evolutionarily Significant Units" for conservation. Trends Ecol Evol 9:373–375

Morris MJ (1991) The use of plant pathogens for biological weed control in South Africa. Agric Ecosyst Environ 37:239–255

Murphy DD, Weiss SB (1988) Ecological studies and the conservation of the bay checkerspot butterfly, *Euphydryas editha bayensis*. Biol Conserv 46:183–200

Naeem S, Thompson LJ, Lawler SP, Lawton JH, Woodfin RM (1994) Declining biodiversity can alter the performance of ecosystems. Nature 368:734–737

Nazzi F, Paoletti MG, Lorenzoni GG (1989) Soil invertebrate dynamics of soybean agroecosystems encircled by hedgerows or not in Friuli, Italy. First Data. Agric Ecosyst Environ 27:163–176

New TR (1984) Insect conservation – an Australian perspective. Junk, The Hague

New TR (1993) Effects of exotic species on Australian native insects. In: Gaston KJ, New TR, Samways MJ (eds) Perspectives on insect conservation. Intercept, Andover, pp 155–169

New TR (1994) Exotic insects in Australia. Gleneagles, Adelaide

New TR, Collins NM (1991) Swallowtail butterflies: an action plan for their conservation. IUCN, Gland, Switzerland

New Tr, Pyle RM, Thomas JA, Thomas CD, Hammond PC (1995) Butterfly conservation management. Ann Rev Entomol 40:57–83

Noonan GR (1985) The influence of dispersal, vicariance and refugia patterns of biogeographical distribution of the beetle family Carabidae. In: Ball GE (ed). Taxonomy, phylogeny and zoogeography of beetles and ants. Junk, The Hague, pp 322–349

Noss RF (1990) Indictors for monitoring biodiversity: a hierarchical approach. Conserv Biol 4:355–364

Pearson DL, Cassola F (1992) World-wide species richness patterns of tiger beetles (Coleoptera: Cicindellidae): indicator taxon for biodiversity and conservation studies. Conserv Biol 6:376–391

Picker MD and Samways MJ (1996) Faunal diversity and endemicity of the Cape Peninsula, South Africa. Biodiv Conserv 5:591–606

Pizzolotto R, Brandmayr P (1990) The carabid groupings of the Nebrodi Mountains in Sicily: Ecological and Historical Indicators. In: Stork NE (ed). The role of ground beetles in ecological and environmental studies. Intercept, Andover, pp 201–207
Platnick NI (1991) Patterns of biodiversity: tropical v temperate. J Nat Hist 25:1083–1088
Plattnick NI (1992) Patterns of biodiversity. In: Eldridge N (ed) Systematics, ecology and the biodiversity crisis. Columbia University Press, New York, pp 15–24
Price PW (1984) Insect ecology. 2nd edn. Wiley, New York
Rebelo AG (1992) Red data book species in the Cape Floristic Region: threats, priorities and target species. Trans R Soc S Afr 48:55–86
Rentz CF (1993) Orthopteroid insects in threatened habitats in Australia. In: Gaston KJ, New TR, Samways MJ (eds) Perspectives on insect conservation. Intercept, Andover, pp 125–138
Richardson DM, Macdonald IAW, Holmes PM, Cowling RM (1992) Plant and animal invasions. In: Cowling RM (ed). The ecology of fynbos: nutrients, fire and diversity. Oxford University Press, Cape Town, pp 271–308
Samways MJ (1988) Classical biological control and insect conservation: Are they compatible? Environ Conserv 15:349–354
Samways MJ (1989) Insect conservation and landscape ecology: a case-history of bush crickets (Tettigoniidae) in southern France. Environ Conserv 16:217–226
Samways MJ (1992) Some comparative insect conservation issues of north temperate, tropical, and south temperate landscape. Agric Ecosyst Environ 40:137–154
Samways MJ (1993) Dragonflies (Odonata) in taxic overlays and biodiversity conservation. In: Gaston KJ, New TR, Samways MJ (eds) Perspectives on insect conservation. Intercept, Andover, pp 111–123
Samways MJ (1994) Insect conservation biology. Chapman and Hall, London
Samways MJ (1995) Conservation of the threatened endemic dragonflies of South Africa. In: Corbet PS, Dunkle SW, Ubukata H (eds) Proc Int Symp on the Conservation of dragonflies and their habitats. Japanese Society for the Preservation of Birds, Kushiro, pp 8–15
Samways MJ, Harz K (1982) Biogeography of intraspecific morphological variation in the bush crickets *Decticus verrucivorus* (L.) and *D. albifrons* (F.) (Orthoptera: Tettigoniidae). J Biogeogr 9:243–254
Samways MJ, Caldwell PM, Osborn R (1996) Ground-living invertebrate assemblages in native, planted and invasive vegetation in South Africa. Agric Ecosyst Environ 59:19–32
Samways MJ, Moore SD (1991) Influence of exotic conifer patches on grasshopper (Orthoptera) assemblages in a grassland matrix at a recreational resort, Natal, South Africa. Biol Conserv 57:205–219
Sapp J (1994) Evolution by association. Oxford University Press, Oxford
Slingsby P, Bond WJ (1985) The influence of ants on the dispersal distance and seedling recruitment of *Leucospermum canacarpodendron* (Proteaceae). S Afr J Bot 51:30–34
Soulé ME, Kohm KA (eds) (1989) Research priorities for conservation biology. Island Press, Washington, DC
Stork NE (1993) How many species are there? Biodiv Conserv 2:215–232
Tilman D, Downing JA (1994) Biodiversity and stability in grasslands. Nature 367:363–365
Tilman D, May RM, Lehman CL, Nowak MA (1994) Habitat destruction and the extinction debt. Nature 371:65–66
Tribe GD, Richardson DM (1994) The European wasp, *Vespula germanica* (Fabricius) (Hymenoptera: Vespidae), in southern Africa and its potential distribution as predicted by ecoclimatic matching. Afr Entomol 2:1–6
Vane-Wright RI (1986) White monarchs. Antenna 10:117–120
Vogler AP, DeSalle R (1994) Diagnosing units of conservation management. Conserv Biol 8:354–363

Williams KS (1993) Use of terrestrial arthropods to evaluate restored riparian woodlands. Restor Ecol 1:107–116

Wood PA, Samways MJ (1991) Landscape element pattern and continuity of butterfly flight paths in an ecologically landscaped botanic garden, Natal, South Africa. Biol Consev 58:149–166

Wright MG (1993) Insect conservation in the African Cape fynbos, with special reference to endophagous insects. In: Gaston KJ, New TR, Samways MJ (eds) Perspectives on insect conservation. Intercept Press, Andover, pp 97–110

Yen AL, Butcher RJ (1994) An overview of the conservation status of non-marine invertebrates in Australia. Australian Nature Conservation Agency, Canberra

19 Loss of Vertebrate Diversity Following European Settlement of Australian Mediterranean Regions

B.J. Fox

19.1 Introduction

The loss of vertebrate diversity in Australia has been dramatic in the two centuries since the European occupation. It has been possible to document this decline because an accurate assessment of the pre-European fauna was available. The information enabling the reconstruction of the mammal fauna was obtained from many sources including owl pellets from roost sites and subfossil cave deposits (Morton and Baynes 1985). Burbidge et al. (1988) were able to make use of an extremely important source of information by gathering the knowledge of the mammals that occupied the central deserts of Australia from memories of the aboriginal inhabitants. Museum collections of skins and preserved specimens were used to obtain positive identification in each local area for each species that had lived in that area, together with the name used by that aboriginal language group. As almost 500 distinct language groups have been identified in Australia this provided reliable and detailed information for localised areas. In addition to the work of Burbidge et al. (1988) in semi-arid and arid regions in central western Australia (see Fig. 19.1), a similar technique has also been used in the Flinders Ranges in the eastern mediterranean region of southern Australia (Tunbridge 1991).

19.2 Pre-European Mammal Fauna

In 1985, Morton and Baynes provided a detailed assessment of how the present day mammal fauna differed from that for fossil and sub-fossil sites. They compared faunas established from present day censuses with data obtained from fossil deposits and subfossil cave deposits from the same localities. They included data for the western half of the Australian continent, with some sites from three climate regions: the mediterranean region

Ecological Studies, Vol. 136
Rundel et al. (eds.) Landscape Degradation and Biodiversity in Mediterranean-Type Ecosystems

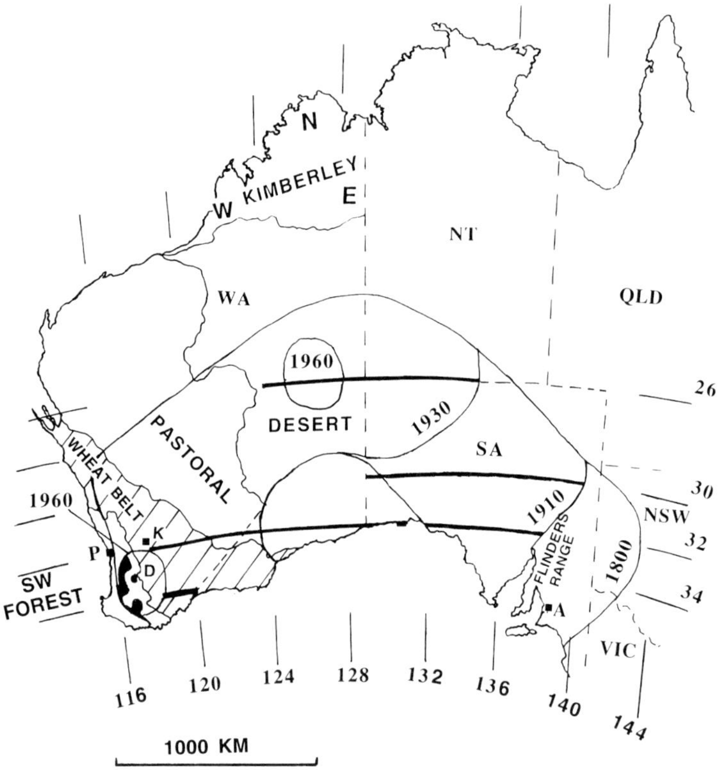

Fig. 19.1. Map of part of Australia showing distribution of the numbat. The former distributions are shown as *solid lines* associated with their appropriate date, and the distribution at 1980 filled in solid black. The Kimberley, desert, pastoral and SW forest regions are indicated, with the wheat belt (mediterranean region) hatched. State boundaries are shown as *dotted lines* with abbreviations for states, while the positions of Adelaide (*A*), Dryandra forest (*D*), Kellerberrin (*K*) and Perth (*P*) are indicated, together with the Flinders Ranges. Latitudinal transects are shown as *heavy lines* at 26, 30, 32, 34° S

(wheatbelt), the arid region, and two sites from the tropical region to the north (see Fig. 19.1). For arid sites, 41% of polyprotodont marsupial species remain (insectivorous dasyurids and bandicoots) and 48% of rodents species remain. For the mediterranean sites 51% of marsupial species and 56% rodent species remain. This conflicts markedly with the tropical region further north where no loss of species was recorded.

19.3 Critical Weight Range

A later and more extensive comparison by Burbidge and McKenzie (1989) shows a similar trend as one moves from the desert regions to higher

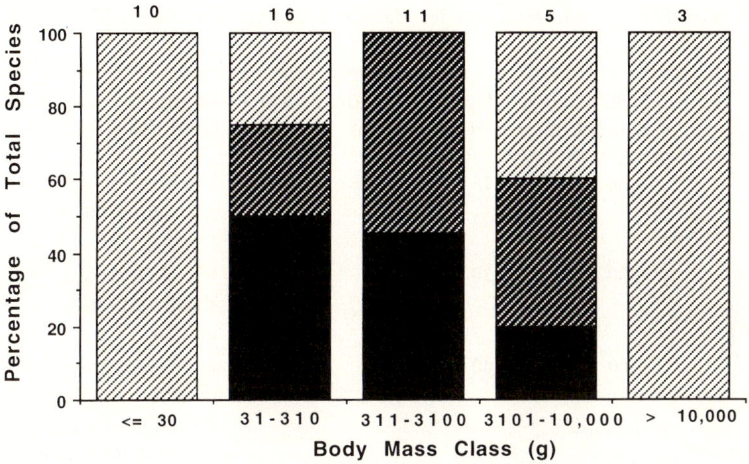

Fig. 19.2. Percentage decrease (*dark hatching*) and percentage extinction (*solid*) for mammal body weight classes. Data extracted from Burbidge and McKenzie (1989)

rainfall areas. They report an interesting pattern in the number of species of mammals going extinct or declining, in relation to the body weight of the species. This information is best summarised for the wheatbelt region (see Fig. 19.1) by graphing these percentages for five weight classes, where I have combined classes from the data of Burbidge and McKenzie (1989) and indicated the number of species in each weight class (Fig. 19.2). The pattern is clear: none of the ten species less than 30 g, nor the three species above 10 kg, have been affected. This pattern is repeated for all of the separate regions and for Western Australia overall, so that on closer examination Burbidge and McKenzie (1989) found that all species suffering decline or extinction lay within the range 35 g to 4.2 kg, which they termed the critical weight range (CWR). Expansion of their idea to cover the whole of Australia added one more species and raised the upper limit to 5.5 kg. A summary of their results for the effects on CWR species in four of the regions is shown in Table 19.1. In this way the role of body mass as a factor in the loss of species (as it determines which species lie in the critical weight range) is included in the summary table of their factors shown below.

19.4 Regional Losses

For 22 desert sites their figures demonstrate that 73% of species have gone extinct and 15% declined, with 50% extinct and 20% declined from 28

Table 19.1. Fate of critical weight range species. Data collated from Burbidge and McKenzie (1989)

Region	Mean Annual Rainfall (mm)	Declined (%)	Extinct (%)	Total (%)	Changed Land Use
Desert (arid)	250	15	73	88	Unchanged except for Aborigines move to towns
Pastoral (arid)	250	20	50	70	Extensive pastoralism sheep and cattle grazing
Wheat Belt (mediterranean)	400	40	50	90	Extensive clearing (>65%) Agriculture, small remnants of native vegetation
Extreme SW (mesic temperate)	700	30	20	50	Forestry, agriculture, some clearing (35%), much native vegetation remaining
E and W Kimberley (mesic tropical)	600	26	21	47	Extensive pastoralism
North Kimberley (mesic tropical)	1100	0	0	0	Some pastoralism, mostly unchanged

pastoral sites. For 29 sites in the wheatbelt region with mediterranean climate, these figures are 50% and 40% respectively. These figures are reduced in regions with greater rainfall. The extreme southwest region exhibits 20 and 30% respectively from 16 sites. Further north in the Kimberley region within the tropics, 21% went extinct and 26% declined from 19 sites in the east and west regions where extensive pastoralism occurs. However, there have been no extinctions or declines from any of the 21 sites surveyed in the North Kimberley region where there is little pastoralism and the habitats are unchanged.

Burbidge and McKenzie (1989) define a Fauna Attrition Index (FAI) in terms of the number of CWR species, the number of extinct species and the number of stable species. They demonstrate that 96% of the variance in FAI can be explained by a negative relationship with rainfall alone, although from the above percentages it is also clear that the degree of disturbance (agriculture, clearing, pastoralism) is also important. For the wheatbelt region the effect on different vertebrate groups has been summarised in Table 19.2 from the data presented by Burbidge and McKenzie (1989). The losses have been greatest for the 137 species of non-flying mammals

Table 19.2. Species loss in different groups for Western Australia. Data from Burbidge and McKenzie (1989)

Group	Total No. species	Declined species	Extinct species	Total affected
Bats	35	0	0	0
Reptiles	400	2 (0.5%)	0	2 (0.5%)
Birds	425	10 (2.4%)	3 (0.7%)	13 (3.1%)
Mammals (Non-flying)	137	24 (17.5%)	17 (12.4%)	41 (29.9%)

recorded with 17 (12.4%) extinct and 24 (17.5%) declining while for the 425 species of birds only 3 (0.7%) have gone extinct and 10 (2.4%) have declined. The 400 species of reptiles have shown even less effect with no extinctions and only 2 (0.5%) species declining, while bats have not been affected at all.

Saunders and Curry (1990) provided an interesting very detailed contrast, to the more extensive study by Burbidge and McKenzie (1989), when they examined the bird fauna at two specific sites: Kellerberrin from the wheatbelt region of Western Australia with a mediterranean climate; and Murchison River from the pastoral region that abuts the wheatbelt and experiences a semi-arid climate. They present detailed data for passerines and non-passerines separately and I have illustrated their results as histograms of the number of species extinct, decreased, unchanged, or increased (Fig. 19.3). With a total of 7% of the 118 bird species affected detrimentally, the Murchison River has lost a similar number of species to that quoted by Burbidge and McKenzie (1989) for the whole of Western Australia, but from a bird fauna one quarter the size. However, with a total of 29% of the 131 species affected the bird fauna at Kellerberrin has lost a far greater number of species and is more similar to the results indicated for mammals in the total wheatbelt region (Table 19.2).

Part of the explanation for the disagreement in the percentages for birds affected relates to the fact that one is much more likely to record decreases and extinctions from a single small site (like Kellerberrin) than from an entire region and it is most likely that more data is available for specific sites than for a whole region. Saunders and Curry (1990) go to considerable lengths to establish the validity of the direct comparison between their two sites. The differences between their sites relate to the increased rainfall at Kellerberrin and the massive amount of clearing and habitat fragmentation that has occurred there (see below) in comparison to the grazing (but not clearing) that has taken place at Murchison River. Given that the increase in detrimental effects of Kellerberrin is in the opposite direction to the

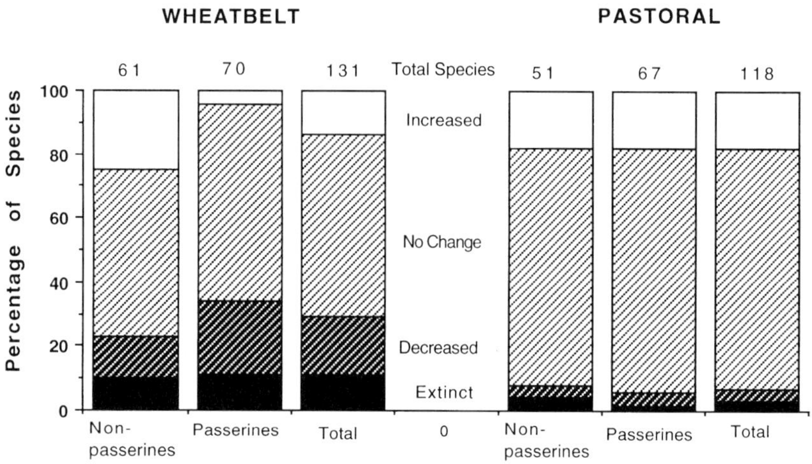

Fig. 19.3. Birds' percentage decrease and percentage extinction in the wheatbelt. Data collated from Saunders and Curry (1990)

trend with rainfall shown by Burbidge and McKenzie (1989) it would appear that the major effect here should be attributed to the habitat loss and fragmentation (see below).

19.5 Factors Causing Loss of Biodiversity

Burbidge and McKenzie (1989) documented and discussed a number of factors that might contribute to the loss of Australia's vertebrate fauna, and I have pulled these together into a summary (Table 19.3) with an indication of the role of human intervention as direct, indirect or independent.

19.5.1 Fragmentation and Loss of Habitat

The effect of habitat loss by conversion to agricultural production is obvious and has been dramatically illustrated by Merriam and Saunders (1993), based on data from Hobbs and Saunders (1992). They present a graphic picture of progressive fragmentation of native vegetation in the Kellerberrin area at four time periods: 1920, 1940, 1960 and 1984. I have attempted to quantify this fragmentation by estimating the number of fragments in five size classes that approximate to >100, 20, 10, 5 and 1 km^2 with the large gap between the very large and large categories purposefully chosen to emphasise how extreme was the degree of fragmentation that

Table 19.3. Summary of factors causing loss of Australia's vertebrate fauna

Factor[a]	Role of human intervention[b]		
	Direct	Indirect	Independent
Loss of habitat	Yes		
Fragmentation of habitat	Yes		
Competition from introduced predators		Yes	
Predation from introduced predators		Yes	
Diversion of environmental resources	Yes		
Reduction in vegetative cover by introduced herbivores		Yes	
Changed fire regimes	Yes		
Body mass (critical weight range species)			Yes

[a] Factors after Burbidge and McKenzie (1989).
[b] My assessment of role of human intervention.

Table 19.4. Approximate number of fragments in each of five rough size classes at Kellerberrin. (Counts based on maps presented in Hobbs and Saunders 1992: Fig. 2)

Area (km^2)	Size class	1920	1940	1960	1984
1	Tiny	87	190	210	236
5	Small	5	9	8	13
10	Medium	0	15	5	3
20	Large	3	5	0	0
>100	Very large	1	0	0	0
	Total	96	219	223	252

took place between 1920 and 1940 (Table 19.4). In 1920 there was one contiguous area of native vegetation well over 100 km^2 that covered much of the study area although its shape and margins were most irregular, because of vegetation clearing already underway. Twenty years later the few large areas remaining are almost an order of magnitude smaller. There is also a doubling of the number of fragments between 1920 and 1940. However, while complete loss of habitat would virtually ensure complete loss of species, there is not a linear decline and there can be a substantial loss of habitat in the short term before there is any marked loss of species recorded. Loss of habitat is also generally associated with fragmentation. The effects of fragmentation in this mediterranean landscape and the impact, of the degree of clearing and fragmentation that has occurred up to 1984, on all vertebrate groups is obvious and has already been documented for birds (see Fig. 19.3; Saunders and Curry 1990). The direct relationship between fragment area and the number of species able to be supported is well known. However, an additional negative relationship between the number of species, the degree of disturbance to which a fragment has been subjected, and interaction effects between size and degree of disturbance has also been demonstrated for ground-dwelling small mammals in rainforest fragments (Dunstan and Fox 1996).

19.5.2 Diversion of Resources and Loss of Vegetative Cover

Burbidge and McKenzie (1989) also considered the diversion of environmental resources from the native species to those species introduced by humans, most notably to sheep and cattle and the crops grown on land that had once supported habitat for native animals. As well as the diversion of food resources to introduced herbivores there was also the simple loss of vegetative cover, eaten or trampled by these introduced herbivores.

19.5.3 Changes to Fire Regimes

Another factor was changes to the fire regimes, brought about by the shift away from 'aboriginal care of the land', which included frequent intentional burning but restricted to small areas. European land management practices involved many fewer fires, usually wildfires that burned extensive areas with greater intensity. The result was a shift from a patchwork quilt of habitats with different fire histories, to large areas with uniform fire histories and the associated loss of protective cover in these habitats for vulnerable species.

Evidence to support the impact of changed fire regimes on the loss of CWR species has been put forward by Burbidge et al. (1988), and this ties in nicely with evidence from the map of past numbat distribution (Friend 1990), where an isolated population of numbats in the Gibson desert in Western Australia hung on until the 1960s. This area was one of the last areas subjected to the traditional burning practices of the aboriginal traditional owners of the area before they abandoned their traditional lifestyle in the late 1960s. In the Dryandra Forest the loss of shrub cover and log refuges caused by changes in the fire frequency has also been suggested as one factor contributing to the sharp decline in numbat numbers in that area in the late 1970s.

19.5.4 Introduced Species

Here I wish to examine specific examples of the effect of an introduced herbivore (European rabbit *Oryctolagus cuniculus*) and an introduced predator (European red fox *Vulpes vulpes*) on one native species in the CWR that has suffered drastic reduction in its range since 1800 (numbat, *Myrmecobius fasciatus*). I have chosen this species as an example as its decline is particularly well documented because of its very distinctive appearance that made observations of this species reliable as there was no possibility of mistaken identifications, and good information on past distribution has been collated (Friend 1987, 1990). The fox feeds on its accustomed prey the rabbit, but also eats alternative native prey such as the numbat. Detailed maps are available for the changing distributions of these three species: the shrinking distribution of the numbat from 1800 to 1980 (Friend 1990), the pattern of spread for the rabbit from 1870 to 1960 (Myers 1971, 1986) and probable dates of the first arrival of the red fox in local districts from Redhead et al. (1991, using data adapted from Jarman 1986).

19.6 Time Lines

I transferred these data from the three maps described above onto one common map and then selected a number of latitudinal bands or transects for which there were available sufficient common points for the three distributions to allow estimation of the time that each of the three waves of invasion or disappearance passed that particular geographical point. It was necessary to interpolate from the original maps to derive the appropriate date at some of these points. Each such "intersection point" would then have five values: latitude and longitude, plus a year for each of the three species. I plotted these values for one latitude band onto a graph with longitude on the x-axis and time on the y-axis and joined up the points for each species into a time line for that species that represented a trajectory through time and space to describe either the pattern of its demise or invasion (Fig. 19.4). For instance, along latitude 30° S (see Figs. 19.1, 19.4A) at longitude 140° E the wave of invading rabbits passed in 1888, followed by the first record of foxes in 1908 and the last record of numbats two years later in 1910. In this case there is a time lag of 22 years between the arrival of the rabbit and the loss of the numbat, with a 2-year time lag for the fox. It is noticeable that these time lags increase as the time path moves west to longitude 129° E, but the time lag is always much less for the fox than for the rabbit. Figure 19.4B shows the time path along a band just north of 32° S, the rabbit and fox time paths proceed in parallel at very similar rates, while the time path for the numbat diverges further as it moves west with the most marked divergence between 118° S and 120° S, the region that contains the Kellerberrin study area north of Albany. I have considered two

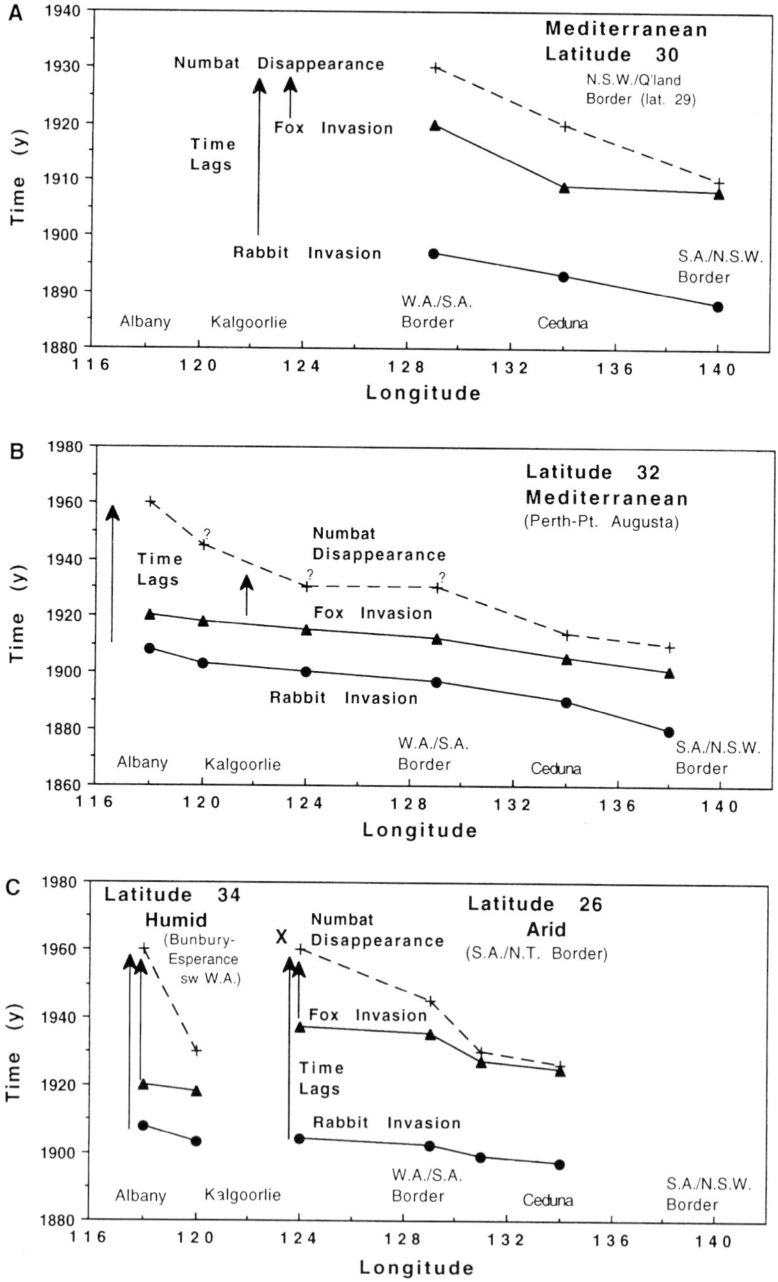

Fig. 19.4a–c. East to west time path for the numbat, the fox and the rabbit at: *A* latitude 30°, *B* latitude 32°, and *C* latitudes 26° and 34°

other latitudinal bands (Fig. 19.4C), one along 26°S entirely within the arid zone, and a second at latitude 34°S which is in the humid climatic zone in the southwest of Western Australia (see Fig. 19.1). For the arid zone 26°S band, the numbat disappearance remains relatively close to the fox arrival as far as the Western Australia/South Australia border, but there is a marked divergence further west. It was in this region of the Gibson Desert that a small isolated area continued to support a population of numbats in 1960. There is also a marked divergence seen for the 34°S band in the humid zone.

19.7 Experimental Evidence for Predation

Friend (1990) provided a map of the isolated areas where the numbat was still extant in 1980 and where it is still present today. These areas (Perup Nature Reserve, Dryandra Forest) lie between and just west of the ends of the 32°S and 34°S transects. Information on the rate of numbat sightings per 100 km of survey in the Dryandra Forest is available at different times over the period from 1954 to 1988 (Friend 1990). Sightings by John Calaby over three years from 1954–1956 were at the mean rate of 3.5 per 100 km. There were indications of a marked drop in numbat numbers in the late 1970s and by 1979 when another survey was conducted by J. Turner there was only a single sighting in over 700 km of driving. Further surveys in 1981, 1982 and 1983 covering a total of over 2200 km revealed a mean rate of 0.6 sightings per 100 km (ranging from 0.4 to 1.2). The laying of poison baits to reduce fox populations in the Dryandra forest began in late 1982 and surveys in 1984, 1985, 1987 and 1988 had a mean sighting rate of 3.3 per 100 km with 42 sightings in 1265 km of survey. It seems clear that the reduction in the predation pressure from foxes has allowed the numbat population to recover to the 1950s level, and more supporting evidence is available. Analysis of the sighting data per 100 km provided by (Friend 1990) for three surveys in 1984–85 indicates a marginally significant difference ($p = 0.0515$, one-tailed) between areas of forest that were baited (mean sighting rate 3.71 ± 1.05 SE) and those unbaited (1.42 ± 0.27 SE).

Kinnear et al. (1988) have provided additional evidence for the detrimental effect of fox predation, in this case for rock wallabies (which also fall within the critical weight range). Using these data I have taken account of the different numbers of rock wallabies present at their four sites by standardizing them to have their mean value between 1980 and 1982 set at a value of 1.0, so that for later time periods values greater than 1.0 represented increases in rock wallaby numbers and values less than 1.0 represented decreases. I have presented these values together with error bars for this mean value in Fig. 19.5. Fox control began in May 1982 and I have graphed the relative abundance data for rock wallabies from 1984 and 1986 (in terms of the pre-control mean = 1.0). At the sites where foxes had been removed there were threefold increases in rock wallaby numbers while numbers on the control sites remained within the range of the error bars for the pre-control period.

19.8 Model for the Impact of European Invasion

In 1990, Stephen Morton put forward a conceptual model for the impact of European settlement on vertebrate animals in arid Australia and I have used it as the basis for a model for the mediterranean climate zone in Australia. Many of the same factors are involved in the mediterranean climate areas. However, there are additional factors such as: habitat loss and degradation associated with cropping, and the impact of fragmentation and disturbance, in addition to the grazing effects observed in arid areas. My adaptation of Morton's model for a mediterranean climate is shown in Fig. 19.6.

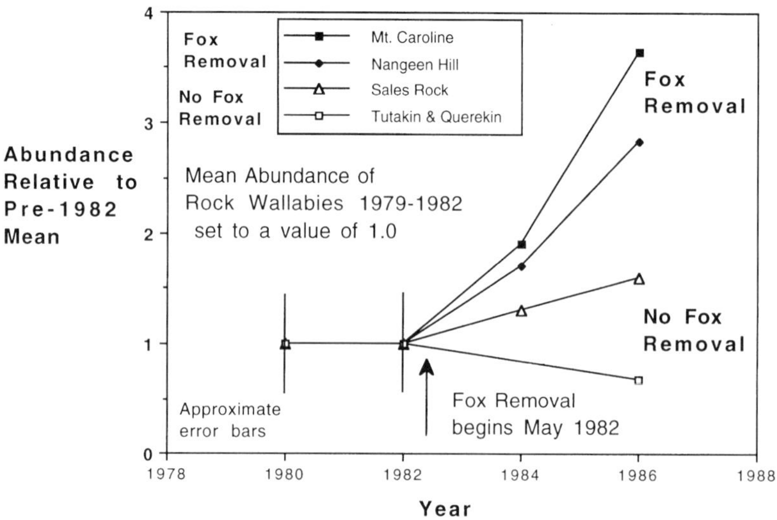

Fig. 19.5. The effect of fox removal on rock wallaby abundance compared with control sites. Data extracted from Kinnear et al. (1988)

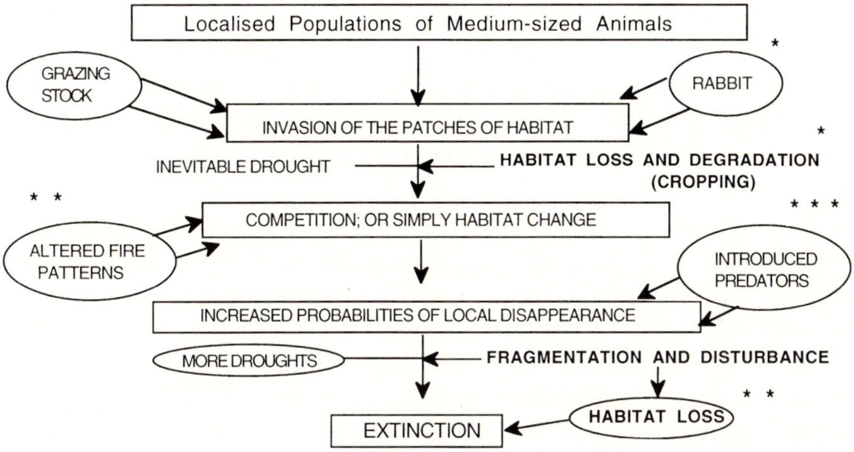

Fig. 19.6. Model of the impact of European settlement on vertebrate animals in the mediterranean-climate region of Western Australia. Number of *asterisks* and number of *arrows* indicate relative importance of factors in the mediterranean region. (Adapted from Morton 1990, amendments for mediterranean-climate areas are shown in *boldface type*)

19.9 Conclusions

Australian ecosystems in general, but mediterranean ecosystems in particular, have suffered an unprecedented loss of animal biodiversity in the period following European invasion. Mammals have suffered most from this loss of biodiversity, most noticeably in the critical weight range (0.1–5.5 kg). Detailed information for rodents, small dasyurid marsupials and bandicoots indicates that only 44% of the original fauna remain in the mediterranean region that abuts the southern boundary of the arid zone. However, in the less developed tropical region that abuts the northern boundary of the arid zone there has been much less loss of species. While human involvement in this loss must be regarded as the ultimate cause, a number of different proximal causes have been implicated. Not least of these is the European rabbit (*Oryctolagus cuniculus*), a Mediterranean immigrant that was introduced to Geelong in Victoria from Spanish stock in the late 1850s, and then spread across the continent. Subsequent introductions to the mediterranean regions in Western Australia were made at Geraldton in the 1890s and the invading wave from the east merged with these by 1910. The introduction of another European immigrant, the red fox (*Vulpes vulpes*) as a predator on the rabbit, has also had a major impact on native mammal species. The hard hooves of human's grazing animals, primarily sheep in mediterranean southern Australia, have played a major role in habitat destruction through grazing and trampling of shrubland habitats. The final, and probably most major impact is massive clearing of native habitats for the cultivation of the crops of Mediterranean origin, with wheat being the most extensive. The habitat fragmentation and loss have been well documented in southern Australia as has the attendant loss in biodiversity.

Study of temporal and spatial scales to produce time lines for invasions and local extinction of species illustrates that the European red fox is a major factor in loss of the numbat from much of the mediterranean climate zone in Australia. Study of time and spatial scales also illustrates that other factors are operating, with fragmentation and habitat loss related to cropping and changed fire regimes as the next most important. No single factor is solely responsible for the loss of biodiversity. Different factors have their greatest impact on different groups. The only factor common to all groups is the almost universally detrimental role played by humans.

References

Burbidge AA, McKenzie NL (1989) Patterns in the modern decline of Western Australia's vertebrate fauna: causes and conservation implications. Biol Conserv 50:143–198

Burbidge AA, Johnson KA, Fuller PJ, Southgate RI (1988) Aboriginal knowledge of the mammals of the central deserts of Australia. Aust Wildl Res 15:9–39

Dunstan CE, Fox BJ (1996) The effects of fragmentation and disturbance of rainforests on ground-dwelling small mammals on the Robertson Plateau, New South Wales, Australia. J Biogeogr 23:187–201

Friend JA (1987) Local decline, extinction and recovery: relevance to mammal populations in vegetation remnants. In: Saunders DA, Arnold GW, Burbidge AA, Hopkins AJM (eds) Nature conservation: the role of remnants of native vegetation. Surrey Beatty, Sydney

Friend JA (1990) The numbat *Myrmecobius fasciatus* (Myrmecobiidae): history of decline and potential for recovery. Proc Ecol Soc Aust 16:369–377

Hobbs RJ, Saunders DA (eds) (1992) Reintegrating fragmented landscapes: towards sustainable production and nature conservation. Springer, Berlin Heidelberg New York

Jarman P (1986) The red fox – an exotic, large predator. In: Kitching RL (ed) The ecology of exotic animals and plants. Wiley, Brisbane, pp 43–61

Kinnear JE, Onus ML, Bromilow RN (1988) Fox control and rock wallaby population dynamics. Aust Wildl Res 15:435–447

Merriam G, Saunders DA (1993) Corridors in restoration of fragmented landscapes. In: Saunders DA, Hobbs RJ, Ehrlich PR (eds) Nature conservation 3: reconstruction of fragmented ecosystems. Surrey Beatty, Chipping Norton, New South Wales

Morton SR (1990) The impact of European settlement on the vertebrate animals of arid Australia: a conceptual model. Proc Ecol Soc Aust 16:201–213

Morton SR, Baynes A (1985) Small mammal assemblages in arid Australia: a reappraisal. Aust Mammal 8:159–169

Myers K (1970) The rabbit in Australia. Proc Adv Study Inst Dynamics Number Popul (Oosterbeek, 1970) Purdoc, Wageningen, pp 478–506

Myers K (1986) Introduced vertebrates in Australia, with emphasis on mammals. In: Groves RH, Burdon JJ (eds) The ecology of biological invasions: An Australian perspective. Australian Academy of Science, Canberra, pp 120–136

Redhead TD, Singleton GR, Myers K, Coman BJ (1991) Mammals introduced to southern Australia. In: Groves RH, di Castri F (eds) Biogeography of mediterranean invasions. Cambridge University Press, Cambridge

Saunders DA, Curry PJ (1990) The impact of agricultural and pastoral industries on birds in the southern half of Western Australia: past, present and future. Proc Ecol Soc Aust 16:303–321

Tunbridge D (1991) The story of the Flinders Ranges mammals. Kangaroo Press, Kenthurst, New South Wales

20 Bird Diversity in a Changing Landscape (Tuscany, Italy)

A. Farina

20.1 The Landscapes of Tuscany

The change of land use occurred in the last five decades in most of the Italian peninsula as well as in other parts of southern Europe have produced severe effects on a broad spectrum of plants and animals (Farina and Naveh 1993). The northern regions of Italy, massively urbanized and industrialized, are particularly exposed to these modifications.

The aim of this contribution is to describe birds diversity in changing valuable rural landscapes of the Tuscany region, pointing out the importance of birds as indicators of the environment "quality" (Furness et al. 1993) and of landscape changes across a broad range of spatial and temporal scales (Noss 1990; Jarvis 1993), focusing on the causes of farmland vanishing, describing the consequences of this decline on many species of birds and providing managing guide lines to be extended to other Mediterranean-type threatened landscapes. An overview of small and large-scale studies on birds carried out recently in Tuscany and pertaining mainly lowland and hilly farmlands and upland ecotones is discussed.

20.1.1 Geomorphology, Climate, and Vegetation

The Tuscany Region, 22 998 km^2 in size, is characterized by a complicate geomorphology ranging from islands of the Tuscan archipelago to the Apennines peaks (Fig. 20.1). About 22% of the area is located from sea level to 100 m, 54% from 100–500 m, 18% from 500–900 and 6% above 900 m. The climate can be divided into eight classes, according to the Thornthwaite's classification ranging from perhumid to semiarid (Bigi and Rustici 1984). The perhumid type is distributed along the Apennines chain and surrounds the summit of the isolate Monte Amiata as a small ring. The semiarid type occurs in few localities long the coast (e.g. the Ombrone estuary) and in the northern coast of Elba island, close to Portoferraio harbor.

Ecological Studies, Vol. 136
Rundel et al. (eds.) Landscape Degradation and Biodiversity in Mediterranean-Type Ecosystems

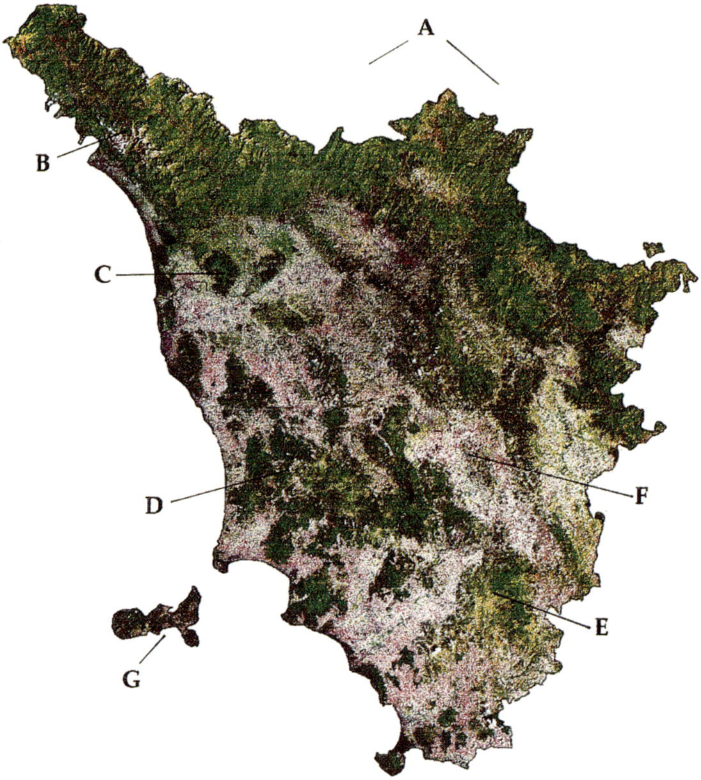

Fig. 20.1. Landsat image of Tuscany. Woodland and grassland from dark green to pale green; cultivation pale brown to pink; urban settlements blue to violet: A- Tosco-emiliano Apennines; B- Apuane Alps; C- Monte Pisano; D- Metallifere Hills; E- Monte Amiata, F- Val di Chiana; G- Elba Island

The vegetation of Tuscany ranges from maquis shrublands dominated by *Quercus ilex*, *Phillirea* sp., *Arbutus unedo*, and *Myrtus communis*, to mountainous beech forest (*Fagus sylvatica*) and moorlands (*Vaccinium* spp.) through many intermediate mesic vegetation types (e.g. *Ostrya carpinifolia*, *Quercus pubescens*, *Q. cerris*, *Fraxinus ornus*, *Carpinus betulus*, and *Acer campestre*).

20.1.2 The Landscape

According to Pignatti (1994), two main types of landscapes can be recognized in Tuscany: the northern Apennine landscape along the Tosco-Emiliano Apennines and the Tuscany landscape (*sensu strictu*) in the central and coastal parts. But Rossi et al. (1994) at a finer scale include morphological criteria and describe nine main landscapes: Apennines, Alpi Apuane, hills and small mountains, Pliocene hills, tuffaceous plains, intermontane basins, flood plains, coastal plains, islands and promontories. From the land use inventory carried out in 1991, woodlands are the most common land cover (47%) concentrated mostly in the northern and eastern parts, then cultivations (34.9%), pastures (10%) and urbanized areas (5.3%) (Rossi et al. 1994). As in other Mediterraean countries, Tuscany landscape can be considered a cultural landscape in which nature and humans have been reciprocally conditioned by a feed-back lasting several millennia (see also the World Heritage Center of UNESCO).

The variety of the land mosaic is large. High mountains (Tosco-emiliano Apennines), with peaks on average above 1500 m, border the east and north side of the region. A coastal range (Metallifere Hills) divides the coastal *Maremma* from the intermontane basins (Arno and Chiana Valleys). Isolated hills (Alpi Apuane and Monte Pisano in the north and Monte Amiata in the south) increase the complexity of this unique landscape. Pliocene lacustrine deposits, deeply eroded and shaped into rounded hills, occupy the central part of the Tuscany from Pisa to Arezzo provinces. Rocky coasts, especially on the Tuscan archipelago, alternate with sandy dune systems of mainland. Slope terracing and coastal wetland reclamation are the result of a long history of human-nature feedback that has created the cultural landscape of Tuscany.

This Tuscan landscape shows a broad variety of distinctive types according to the prevailing land use: coastal pastures (*maremme*), coastal farmlands, hilly and sub-montane farmland (*coltura mista*), chestnut orchards and olive orchards. Often these different cultivations are interdispersed in a fine-grained mosaic. The complicated mosaic of fields, pastures, plantations and woodlands favors the development of ecotones enhancing the biodiversity at different spatial and temporal scale.

20.1.3 The Recent Landscape Changes and the Causes of Land Degradation

The Tuscan landscape, immortalized by renaissance painters, had conserved most of its unique characters until the beginning of this century. It is after the Second World War that many patterns and land uses of this landscape changed markedly as a result of a sharp decline in agriculture and livestock rearing. Industrial development of the Arno valley and the growing demand for space for houses and development along the coast have permanently modified large areas of Tuscany. The increase development of a road network has produced further habitat fragmentation. Introducing irrigated crops has modified the diversity and quality of agriculture products. In addition, extensive land abandonment along the mountain ranges and the expansion of pine plantations have encouraged widespread natural and human-induced fires.

The land abandonment in the mountains and the intensification of agriculture in hills and lowlands have produced the reduction of crop diversity and an increase of field size reducing respectively fence rows and ecotonal areas. The highly integrated system of the traditional agriculture has been replaced gradually by a system in which agriculture practices had patterns and processes separate from the livestock farming and forestry practices (Farina 1991).

As in many other Mediterranean countries (Baldock 1994), seasonal grazing has been abandoned in mountains and low wetlands and replaced by resident livestock rearing. Broad differences between Tuscan administrative provinces arise when quality and quantity of agricultural products and livestock abundance are compared. For instance in the north (Massa Carrara , Lucca and Pistoia provinces) the size of farms is small (70 to 80% of all farms are less then 10 ha in size), and in these provinces agriculture decline has been greater. In Grosseto and Siena provinces where farms are larger, an increase of the agriculture activity has taken place (Dipartimento Statistica, Elaborazione Dati e Documentazione 1993).

20.2 Ongoing Studies

Data used in this chapter come from two main sources: local and large scale studies.

20.2.1 Small-Scale Studies

Data on changes in bird diversity in Tuscany has come from different researchers, and has been collected in a series of short term censuses in different periods and localities: in Maremma Regional Park (Farina 1979), in sub-montane *coltura mista* (Tellini 1988), in Sub-mountain grasslands of Lunigiana (Farina et al. 1989), in chestnut orchards of northern Tuscany (Puglisi 1992), in upland ecotones (Farina 1994), and in *colturamista* of Lunigiana (Illner et al. 1992; Farina 1995). Other investigations have been carried out recently in beech forest of the Northern Apennines, in tall maquis shrublands of the central Tuscan hills, and in mixed woodlands of Monte Amiata in the southern Tuscany (Fig. 20.2) (Farina & Brogi, 1995).

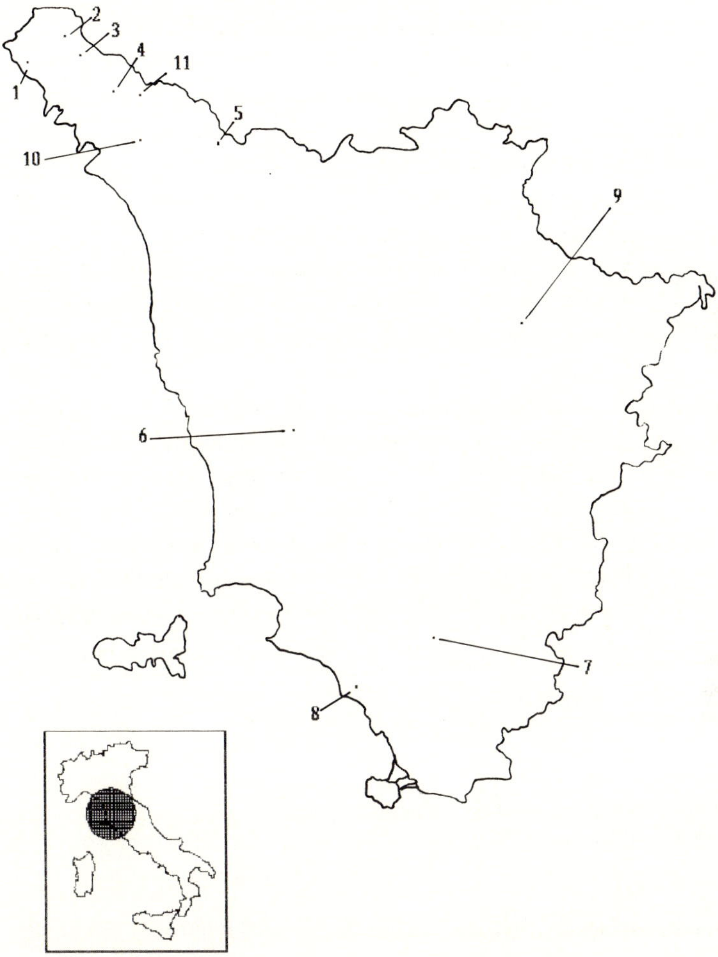

Fig. 20.2. Localization of the small scale study areas, as described in Farina et al. (1989), Farina (1979, 1995), Farina & Brogi (1995), Tellini (1988), Puglisi (1992), and Illner et al. (1994): 1. Zeri sub-mountain grasslands; 2. Logarghena sub-mountain grasslands; 3. Comano sub-mountain grasslands; 4. *Coltura mista*; 5. M. Tondo upland ecotone; 6. Monterufoli upland ecotone; 7. Monte Amiata ecotone; 8. Maremma Regional Park; 9. Val di Chiana lowland; 10. Alpi Apuane chestnut orchards; 11. *Coltura mista*

20.2.2 Large-Scale Studies

Bird distribution and abundance in Tuscany are becoming well studied through recent investigations carried out between 1982 and 1986: BBS (breeding bird survey) 1984–85 (Farina and Meschini 1987) (Fig. 20.3) and the Tuscany Atlas Project of breeding birds (unpublished data). From 1982 to 1986, an atlas project was organized in Tuscany employing amateur bird

Fig. 20.3. Distribution of roads used to count birds in 1984–1985 period

Table 20.1. Relative importance (%) of primary land uses along the 1984–85 Tuscany Breeding Bird Survey roads. Data from Farina and Meschini (1987)

Land use	Relative importance (%)
Mountain grassland	0.65
Open farmland	28.80
Wooded farmland (*coltura mista*)	28.62
Woodland and shrubland	25.39
Forest	16.32
Wetland	0.11

watchers to help map bird distributions over a 10 × 10 km grid covering 276 units (maps of 1:25 000 scale). Counts on 60 species selected among the most common birds were carried out during the breeding seasons of 1984 and 1985 along 177 roads, each of 40 km in length. A total of 5607 point counts were made, each 3 minutes in duration, at sites 1 km apart over a range of different types of landscape (Table 20.1).

20.3 The Importance of the Tuscany Region for Birds

In Tuscany, birds are important components of the landscape throughout the year. According to the season, the importance of habitats changes following the bird movements. A complex mosaic of habitats due to natural (morphological, climatic, and vegetational) as well as human constraints allows such a dynamic situation to exist.

The geographic position of Tuscany in the center of the Italian peninsula and favourable microclimates offer a high variety of habitats for many species of terrestrial and aquatic birds, encouraged by a traditional low intensity agriculture. These habitats and landscapes, ranging from coastal and inland wetlands, lagoons, sea cliffs, maquis, grassy hills, mountain forests and summit grasslands, sustain permanent resident populations of birds and attract migratory, erratic, and transient birds (Farina 1988). *Maremma* low-lands, hilly *coltura mista* and mountain farmlands have a primary role as wintering and stop-over foraging sites for many species of water and terrestrial birds as well recognized by ornithologists of the last century as Paolo Savi (Ornitologia Toscana 1827–31).

Particularly important are the distributions of larks and shrikes, two groups of birds in decline in many regions of Europe where agriculture has been intensified (Tucker and Heath 1994; Gates et al. 1993). From Atlas data and the BBS data, these birds appear to remain widespread in many parts of the Tuscany (Table 20.2, Figs. 20.4 and 20.5).

20.4 The Importance of Coastal Farmlands and Pine Plantations

The reclamation of most of the *maremma* wetlands has modified the coastal area, reducing and isolating lagoons and marshes. Nevertheless, in some areas, as at the Ombrone estuary, a peculiar landscape can be observed. Reclaimed areas are only partially plowed and permanent pastures

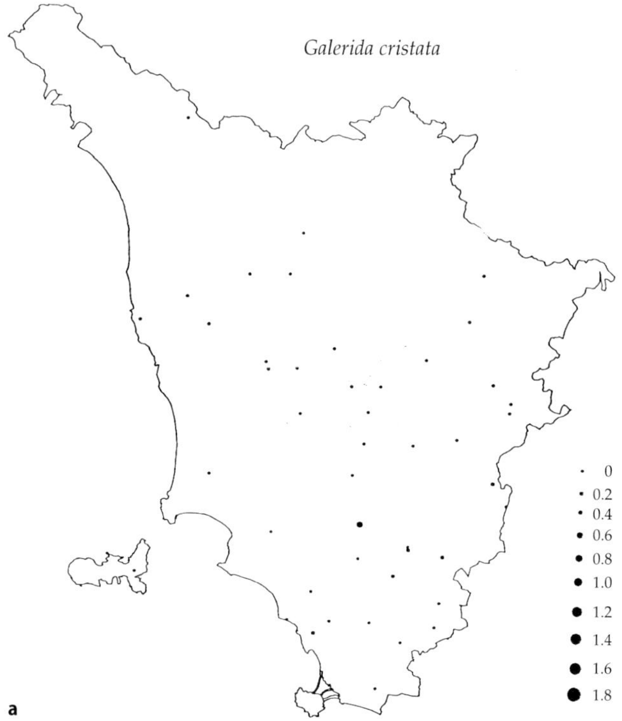

a

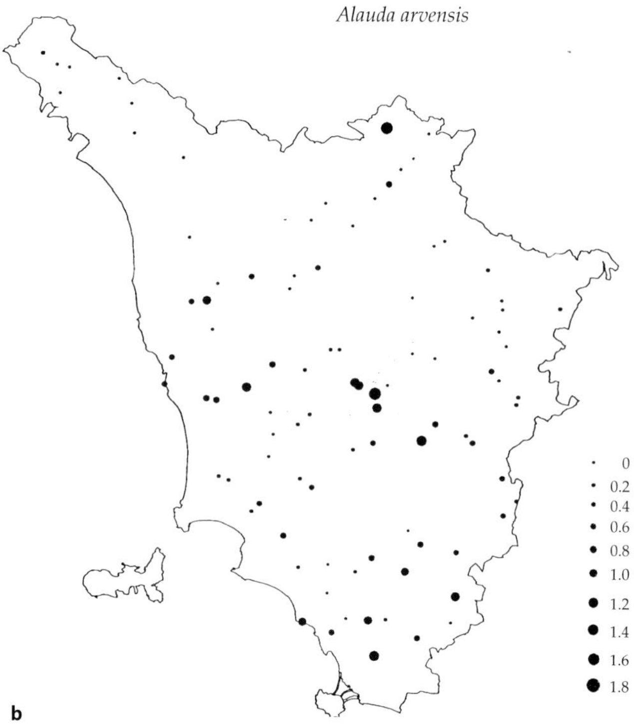

b

Fig. 20.4a–c. Distribution and mean abundance of larks (*Alauda arvensis, Lullula arborea* and *Galerida cristata)* in Tuscany. Data from Farina and Meschini (1987)

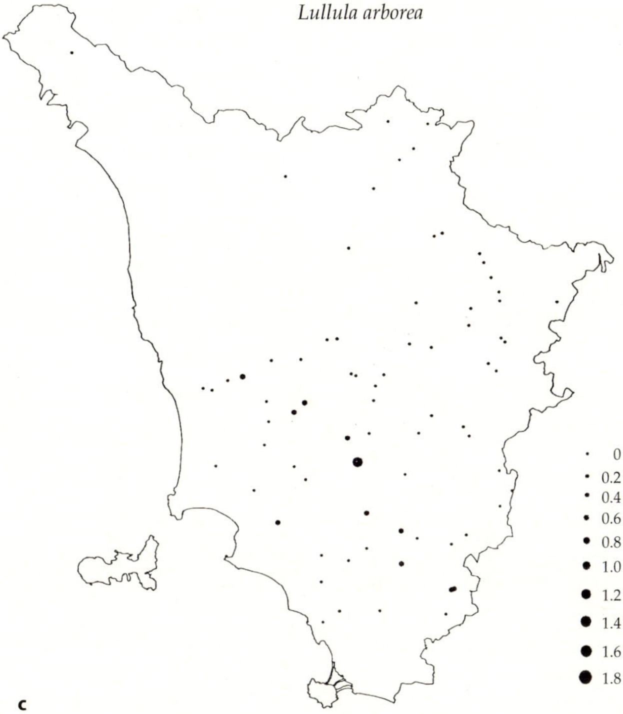

Fig. 20.4a–c. (*Continued*)

Table 20.2. Number of 10 × 10 km sections (n) and percentage (%) in which larks and shrikes were recorded in Tuscany (data from Atlas 1982–86)

	n	%
Larks		
– *Melanocorypha calandra*,	3	1.08
– *Calandrella brachydactyla*,	36	13.04
– *Galerida cristata*	112	40.57
– *Lullula arborea*,	146	52.89
– *Alauda arvensis*,	215	77.89
Shrikes		
– *Lanius collurio*	252	91.30
– *Lanius minor*	38	13.76
– *Lanius senator*	109	39.49

with scattered pines (*Pinus pinea*) persist. Dense pine plantations (*Pinus pinaster* and *P. pinea*) occur as a buffer between coastal dunes and salt marshes and the permanent pastures and arable fields. The surrounding hills are covered by olive orchards, tall maquis, and mature woodlands. In this landscape the role of grazing by herds of *maremmana* cattle and horses is very important to maintain a high ecodiversity (sensu Naveh 1995).

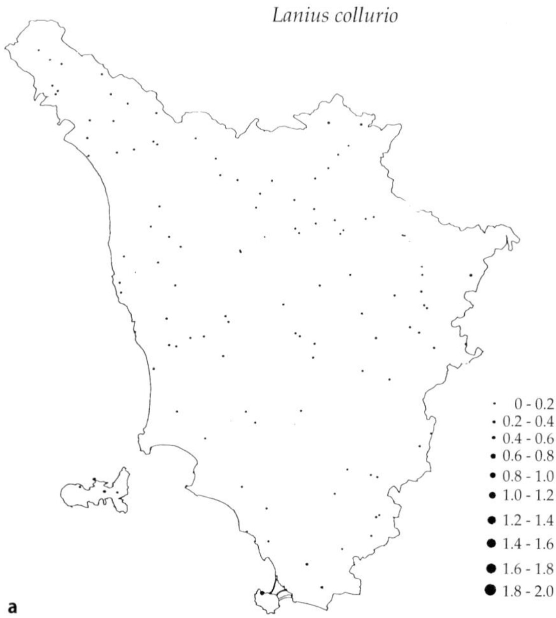

a

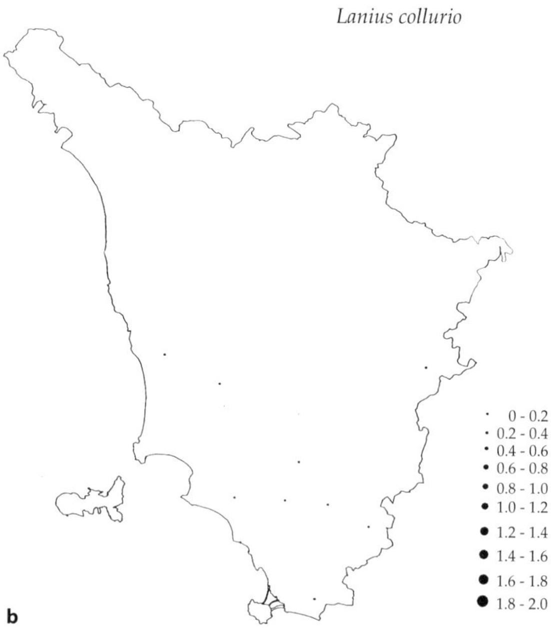

b

Fig. 20.5a–c. Distribution and mean abundance shrikes (*Lanius collurio*, *L. minor* and *L. senator*) in Tuscany. Data from Farina and Meschini (1987)

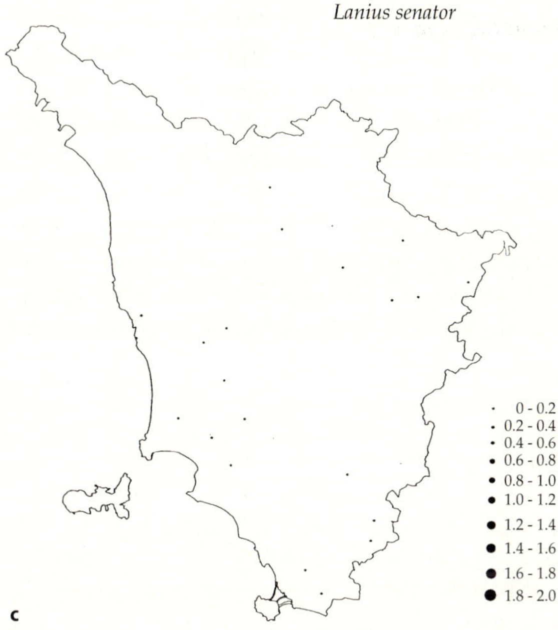

Fig. 20.5a–c. (*Continued*)

In farmland and in pine plantations of the Maremma Regional Park, birds were censused during the breeding period by Farina (1979). This landscape is particularly rich in species (71 were recorded from May to June 79) and some (*Merops apiaster, Burhinus oedicnemus* and *Coracias garrulus*) have been recently included in the Tuscany red list (Sposimo and Tellini 1995). A group of declining species including *Alauda arvensis, Galerida cristata, Anthus campestris* and *Lanius minor* are common breeders. Mass tourism, changing agriculture practices, and livestock reduction are the more direct threats that endanger this valuable landscape that requires traditional agricultural practices to maintain its emergent characteristics.

20.5 The Importance of Hilly and Submontane Rural Areas

20.5.1 *Coltura Mista* Farmland

The hilly rural landscape of Tuscany (*coltura mista*) is a widespread agro-ecosystem along the Apennines. This system is composed of small, plowed and/or grazed fields bordered by wood or stone fences, tree borders and woodlots. In well exposed slopes vineyard and olive orchards are the dominant cultivated crops. The disappearance of this agricultural system has been described for the Magra Valley in northern Tuscany (Farina 1991) and in the Solano Basin of north-central Tuscany (Vos and Stortelder 1992). The *coltura mista* plays a significant role in the conservation of bird diversity in different periods of the year (Farina 1984a, 1988), but it is in spring that bird diversity is highest (Table 20.3). Bird diversity and abundance are particularly high in the traditionally cultivated areas close to the villages (Farina 1995) in which the seasonality of available resources plays an important role in attracting wintering and trans-Saharan migratory birds (Farina 1988).

20.5.2 Chestnut Orchards

In Tuscany, chestnut (*Castanea sativa*) orchards are clearly in decline everywhere. Although planted by farmers in most of Tuscany, chestnut orchards represent important habitats for many species of birds, especially tits and woodpeckers, because these Plantations are structurally very close to mature natural forests (Puglisi 1992) (Table 20.4). Chestnut orchards are

Table 20.3. Richness (S) and bird diversity (H′) in an area of *coltura mista* in northern Tuscany in different seasons. Data from Farina (1995)

	S	H′
February	45	2.77
April	65	3.17
July	59	3.07
October	55	2.62

Table 20.4. Richness (S), diversity (H′), evenness (J′) and mean abundance (a) of birds in four chestnut orchards in Alpi Apuane chain (Northern Tuscany). Data from Puglisi (1992)

	S	H′	J′	a
Roggio	36	2.85	0.79	33.5
Campolemisi	37	2.82	0.78	26.4
Antona	36	2.77	0.77	25.4
Vinca	34	2.64	0.75	32.8
Total	49	2.90	0.74	29.5

especially widespread in low mountain farmlands and generally are locally of modest size. However, inserted in the landscape they acquire an important ecological role. Some of the causes of the decline of chestnut orchards are on the loss of economic value of chestnut fruits, pathogens that accelerate the death of mature plants, secondary succession which occurs when the orchard are abandoned, and finally the high fire risk due to the persistence of dead leaves in the understory of the orchards. Most of the birds that breed in this plantation moves into the surrounding agricultural mosaic at the end of the breeding season. As pointed out by Puglisi (1992), the size of the orchards can affect diversity and abundance of these species.

20.6 Evidence of Bird Decline in Lowland Farmlands

The degradation of natural and cultural landscape and the vanishing of many keystone habitats within the Tuscany Region are producing a broad decline in many species of birds. As recently stressed by Sposimo and Tellini (1995), this decline is particularly evident for birds of open spaces caused by the reduction of open fields cultivated in a traditional way and by the abandonment of pastures. This is a trend also common to other European countries (Gates et al. 1994).

In Val di Chiana in eastern Tuscany, a general decrease of abundance of many species of resident and migratory birds was demonstrated by Tellini (1988) who compared a traditional low intensity agricultural system with an intensive modern monocultural system (Tables 20.5 and 20.6). Of the 22 species common to the two systems, 19 species were decreasing in the latter system and only three (*Motacilla flava, Alauda arvensis,* and *Emberiza calandra*) were increasing in the modern agriculture system. Tellini argued that most of this trend depends on a variety of converging factors such as

Table 20.5. Ecological characteristics of bird communities in traditional versus modern agrosystems along 14 transects in two different agricultural system of Val di Chiana. Data from Tellini (1988)

Bird community characteristic	Traditional	Modern	Ratio (%)
mean richness	21.43	8.71	47%
mean diversity	2.33	1.27	45%
mean evenness	0.76	0.63	16%
mean biomass (gr/10 ha)	955.53	575.75	38.4%
mean consuming biomass (gr/10 ha)	279.20	132.29	52.1%

Table 20.6. Bird species that decrease from traditional to modern agriculture along 14 transects in Val di Chiana. Data from Tellini (1988)

Parus major	*Sylvia melanocephala*
Troglodytes troglodytes	*Serinus serinus*
Aegithalos caudatus	*Sylvia atricapilla*
Carduelis chloris	*Erithacus rubecula*
Galerida cristata	*Passer domesticus italiae*
Carduelis carduelis	*Fringilla coelebs*
Phylloscopus collybita	*Passer montanus*
Vanellus vanellus	*Motacilla alba*
Prunella modularis	*Saxicola torquata*
Pica pica	

a loss of forest edges, lower crop diversity, abandonment of livestock production, and a reduction of agricultural residue in the fields. The rarefaction of the bird density in modern versus traditional farming systems is especially evident in winter when food shortage is more significant.

20.7 The Importance of Forest-Grassland Ecotones

The Tuscany Region has high forest coverage (47%) when compared with other Italian regions. The forest coverage is increasing (ranging from 17–25% annually depending on the localities) to the detriment of open areas (Farina 1994b) and has been occurring very dramatically in submontane clearings. This process has negative effects on bird populations, especially if forested areas are adjacent to open cultivated or grazed ranges. The forest expansion reduces the ecotonal areas, as shown in three sites recently studied by Farina and Brogi (1995) across Tuscany (Table 20.7). In these sites, a rapid forest expansion, the reduction of clearings, the decline of agro-pastoral system, and the plantation of pines are expected to produce a sharp reduction of bird abundance and diversity in the coming decades.

Recent investigations carried out at Monte Cavalbianco (Farina 1994a) and other selected mountain ecotones (Farina et al. 1989) have confirmed the importance of forest-grassland ecotones. In the Monte Cavalbianco site, the edge effect was significant and easily detected. Bird density decreased from 180 birds $10\,ha^{-1}$ at the forest edge to 25 birds $10\,ha^{-1}$ at 450–600 m from the edge.

Mountain ecotones are important during the breeding season because are rare habitats and some species as *Anthus trivialis*, *Lanius collurio*,

Table 20.7. Number of wintering (W), spring transient (Sp), Summer (Su) breeding and Autumn (Au) transient birds, total species richness (Stot), species diversity (H′) and evenness (J′) at three different sites of Tuscany Region (from Farina and Brogi 1995)

	W	Sp	Su	Au	Stot	H′	J′
M. Tondo	26	49	45	39	64	2.71	0.65
Madonna Q.	41	46	53	46	73	3.00	0.73
Monterufoli	37	42	42	41	57	2.91	0.72

Table 20.8. Richness (S) and diversity (H′) in five upland ecotones of north Tuscany (Farina et al. 1989)

	S	H′
Comano	47	3.34
Logarghena	54	4.36
Zeri	44	3.32
Varese 1	45	3.27
Varese 2	37	3.47

Oenanthe oenanthe, *Lullula arborea*, and *Sylvia communis* find suitable breeding habitats. In autumn and in spring many species of birds such as thrushes and pipits use these areas as foraging sites (Farina et al. 1989) (Table 20.8).

20.8 Guidelines to Conserve Mediterranean Landscapes and their Fauna

The conservation of the highest variety of habitats is one of the most important strategies to help preserve declining bird populations in the Mediterranean region. How to achieve this goal is not an easy task. First of all, most of the lands are private and thus it is often difficult to convince owners to modify their land-use practices which are strongly driven by economic considerations.

Coastal lagoons, marshes, rivers, mature forests, open forests and shrublands have to be considered endangered valuable landscapes in Tuscany and priorities in conservation should be assigned.

The ongoing landscape changes are producing a simplification of the land mosaic and restricting the diversity in two main habitats: open intensive agriculture and dense woodland. To counter this trend, there is a need for a land use policy to favor the conservation of a fine scale of landscape heterogeneity and protect natural landscape remnants. Scientific discussions on the ideal size and shape of natural reserves has becoming common in the US and in Europe, with studies which have show the importance of small remnant plots (Loman and Von Schantz 1991).

Agro-pastoral systems should be supported. Grazing has positive effects in maintaining rare habitats such as mountain grasslands. In Mediterranean uplands grazing is critical for the survival of many species of birds such as thrushes, pipits and larks (Farina, unpublished data, Sposimo and Tellini 1995).

Recently, increased fire frequency is creating concern around the Mediterranean Basin, but far from human permanent settlements wild fires could represent a benefit for some birds. Fires create temporary habitats as short and scattered shrublands and some species of buntings and warblers select these burned areas as breeding places.

To maintain the complexity of the landscape mosaic, prescribing burns should be a good but not sufficient strategy. Also important is the traditional agriculture strongly need to maintain the *coltura mista* of the Mediterranean uplands. In lowlands, the decrease use of irrigation and more consideration for edges, woodlots and marshes should be important resource management issues. These recommendations to preserve diversity in the Mediterranean countryside must be coupled to ad hoc local plans to integrate local needs and global perspectives for the conservation of biodiversity.

Recommendations for land use protocols such as described in the red books of threatened landscapes (Naveh and Lieberman 1984, CESP-IUCN 1993) should be incorporated and locally tested more widely as a tool in land use management. An existing on the western Crete and the management-threatened landscape (Grove et al. 1993, Chapter 7 in this volume) is a good example of this procedure. The approach used in this case study has allowed resource managers to describe and focus landscape problems so as to center attention on the future possible scenarios of economic and environmental conflicts. Such approaches are now being employed to evaluate the effect of the decline of pine plantations along the coast of Tuscany in relation to issues of expanding tourism, irrigated agriculture, urban expansion, grazing reduction and growing concerns over problems of biodiversity and habitat conservation.

Multidisciplinary ecological knowledge, an efficient educational program, and public awareness seem to be the key factors to be considered for more effective conservation action in this cultural landscape.

20.9 Summary

In the last five decades changes of land use have modified structure, spatial patterns and functioning of most of the Mediterranean landscapes. The intensification of agriculture and the urbanization in lowlands, and the land abandonment on hilly and mountain ranges have restricted available habitats for many species of animals reducing the biological diversity.

The distribution of birds has been studied in Tuscany (North Italy), a region facing important landscape modifications, where bird diversity is particularly high. This region has a variety of geomorphic, microclimatic, biological and cultural systems such as coastal wetlands (*maremma*), maquis, extensive croplands on both lowlands and hill slopes, mountain forest and terraced fields, summit grasslands and pasturelands. Investigations carried out in different sites and in different years have allowed to evaluate the effects of landscape modification on bird populations.

Coastal farmlands and pine plantations, *coltura mista*, chestnut orchards and upland ecotones seem very important habitats for many species of resident and migratory birds. But mass tourism, coastal urbanization and industrial development combined with the abandonment of upland farmland have severely affected distribution and abundance of many birds. Those species living in open cultivated spaces seem particularly menaced.

In fertile lowlands, the intensification of agriculture is increasing habitat fragmentation, reducing the possibility of settlement even to many common birds. In the past the traditional cultivation practices allowed the presence of a great number of birds especially in winter. Today the industrial and intensive agriculture is reducing the sustainable capacity of these landscapes.

The maintenance of traditional agriculture and pastoralism, coupled with prescribing burning and low intensity farming, appear to contribute to strategies to conserve habitats and bird diversity in Tuscany as well as in other Mediterranean Basin countries.

References

Baldock OD (1994) The nature of farming: low intensity farming systems in nine Europe countries. IEEP, London

Bigi L, Rustici L (1984) Regime idrico dei suoli e tipi climatici in Toscana. Regione Toscana, Dipartimento Agricoltura e Foreste, Firenze

Commission on Environmental Strategy and Planning of IUCN (1993) Red Books for Threatened Landscapes. Proceedings of a Symposium and Workshp. Montecatini, Italy

Dipartimento Statistic, Elaborazione Dati e Dcumentazione (1993) 4° censimento generale dell'agricoltura 21 ottobre 1990. Quadro generale e prime valutazioni. Toscana, Dati definitivi. Regione Toscana, Giunta Regionale, IRPET. Informazione Statistica 9, Firenze
Farina A (1979) Breeding census of an Italian Mediterranean habitat, the Parco Naturale della Maremma. VII Cong Bird Census Work, Gottingen, pp 129–135
Farina A (1985) Habitat use and structure of a bird community in a farmland. In: Moroni, A, Anelli A, Ravera O (eds) Ecologia. Atti del secondo congresso nazionale della Società Italiana di Ecologia, 25–28 June 1984, Padova, pp 679–696
Farina A (1988) Bird community structure and dynamics during spring migration in selected habitats of northern Italy. Bull Zool 55:327–336
Farina A (1991) Recent changes of the mosaic pattern in a montane landscape (North Italy) and consequences on vertebrate fauna. Option Mediterreaneenes – Serie sem, 15:121–134
Farina A (1994a) Birds in a mountain landscape. In: Dover J (ed) Fragmentation in agricultural landscapes. Proc 3rd ann conf of IALE (UK). Myerscough College, Bilsborrow, Preston. pp 153–160
Farina, A (1994b) Effect of recent changes of the summit landscape on vertebrate fauna of the northern Apennines. Fitosociologia 26:171–175
Farina A (1995) Distribution and dynamics of birds in a rural sub-Mediterranean landscape. Lands Urban Plan 31:269–280
Farina A, Meschini E (1987) The Tuscany breeding bird survey and the use of bird habitat description. Proc IX Int Conf Bird Census and Atlas Work, Dijon, 1985. Acta Ecol, Oecol Gen 8:145–155
Farina A, Giacomazzi F, Andreotti S (1989) Struttura e dinamica delle comunità di uccelli nelle praterie montane secondarie dell'Appennino settentrionale. Economia Montana. Linea Ecologica 22:19–24
Farina A, Naveh Z, eds (1993) Landscape approach to regional planning: The Future of the Mediterranean landscapes. Lands Urb Plan 24
Farina A, Brogi L (1995) Struttura e dinamica delle comunità di uccelli in tre aree del demanio forestale della Regione Toscana. Regione Toscana
Farina A, Naveh Z (1997) The green book of the coastal Tuscany pine plantations. In prep
Furness RW, Greenwood JJD, Jarvis PJ (1993) Can birds be used to monitor the environments? In: Birds as monitors of environmntal change. Chapman & Hall, London, pp 1–41
Gates S, Gibbons DW, Lack PC, Fuller RJ (1993) Declining farmland bird species: modelling geographical patterns of abundance in Britain. In: Edwards PJ, May RM, Webb NR (eds) Large-scale ecology and conservation biology. Blackwell Science pp 153–178
Groves AT, Ispikoudis J, Kazaklis A, Moody JA, Papanastasis V, Rackam O (1993) Threatened Mediterranean landscapes: West Crete. Final Report
Illner H, Judas M, Juneman K, Oelke H (1994) Italian summer bird population (Lunigiana, NW Tuscany, Apennines). Boll Mus Sci Nat Lunigiana 8:13–94
Jarvis PJ (1994) Environmental changes. In: Birds as monitors of environmental change. Chapman & Hall, London, pp 42–85
Loman J, Von Schantz T (1991) Birds in a farmland – more species in small than in large habitat island. Conserv Biol 5:176–188
Naveh Z (1995) From biodiversity to ecodiversity: new tools for holistic landscape conservation. Int J Ecol Environ Sci 21 (in press)
Naveh Z, Lieberman AS (1984) Landscape ecology; theory and application. Springer-Verlag, Berlin
Noss RF (1990) Indicators for monitoring biodiversity: a hierarchical approach. Conserv Biol 4:355–364
Pignatti S (1994) Ecologia del paesaggio. Utet, Torino

Puglisi L (1992) Gli uccelli dei castagneti apuani: struttura e dinamica dei popolamenti. Dissertation

Rossi R, Merendi GA, Vinci A (1994) I sistemi di paesaggio della Tocana. Regione Toscana, Giunta Regionale, Firenze

Sposimo P, Tellini G (1995) L'avifauna in Toscana. Lista rossa degli uccelli nidificanti. Regione Toscana, Giunta Regionale, Firenze

Tellini G (1988) L'influenza di differenti ordinamenti colturali sulle comunità ornitiche presenti in una zona agricola della Toscana (Valdichiana). Unpubl Dissertation

Tucker GM, Heath MF (1994) Birds in Europe: their conservation status. Birdlife Conservation Series no. 3. Birdlife International, Cambridge

Vos W, Stortelder A (1992) Vanishing Tuscan landscapes: landscape ecology of a submediterranean-montane area (Solano Basin, Tuscany, Italy). Pudoc, Wageningen

21 Patterns of Mammalian Biodiversity, Urbanization, and Land Use in Southern California

R.D. Quinn

21.1 Introduction

This chapter reviews causes of changing patterns of mammalian biodiversity in southern California, and discusses measures that could be taken to preserve diversity while accommodating continuing urbanization.

Of the five areas of the world with Mediterranean climates, California is well suited for an evaluation of anthropogenic landscape degradation because its human population has been growing exponentially for almost a century and a half. As immigrants began arriving in large numbers in the mid-nineteenth century, a relatively stable landscape lightly populated by indigenous peoples and Mexican ranchers was quickly swept away. It was replaced by commercial agriculture and industries extracting natural resources, accompanied by continuously expanding urban areas. The pattern of this growth has been, and continues to be, one that engenders landscape degradation – decentralized, low density, and uncoordinated. This pattern has long been described as "urban sprawl." Changes in land use connected with human economic activities that developed gradually over millennia in the Mediterranean Basin were compressed into a few decades in California. The state offers an extreme example of the consequences of continuing decentralized, post-industrial, urban growth.

The mammals of California can be rather easily used as accurate indicators of ecosystem change. They are well studied over most of the period when changes have been most rapid, and their ranges are relatively well known, as are their associations with particular plant communities and physical features. The first comprehensive review of the mammalian fauna of California appeared almost a century ago (Stephens 1906), closely followed by careful surveys focused on particular parts of the state by Joseph Grinnell and colleagues (Grinnell 1908, Grinnell and Swarth 1913). Since those early efforts numerous investigators have examined various segments of the mammalian fauna, using taxonomic or geographic boundaries (Grinnell 1922, Grinnell and Storer 1924, Grinnell 1933,

Ecological Studies, Vol. 136
Rundel et al. (eds.) Landscape Degradation and Biodiversity in Mediterranean-Type Ecosystems

Grinnell et al. 1937, Ingles 1965, Hall 1981, Jameson and Peeters 1988). In recent decades this wealth of information as it applies to endangered species has been compiled in a variety of volumes (Steinhart 1990, Theilander et al. 1994), and the State of California has undertaken a thorough and comprehensive compilation of data about the distribution and status of Californian mammals (Zeiner et al. 1990, California Department of Fish and Game 1995).

Fortunately California provides some positive, albeit limited, examples of courses of action that can be taken to maintain and enhance the biological diversity of changing landscapes. The common elements of these examples are planning for growth at the scale of ecological regions, application of the principles of conservation biology for determining the geometry and function of areas to be preserved, and accommodating rather than simply opposing the economic and social forces that propel development.

21.2 Study Area

The study area of 121 650 km^2 includes all of the southern half of the state of California with a Mediterranean climate (Fig. 21.1). The study area encompasses all terrestrial habitats, including the 6 largest Channel Islands west of the southern California coast. The deserts of eastern and east central California are excluded. Beginning at the southern boundary with Mexico, the eastern boundary of the study area coincides with the eastern edge of mediterranean-climate vegetation types, following the interior edge of the Peninsular and Transverse Mountain Ranges. The northern half of the interior boundary coincides with the crest of the Sierra Nevada Mountain Range. The northern boundary crosses the Central Valley and Coast Ranges, intersecting the Pacific Ocean at the center of Monterey Bay. The area is 640 km long, extending from 32° to 37° N latitude, and 64–280 km wide, between 116° and 122° E longitude. The physiographic provinces included within the study area are the Coast Ranges south of Monterey Bay (elevations 0–1780 m), the Transverse Mountain Ranges of southern California (elevations 0–3500 m), the Peninsular Mountain Ranges of the extreme south (elevations 300–1985 m), the southern Sierra Nevada (elevations 180–4420 m), the San Joaquin Valley (elevations 40–180 m) situated between the Sierra Nevada and Coast Ranges, and numerous smaller valleys and plains located between the mountains and the sea (Jones and Stokes, 1987). Three of California's 11 bioregions (south coast, south-central coast, San Joaquin Valley), and the cismontane portion of a fourth

Fig. 21.1. Study area

(southern Sierra Nevada) fall within the study area boundaries (Bakker 1994). The Mediterranean climate area south of the study area in Mexico has been excluded, because information about its mammalian fauna is less complete.

21.3 Human Population Growth

The human population of California has been growing exponentially for 140 years, increasing over that period by about two orders of magnitude (Fig. 21.2). The study area includes Los Angeles, which, with more than nine million people, is the second largest metropolitan area in the United States (Anonymous, 1992). Between the years of 1970 and 1991 the population of the study area grew from 13176600 to 20039000, an increase of 55.6%. During the 12 years from 1980 to 1991, the population of California grew at an annual rate of 2.33%, which corresponds to a population doubling time of approximately 30 years. This rate of growth is higher than that of the world as a whole, and far above rates of most of the remainder of the United States. Increasing the population density of areas that were already settled has accommodated some of this growth, but much of it has not. Each year in the decade of the 1990s the growing population of California is expected to urbanize an area more than three times the size of San Francisco, a city of more than 700000 people (Jensen et al. 1993). Most of this growth is occurring on agricultural lands, and on lands with value as wildlife habitat such as coastal sage shrublands and oak woodlands (Jones and Stokes Associates 1987). If past and present patterns of urbanization

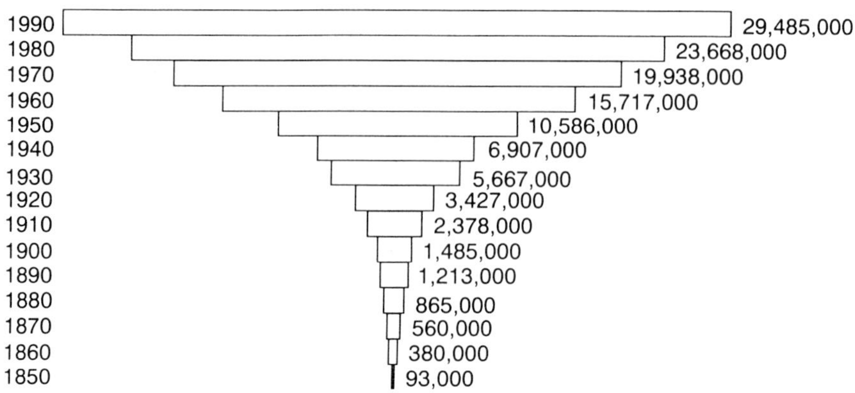

Fig. 21.2. Human population growth in California, 1850–1990

continue, most future growth will be unnecessarily destructive to wildlife habitat, and all other natural values, because it will be scattered and disorganized. In California, most decisions about urban development are initiated by private interests, and regulated by local governments. There is almost nothing in this process that requires, or even provides an incentive, that participants consider the consequences of development at the landscape level in an ecological context, or at a regional level in a political context. Since human development in southern California lacks large-scale coherency or any overall plan, it is inescapable that wildlife habitats steadily shrink, and the value of that which remains is diminished by fragmentation, loss of critical areas, and general disregard for principles of conservation biology.

21.4 The Mammals

Landscape degradation, the general theme of this volume, is invariably accompanied by a loss and simplification of natural ecosystems. In terms of mammalian species richness, landscape degradation is accompanied by the loss of some species, range reductions of many others, while a few become more numerous and widespread. Species that experience increases of ranges and numbers are often alien species that become naturalized on a disturbed and rapidly changing landscape.

The study area is biologically diverse, because it is physically diverse. It contains 125 of the 186 species of terrestrial mammals found in the State of California, and more than one-quarter of all non-marine species of mammals found in the United States. Among these 125 species, 58 species and subspecies are listed by the State of California as endangered, threatened, or as species of special concern (Table 21.1). An endangered species is a taxon (species or subspecies) that is in serious danger of becoming extinct throughout all or a significant portion of its range. A threatened species is one that is likely to become endangered in the near future unless protective measures are taken. Species of special concern, a designation used by California but not by the Federal Government, are species that merit particular attention but have no legal status. They may be considered threatened or endangered under certain circumstances (Sachse and Andrews 1993). They are species that are sufficiently rare, or about which so little is known, that they may possibly be rare or endangered, or could become so in the future. Within the study area the Federal Government has seven taxa of mammals categorized as endangered, and an additional four as candidates or proposed for listing (Table 21.1; Federal Register 1996). The Fed-

Table 21.1. Status of rare and declining taxa of mammals of southern California

Species	Status[a]	Notes[b]
INSECTIVORA (Insectivores)		
Soricidae (Shrews)		
1. *Sorex lyelii*	sc	uk
2. *Sorex ornatus relictus*	sc, c	al
2a. *S. ornatus salicornicus*	sc	ul
2b. *S. ornatus willetti*	sc	uk, i
CHIROPTERA (Bats)		
Vespertilionidae (Evening Bats)		
3. *Euderma maculatum*	sc	r
4. *Plectotus townsendii pallescens*	sc	d, rs
4a. *P. townsendii townsendii*	sc, ii	i
5. *Antrozous pallidus*	sc	rs
Molossidae (Free-tailed Bats)		
6. *Nyctinomops macrotis*	sc	r
7. *Eumops perotis californicus*	sc	d
LAGOMORPHA (Rabbits, Hares, Pikas)		
Leporidae (Rabbits and Hares)		
8. *Lepus americanus*	sc	uk
8a. *L. americanus tahoensis*	sc	uk
9. *Lepus townsendii townsendii*	sc	d
10. *Lepus californicus bennettii*	sc	ul
RODENTIA (Rodents)		
Aplodontida (Mountain Beaver)		
11. *Aplondontia rufa californica*	sc	r
Sciuridae (Squirrels and Chipmunks)		
12. *Tamias speciosus callipeplas*	fss	mi
13. *Ammospermophilus nelsoni*	ct	al
Heteromyidae (Pocket Mice & Kangaroo Rats)		
14. *Perognathus longimembris brevinasus*	sc, ie	ul
14a. *P. longimembris pacificus*	sc, fe	ul
14b. *P. longimembris internationalis*	sc	ul
15. *Perognathus inornatus inornatus*	sc	al
15a. *P. inornatus psammophilus*	sc, ie	al
16. *Perognathus alticola alticola*	sc, iv	uk, mi
16a. *P. alticola inexpectatus*	sc	uk, mi
17. *Chaetodipus fallax fallax*	sc	ul
18. *Chaetodipus californicus femoralis*	sc	ul
19. *Dipodomys elephantinus*	sc	uk
20. *Dipodomys heermanii morroensis*	ce, fe, ie	ul
20a. *D. heermanii dixoni*	nddb	al
21. *Dipodomys ingens*	ce, fe, ii	al
22. *Dipodomys stephensi*	ct, fe, ie	al, ul
23. *Dipodomys merriami parvus*	sc, c	ul
24. *Dipodomys nitratoides exilis*	ce, fe, ii	al

Table 21.1. (*Continued*)

Species	Status[a]	Notes[b]
24a. *D. nitratoides nitratoides*	ce, fe, ie	al
24b. *D. nitratoides brevinasus*	sc, c	al
Cricetidae (Native Mice, Rats, Moles)		
25. *Reithrodontomys megalotis limicola*	sc, iv	ul
25a. *R. megalotis santacruzae*	sc	i
26. *Peromyscus maniculatus anacapae*	sc	i
26a. *P. maniculatus clementis*	sc	i
27. *Onychomys torridus tularensis*	sc	al
27a. *O. torridus ramona*	sc	ul
28. *Neotoma lepida intermedia*	sc	ul
29. *Neotoma fuscipes luciana*	sc	al
30. *Microtus californicus stephensi*	sc	ul
CARNIVORA (Carnivores)		
Canidae (Canids)		
31. *Vulpes vulpes necator*	ct, fss	d
32. *Vulpes macrotis mutica*	ct, fe	al
33. *Urocyon littoralis*	ct, ir	i
33a. *U. littoralis catalinae*	ct	i
33b. *U. littoralis clementae*	ct	i
33c. *U. littoralis dickeyi*	ct	i
33d. *U. littoralis littoralis*	ct	i
33e. *U. littoralis santacruzae*	ct	i
33f. *U. littoralis santarosae*	ct	i
Mustelidae (Weasels, Badgers, Relatives)		
34. *Martes pennanti pacifica*	sc	d
35. *Gulo gulo luteus*	ct, cp	al
36. *Taxidea taxus*	sc	al, d
37. *Spilogale gracilis amphiala*	sc	i
BOVIDAE (Sheep, Goats, and Relatives)		
38. *Ovis canadensis californiana*	ct, cp, fss	al, d
38a. *O. canadensis cremnobates*	ct, cp, pe	d, mi
38b. *O. canadensis nelsoni*	cp, sc	mi, ul

[a] *fe* = Federally listed as endangered; *ce* = California listed as endangered; *ct* = California listed as threatened; *sc* = California listed, special concern; *c* = Federally listed, sufficient information to be classified as endangered or threatened; *pe* = Federally proposed to be listed as endangered; *cp* = California fully protected species; *ie* = International Union for Conservation of Nature (IUCN) listed as endangered; *ii* = IUCN listed as known to be endangered, vulnerable, or rare; *iv* = IUCN listed as likely to become endangered in the near future; *ir* = IUCN listed as having small populations that are at risk; *fss* = Federally listed as sensitive; *nddb* = listed as sensitive by California Natural Diversity Data Base. Data on status from California Department of Fish and Game (1995), Groombridge (1993), and Federal Register (1996).

[b] *uk* = little information available to assess status; *al* = habitat loss due to agriculture; *ul* = habitat loss due to urbanization; *i* = Channel Islands; *r* = rare in study area; *d* = declining populations; *mi* = populations isolated on mountain tops; *rs* = roost sensitive.

eral Government no longer lists species that may be threatened or endangered but for which definitive information is lacking. By either set of standards a significant fraction of the diverse mammalian fauna of southern California is either in danger of extinction, or in the absence of conclusive scientific information, is suspected of being so.

21.4.1 Extinction and Extirpation

The best-known mammalian extinction in California is the California grizzly bear, *Ursus arctos californicus*. It is the official mammal of the State of California, and appears as the central image on the state flag. This once numerous and widespread animal was pursued and killed throughout its range, so that populations declined very quickly as rifle bearing immigrants and hunters arrived in large numbers after the Gold Rush of 1849. The last wild California grizzly in California was killed in 1922 (Storer and Tevis 1955, Hall 1981, 1984).

A second extinct carnivore of the study area is the long-eared kit fox, *Vulpes macrotis macrotis*. This subspecies was known only from subarid valleys of southern California in the late nineteenth and early twentieth centuries (Grinnell et al. 1937). These valleys were fully settled for agriculture by that time. The behavior of kit foxes makes them quite easy to trap or shoot. The last ones near the city of Los Angeles may have been killed for sport, since they were "taken for mounting to taxidermists" (Grinnell et al. 1937).

The earliest documented extinction of a mammal from the study area is that of the giant deer mouse, *Peromyscus nesodytes*. Numerous bones found in cave middens indicate that it lived on San Miguel Island 2000 years ago, but no living specimens have been found during the past 60 years of systematic studies of the island's fauna. Walker (1980) hypothesizes that its extinction is due to the catastrophic degradation of vegetation in the last century from overgrazing by domestic sheep (*Ovis aries*).

The jaguar, *Felis onca*, occurred within the study area within historical times but has long since been extirpated. There are reports of jaguars in the San Jacinto and Santa Rosa Mountains as late as 1860, and from the Tehachapi Mountains and central coast ranges earlier in last century (Merriam 1919, Strong 1926). Some of these animals may have wandered north from the mountains of Mexico, as they still occasionally do along the southern border of Arizona and New Mexico. It is unlikely that jaguars will return to California by means of northern range extension, since they no longer occur near the border between California and Mexico.

21.4.2 Naturalized Species

Fourteen species of mammals have become naturalized in the study area in historical times. Some of these were introduced deliberately, and a few species of domestic mammals have become feral. Three species of cosmopolitan murids, the house mouse (*Mus musculus*), black rat (*Rattus rattus*), and Norway rat (*Rattus norvegicus*) arrived in California in the nineteenth century or earlier. *Rattus* lives in close association with human settlements and agriculture, while house mice are found in wildlands as well. During 1926 and 1927 a population of house mice increased to remarkable densities of as much as 200 000 per ha in a dry lake bed in an agricultural area of the San Joaquin Valley (Hall 1927). Population irruptions of this extraordinary proportion do not occur among native species of rodents, but there have been outbreaks of naturalized house mice in Australia as well (Newsome and Corbett 1975).

The muskrat (*Ondatra zibethicus*) has naturalized in the drainages of the San Joaquin Valley and central coast. It was brought to these places because of its commercially valuable pelt, and was deliberately released or escaped from fur farms. The opossum (*Didelphis virginiana*) was deliberately introduced early in this century from Eastern North America for hunting and for food, and has naturalized at lower elevations throughout the study area.

Feral pigs (*Sus scrofa*) and feral goats (*Capra hircus*) were deliberately released a century or more ago on the Channel Islands. Their heavy grazing, browsing, and soil disturbance have severely impacted island plant communities, and the animals that depend on them (Minnich 1982, Steinhart 1990). The United States Fish and Wildlife Service has developed a recovery plan for all the threatened and endangered species of the Channel Islands. A key element of this plan is the removal of feral mammals from the islands (Thelander et al. 1994). Feral pigs are also widely naturalized on the mainland, in the central coast ranges and the western foothills of the Sierra Nevada. They were deliberately introduced for sport hunting. They are destructive to native vegetation, and compete with native species for food, particularly acorns (Wood and Barrett 1979).

Feral horses (*Equus caballus*) and feral burros (*Equus asinus*) occur in small scattered populations in the study area (Zeiner et al. 1990). They are descended from escaped domestic stock. Larger populations of both species occur on deserts east of the study area. Small populations of free ranging bison (*Bison bison*) occur on one of the Channel Islands, and in the South Coast Range. They are descended from deliberately released animals, and are managed. Feral cats (*Felis cattus*) occur in scattered populations throughout the study area. As predators unfamiliar to native prey they have posed problems for threatened and endangered species of birds on the Channel Islands (Steinhart 1990), and for endangered mammals on the mainland (Zeiner et al. 1990, Palazzo 1994a).

The fox squirrel (*Sciurus niger*) has been introduced in several places within the study area. It favors the fringes of urban and agricultural areas, and seems to be extending its range into foothill woodland and riparian areas (Zeiner et al. 1990). It appears to have entirely replaced the native western gray squirrel (*Sciurus griseus*) in large parts of the suburban San Fernando and San Gabriel Valleys of Los Angeles County, where its aggressive and destructive habits have made it a backyard pest.

21.4.3 Range Expansions

The black bear (*Ursus americanus*) has expanded its range from the southern Sierra Nevada across the Tehachapi Mountains at the southern end of the San Joaquin Valley, and north through the central coast range. In 1933 black bears were deliberately and successfully introduced into the San Bernardino and San Gabriel Mountains of the study area (Burghduff 1934). These range expansions of the largest terrestrial carnivore in California were into areas that were once the exclusive domain of its larger cogener, the California grizzly. Presumably, the black bears are partly occupying the niche vacated by the extinct grizzly.

The California ground squirrel (*Spermophilus beecheyi*) occupies the entire study area. Its numbers and range have expanded in tandem with agriculture and other disturbances of native plant communities, which create habitat for ground squirrel burrows by reducing plant cover. Linear disturbances such as roads and trails are particularly conducive to local range expansion, and have allowed this species to occupy the whole of the Santa Ana Mountains, where they were previously absent (Pequegnat 1951).

A non-native subspecies of red fox (*Vulpes vulpes*) was brought to the study area by fur ranchers in the last century. These have since naturalized in numerous locations in the lowlands of the study area, where populations have rapidly increased (Palazzo 1994b). These spreading populations are having a severe impact on endangered species within the study area, such as the San Joaquin kit fox (*Vulpes macrotis mutica*). In coastal southern California the naturalized red fox has become a threat to endangered species of birds such as the California and light-footed clapper rails (*Rallus longirostris obsoletus*, *R. longirostris levipes*), and California least tern (*Sterna antillarum browni*). The US Fish and Wildlife Service and the California Department of Fish and Game are attempting to remove red foxes from areas where they prey upon, or compete with, sensitive species. They are present in sufficient numbers, and in so many places, that it is likely that most populations will continue to increase. As red fox populations increase their elevational range in the Sierra Nevada there is also the danger that they will meet and hybridize with the threatened Sierra Nevada red fox (*V. vulpes necator*). The Sierran subspecies is confined to the upper reaches of the Sierra Nevada, where it is considered rare, elusive, and with possibly declining populations (Grinnell et al. 1937, Zeiner et al. 1990). It is listed as Federally Sensitive by the United States Forest Service, and as threatened by the State of California (Table 21.1).

The golden beaver (*Castor canadensis subauratus*), which was once widely distributed in the lower drainages of the San Joaquin Valley, had its numbers greatly reduced by trapping (Grinnell et al. 1937, Zeiner 1990). It has subsequently returned to most of its original range. Within the study area populations have been established outside of the historic range by deliberate introduction in the San Bernardino Mountains, the Central Coast ranges, and elsewhere. Beaver populations in the San Bernardino mountains are unstable, tending to increase and decrease under the influence of the drought-flood cycle of the region. During periods of increase they tend to come into conflict with other land uses, necessitating removal of excess animals and undesirable colonies (Pearson 1977).

21.4.4 Direct Conflicts

Mountain lions (*Felis concolor*) occupy the entire study area, with the exception of heavily urbanized coastal valleys and the agricultural-urban center of the San Joaquin Valley (Zeiner et al. 1990, Mansfield 1995). In recent years there have been increasingly frequent encounters between humans and mountain lions, both sightings and attacks. In the past 105 years there have been twelve attacks on humans by mountain lions in California, and nine of those have occurred in the past ten years. More and more animals are appearing in heavily urbanized areas, where no such sightings have occurred in the past. In 1994 a young adult lion was killed in the parking lot of a large shopping mall in the Pomona Valley, 3 km from the nearest wildland. These interactions have been attributed to a steadily growing population of lions, which have not been bountied for 33 years or hunted for sport for 24 years. Lions and people could also be encountering one another more frequently due to the growing human population, and the recent tendency to lightly spread urbanization across wide areas that were previously wildlands. The largest cause of mortality for the population of lions in the Santa Ana Mountains is being struck by vehicles while crossing highways. These mountains are crossed by highways and almost entirely surrounded by solid urbanization, with multi-laned highways along the foothills.

21.5 Species Richness Changes

Considering the rapid ecological changes taking place in the study area, it is remarkable that only four taxa of mammals have been lost. The California grizzly was exterminated deliberately, and the long-eared kit fox fell victim to widespread habitat loss and a trusting nature. The exact cause of the extinction of the giant deer mouse is uncertain, but in a general sense it can be ascribed to the inherent fragility of island ecosystems and the species they contain, particularly when the island is small. The jaguar was at the margin of its range in southern California, and was probably never numerous. In terms of overall species richness, mammals naturalized in the study area in historical times have offset those lost by threefold.

Many of the 58 rare and declining taxa within the study area are categorized in the Notes column of Table 21.1 according to habitat status. Sixteen listed taxa have experienced habitat loss due to agriculture, and 15 taxa habitat loss due to urbanization. These taxa are occupants of the valleys and foothills, where urban and agricultural development are concentrated. Thirteen additional taxa inhabit the Channel Islands, where feral and domestic mammals have caused widespread damage to soils and plant communities.

21.5.1 Mountains

Almost all of the mammals inhabiting the mountains of the study area a century ago continue to do so. This is explained in large part by land ownership patterns. Approximately one-third of the study area is in public ownership, mostly as National Forests that occupy most of the area of all the larger mountain ranges (Fig. 21.1). National Forests have statutory responsibility to manage land for many purposes including wildlife, and to comply with the Federal Endangered Species Act. The Sierra Nevada have two large national parks, and a portion of a third within the study area. A primary responsibility of national parks is to maintain biological diversity. In principle, these federal agencies, together with California State Parks and Reserves, should be able to preserve and restore mammalian biodiversity in all of the mountainous portions of the study area. By geography and by law they possess the ability to adhere to the most important principle of conservation biology, maintenance of large contiguous areas with a wide variety of natural habitats. Even so, mammals that inhabit the mountains of the study area constitute a large group of rare and declining taxa. Five have relatively small populations isolated at the upper elevations of mountains, little is known about another five, and five have declining populations. Three of the mountaintop species in Table 21.1 have populations in only a limited part of one mountain range (*Tamias speciosus callipeplas* – Mt. Pinos chipmunk, *Perognathus alticola alticola* – white-eared pocket mouse, and *P. alticola inexpectatus* – Tehachapi pocket mouse). Very little is known about the two subspecies of *Perognathus*. The other two mountain species listed in Table 21.1, the desert bighorn sheep (*Ovis canadensis nelsoni*) and the peninsular bighorn sheep (*O. canadensis cremnobates*) have small, scattered populations in the mountains of the study area. They are quite sensitive to habitat disturbance of even a small portion of their home range, are susceptible to diseases transmitted by domestic sheep, and may be heavily preyed upon by mountain lions (Zeiner 1990, Torres 1995).

21.5.2 Valleys

Within the study area the largest complement of rare and declining species occurs in the valleys, where agriculture and urbanization have eliminated or altered most habitat for mammals. Approximately 40% of the taxa and species listed in Table 21.1 are inhabitants of valleys and foothills. Seven of these are categorized as endangered. In the coastal and intermediate valleys of the study area, urban expanses have greatly reduced or altogether eliminated most natural habitats, and fragmented and degraded the remainder to the degree that the area retains little of its original value to wildlife. The vast San Joaquin Valley is one of the most productive and valuable agricultural regions in the world. Agricultural development has profoundly altered its landscape, particularly as it relates to water. It was once called the valley of lakes. It was crossed by rivers flowing out of the Sierra Nevada, flanked by wide riparian woodlands, that terminated in large lakes surrounded by wetlands (Dasmann 1994). The Tulare Basin once held more than 3000 km^2 of lakes and wetlands, including one of the largest lakes in the western United States (Cohen 1994). Tulare Lake is gone, and more than 99% of these wetlands have been drained. The water they once contained has been diverted for irrigation. In the entire Central Valley, of which the San Joaquin Valley is the southern half, 89% of riparian forests and 94% of wetlands have been lost in historical times (Jensen et al.

1993). A recent survey of wildlife habitat in the southern San Joaquin Valley floor concluded that only 2.9% remained in good or better natural condition (Anderson et al. 1991).

21.5.3 Grasslands

The native perennial grasslands of the San Joaquin Valley were almost entirely converted to agriculture, and the small remaining fragments have a flora dominated by naturalized annuals. Diminished populations of mammals, from the San Joaquin kit fox (*Vulpes macrotis mutica*) to the Buena Vista Lake shrew (*Sorex ornatus relictus*) must exist on these small, degraded remnants of habitat. The plight of these fragment bound mammals is exemplified by kangaroo rats (*Dipodomys* spp.). Five taxa within this genus have lost more than 95% of their historic range to agriculture in the San Joaquin Valley, and during a recent survey no viable populations of one, the Fresno kangaroo rat (*Dipodomys nitrotoides exilis*) were located (Brylski and Roest 1994). The historic range of these kangaroo rats coincides with the most desirable areas for agriculture, and their only hope for survival in the wild is establishment and expansion of existing ecological reserves by purchase of private lands.

21.6 Multispecies Habitat Conservation Plans

Even in the valleys of the study area, where agriculture and rapid urbanization have had their greatest impact on biodiversity, there is promise of maintaining natural ecosystems on the necessary scale. At the Federal level modifications in the Endangered Species Act in 1982 provided states and local government the option to formulate habitat conservation plans (HCP). These plans are intended to resolve conflicts between human land uses and endangered species, while minimizing delays and legal challenges to the planning process. The key to HCPs is advance planning for the recovery of an endangered species, using the best available scientific information while taking into account the desires of public and private interest groups.

The first HCP to be approved in the United States is on behalf of the endangered Stephens kangaroo rat (*Dipodomys stephensi*), an inhabitant of the annual grasslands of the valleys on the interior side of the coast ranges of southern California. This HCP establishes five reserves totaling 17 400 ha, within a planning area of 216 000 ha, at a cost of $15.3 million US

(Beatley 1994). The amount of Stephens kangaroo rat habitat protected exceeds by 50% that which was estimated through conservation biology population modeling to be necessary for indefinite survival of the species. The HCP allows economic development in the planning area to go forward in an orderly pattern that is compatible with protecting habitat and endangered species on public and privately owned land.

The scope of management for biodiversity was increased to a larger scale in 1991 when a California law authorized the Natural Community Conservation Planning (NCCP) process (Wheeler 1996). The intent of this law is to conserve entire assemblages of naturally occurring species, and the ecological processes that support them, at the landscape level. The first NCCP is nearing completion in the coastal sage scrub habitat of southern California, in the southwestern portion of the study area. The planning area of 15 000 km^2 contains large parts of five counties, and close to half the State's 32 million people (Reid and Murphy 1995). Urban and agricultural development within the NCCP planning area has reduced coastal sage scrub by about 80%, leaving between 160 000 and 200 000 ha. This NCCP contains populations of seven taxa listed in Table 21.1, all heteromyid rodents, plus approximately 80 other potentially threatened or endangered species of plants and animals. Remaining coastal sage scrub is highly fragmented, and in large part privately owned. It falls into numerous regulatory jurisdictions and much of it is degraded to various degrees. Designing an effective reserve system under these circumstances, on some of the most expensive real estate in the country, presents a formidable challenge. From a political and social point of view the most encouraging aspect of the coastal sage NCCP is the consensus that has been built among landowners, environmental organizations, and a wide array of local, state, and federal agencies. From the scientific perspective the most frustrating aspect has been the lack of sufficient data and research, ranging from inadequate maps to basic demographic, genetic, and ecological information about populations and communities (Reid and Murphy 1995).

The plight of rare and declining mammals of the San Joaquin Valley is different from that of mammals of the urban coastal valleys of southern California, in that habitat loss has been caused mostly by agricultural development. The potential solution is similar, NCCPs or HCPs that broker the interests and responsibilities of government, private landowners, and the environmental requirements of listed species. A first step was taken more than a decade ago when the southern San Joaquin Valley was recognized as a place where rare species, particularly mammals, were concentrated. In response to this realization, the Nature Conservancy purchased the Carrizo Plain Natural Area in the southwestern valley, for the purpose of preserving biological diversity at the landscape level. Later in the 1980s urban expansion of the city of Bakersfield ran afoul of local environmental organizations, and later the US Fish and Wildlife Service, because of impingement on the habitat of the endangered San Joaquin kit fox (Beatley 1994). The city responded with a HCP, which raises money from urban land development fees to establish reserve areas and otherwise compensate for the effects of urbanization on kit foxes and other listed species. At about the same time, Kern County, in which Bakersfield lies, began its own HCP over a much larger area. The southern San Joaquin Valley now contains the Carrizo Plain Natural Area, areas set aside by Bakersfield, the Kern National Wildlife Refuge, and at least four other ecological reserves established by the State of California (Beatley 1994, Thelander 1994). These areas could best function to protect rare species and ecosystems if they were brought together under a single NCCP.

21.7 Outlook

The forces that have driven rapid changes in the mammalian fauna of the study area continue to strengthen as the impact of humans on the landscape grows stronger and more pervasive. This chapter brings together what is known about changes in the mammalian assemblage of the study area over the span of about a century and a half. However, it does not begin to fully explain these changes. The most common category in the status column of Table 21.1 is the State of California's "species of special concern". This category is used most often for taxa that are possibly threatened, endangered, or of some other precarious status, but for which scientific information is insufficient to make a determination. The Federal Government recently eliminated a similar designation, Category 2, which was applied until 1996 to 34 of the taxa in Table 21.1 (Federal Register 1996). Lacking a much more detailed understanding of the particular biology of these species, and the functional biology of the ecosystems of which they are members, there is very little hope of formulating successful plans for their long-term survival in nature. The greatest flaw in the federal and California state laws dealing with threatened and endangered species is that taxa slipping toward the brink of extinction are only addressed on a species by species basis. This ad hoc approach, initiated when it may well be too late to reverse or even understand processes of decline set in motion long ago, is destined to fail more often than it succeeds. Statistics from twenty-five years of effort bear this out. The Federal and State Endangered Species Acts have had limited success. At both levels, far more species await consideration for listing than have been listed. In the first 19 years after the passage of the Federal law, of 650 listed species of the United States from all taxa, a total of 16 were removed from the list – 7 because they became extinct, 4 because of errors in the original listing, and 5 because they recovered (Clark 1994). Of the taxa Federally listed in 1990, approximately two-fifths were either stable or improving, but another two-fifths continued to decline. The cost of implementing recovery plans for listed vertebrate taxa is unrealistic, estimated to average millions of US dollars per species (Smith and Maffitt 1994). Other more flexible and imaginative approaches, such as multispecies habitat conservation plans, are needed.

21.8 Conclusion

Very few of the mammalian species that inhabited the study area 150 years ago have been extirpated or become extinct, and naturalized species of mammals have added to the total species richness. Nevertheless, the long range survival of rare and declining taxa of mammals in the study area is by no means assured. The pressures of habitat loss and degradation are relentlessly growing, and many species could easily become extinct in the foreseeable future. The legal protection and recovery procedures accorded to endangered and threatened species by the Federal and State Endangered Species Acts are inadequate to the task of maintaining biodiversity. Setting aside additional large natural areas in the interest of biodiversity, at public expense, is unlikely. The NCCP and HCP approaches that are being pioneered in the study area hold out the greatest practical hope for effective protection and enhancement of biodiversity. These strategies combine interest groups that historically have interfered with one another's intentions, coordinate the planning activities and goals of public agencies that have traditionally worked independently of one another, and work with natural areas on a regional scale. Scientists have a crucial role to play in the process as well, by marshaling available information and the principles of conservation biology, and applying them to regional planning processes. Scientists must also be the ones who decide what additional information is needed to make useful decisions, and act as advocates for funding the necessary research.

In the United States many courses of action are available to conserve biodiversity. The goal of preserving biodiversity is still within reach – even in crowded, rapidly growing southern California. In the end, in California and in the rest of the United States, it will be the collective desires of citizens, as expressed in the political process, which will determine which options are exercised.

Acknowledgments. I thank Fabian Jaksic and Glenn Stewart for helpful comments on an earlier draft of this manuscript. Portions of the work leading to this paper were supported by the Pacific Southwest Forest and Range Experiment Station of the US Forest Service, and by the Cal Poly Kellogg Unit Foundation, Inc.

References

Anonymous (1992) California county fact book. California Association of Counties. Sacramento

Bakker E (1994) California: The great mosaic. In: Thelander C, Pearson D, Olson G (eds) Life on the edge: a guide to California's endangered natural resources: wildlife. BioSystems Books. Santa Cruz, California, pp 6–17

Beatley T (1994) Habitat conservation planning. University of Texas Press, Austin

Brylski P, Roest AI (1994) Fresno kangaroo rat. In: Thelander C, Pearson D, Olson G (eds) Life on the edge: a guide to California's endangered natural resources: wildlife. BioSystems Books. Santa Cruz, California, pp 76–77

Burghduff A (1934) Black bears released in southern California. California Fish and Game 20:83–84

California Department of Fish and Game (1995) California Department of Fish and Game natural diversity data base: endangered and threatened animals of California. Sacramento

Clark JA (1994) The Endangered Species Act: its history, provisions, and effectiveness. In: Clark TW, Reading RP, Clarke AL (eds) Endangered species recovery. Island Press, Washington, DC, pp 2–43

Cohen AN (1994) The hidden costs of California's water. In: Thelander C, Pearson D, Olson G (eds) Life on the edge: a guide to California's endangered natural resources: wildlife. BioSystems Books. Santa Cruz, California, pp 288–301

Dasmann R (1994) California's biodiversity: the impact of development. In: Thelander C, Pearson D, Olson G (eds) Life on the edge: a guide to California's endangered natural resources: wildlife. BioSystems Books. Santa Cruz, California, pp 18–22

Federal Register (1996) Endangered and threatened species, plant and animal taxa; proposed rule. Fed Register 61(40):7596–7608

Grinnell J (1908) The biota of the San Bernardino Mountains. Univ Calif Publ Zool 5:1–170

Grinnell J (1922) A geographical study of the kangaroo rats of California. Univ Calif Publ Zool 24:1–124

Grinnell J (1933) Review of the recent mammal fauna of California. Univ Calif Publ Zool 40:71–234

Grinnell J, Dixon JS, Linsdale JM (1937) Fur-bearing mammals of California: their natural history, systematic status, and relations to man. University of California Press, Berkeley

Grinnell J, Storer TI (1924) Animal life in the Yosemite. University of California Press, Berkeley

Grinnell J, Swarth HS (1913). An account of the birds and mammals of the San Jacinto area of southern California. Univ Calif Publ Zool 10:197–406

Groombridge B (1993) 1994 IUCN red list of threatened animals. IUCN, Gland, Switzerland

Hall, ER (1927) An outbreak of house mice in Kern County, California. Univ Calif Publ Zool 30:189–203

Hall ER (1981) The mammals of North America. John Wiley, New York

Hall ER (1984) Geographic variation among brown and grizzly bears (*Ursus arctos*) in North America. Spec Publ, Mus Nat Hist, Univ Kansas 13:1–16

Ingles LG (1965) Mammals of the Pacific States. Stanford University Press, Stanford
Jameson EW, Peeters HJ (1988) California mammals. University of California Press, Berkeley
Jones & Stokes Associates, Inc. (1987) Sliding toward extinction: the state of California's natural heritage, 1987. Jones and Stokes Associates, Sacramento
Mansfield T (1995) What we've learned about lions. Outdoor Calif 57(3):4–5
Merriam CH (1919) Is the jaguar entitled to a place in the California fauna? J Mammal 1:38–40
Minnich R (1982) Grazing, fire, and the management of vegetation on Santa Catalina Island, California. In: Conrad C, Oechel W (eds) Dynamics and management of Mediterranean-type ecosystems. Pacific Southwest Forest and Range Experiment Station, Berkeley, California, pp 444–449
Newsome A, Corbett L (1975). Outbreaks of rodents in arid and semiarid Australia: causes, preventions and evolutionary considerations. In: Prakash I, Ghosh P (eds) Rodents in desert environments. Junk, The Hague, pp 117–153
Palazzo T (1994a) Keeping out cats. In: Thelander C, Pearson D, Olson, G (1994) Life on the edge: a guide to California's endangered natural resources: wildlife. Biosystems Books, Santa Cruz, California, pp 77
Palazzo T (1994b) Unwelcome predator. In: Thelander C, Pearson D, Olson, G (1994) Life on the edge: a guide to California's endangered natural resources: wildlife. Biosystems Books, Santa Cruz, California, pp 183
Pearson CD (1977) Population and natural history studies of beavers (*Castor canadensis*) on Las Flores Ranch, Summit Valley, California. MS thesis, California State Polytechnic University, Pomona, California
Pequegnat WE (1951) The biota of the Santa Ana Mountains. J Ent Zool 42:1–84
Reid TS, Murphy DD (1995) Providing a regional context for local conservation action. Bioscience 45:S84–S90
Sachse TP, Andrews B (1993) California endangered species resource guide. California Department of Education, Sacramento
Smith KA, Maffitt NT (1994). Wildlife conservation law in California: from bag limits to biodiversity. In: Thelander C, Pearson D, Olson, G (1994) Life on the edge: a guide to California's endangered natural resources: wildlife. Biosystems Books, Santa Cruz, California, pp 32–40
Steinhart P (1990) California's wild heritage: threatened and endangered animals in the Golden State. California Department of Fish and Game, Sacramento
Stephens F (1906) California mammals. West Coast: Publishing Company, San Diego
Storer TI, Tevis LP (1955). California grizzly. University of California Press, Berkeley
Strong WD (1926). Indian records of California carnivores. J Mammal 7:59–60
Thelander C, Pearson D, Olson, G (1994) Life on the edge: a guide to California's endangered natural resources: wildlife. Biosystems Books, Santa Cruz, California
Torres S (1995). Mountain lions – California's largest carnivore. Outdoor Calif 57(3):8–9
Walker P (1980) Archaeological evidence for the recent extinction of three terrestrial mammals on San Miguel Island. In: Power D (ed) The California islands; proceedings of a multidisciplinary symposium. Santa Barbara Museum of Natural History, Santa Barbara, California, pp 703–715
Wheeler DW (1996) Ecosystem management as an organizing principle in land use. Outdoor Calif 57(3):4–10
Wood GW, Barrett RH (1979) Status of wild pigs in the United States. Wildl Soc Bull 7:237–246
Zeiner DC, Laudenslayer WF, Mayer KE, White M (eds) (1990) California's wildlife: volume III, mammals. California Department of Fish and Game, Sacramento

22 Biodiversity in Fluctuating Dry-Land Environments: Basic and Applied Aspects

F.M. Jaksic and P. Feinsinger

22.1 Introduction

Biodiversity, the current buzzword mouthed by everyone from vendors of "rainforest" candies to right-wing politicians or propagandists for petroleum corporations, is difficult to define and even more difficult to link to causal mechanisms (Walker 1992). "Preserving biodiversity" means different things to conservation biologists working at different levels of organization, from gene pools or species to landscapes or ecosystems (Soulé 1991; Franklin 1993; Wilcove 1993; Tracy et al. 1994). To scientists working at the level of species assemblages (i.e., community ecologists), preserving biodiversity means trying to maintain the greatest possible proportion of the native species of interest in a given zone, usually within certain taxonomic boundaries with which the scientists feel comfortable.

The fact that species interactions extend beyond convenient taxonomic boundaries, and that species distributions rarely coincide except at very broad spatial scales or at very sharp habitat boundaries, makes the quantification and preservation of biodiversity an almost impossible task. Because not all species can be listed, monitored, or managed to the same degree of precision, the attention of community ecologists often focuses on so-called "flagship species" that may be used for purposes of publicity and politics. Pleas on behalf of flagship species are often most successful if these organisms are both charismatic (e.g., with feathers or fur, big eyes, and soft or fierce temperaments) and demonstrably vulnerable to human intervention. More objectively, flagship species are often chosen because they are "umbrella species" (whose large ranges indirectly preserve a great number of other, less charismatic species living in the same habitats) and/or "keystone species" (having manifold or cascading effects on the remainder of the species assemblage). By concentrating on preservation of such flagship species, biodiversity by many different definitions may be sustained, even if fortuitously (Noss 1990; Tracy et al. 1994).

Ecological Studies, Vol. 136
Rundel et al. (eds.) Landscape Degradation and Biodiversity in Mediterranean-Type Ecosystems

Of course, the definition and selection of flagship species begs the question: just how important is biodiversity for community function and resilience after disturbances? Paraphrasing Mares (1992), most people faced with the choice of preserving four out of five species – one deer, one rabbit, one rodent, and two hawks – very likely will elect to discard one of the two hawks. Implicitly this decision relates biodiversity to taxonomic differentiation. Taxonomic and ecological diversity are not necessarily equivalent, however. Within any given community, regardless of taxonomic affinity some species may be coupled together in unique ways, or may be major determinants of community processes, while others may play roles that are nearly redundant (Walker 1992). It follows that, to maximize conservation of community function and resilience in the example above, perhaps it would be wisest to focus conservation efforts on the two predators while relinquishing one of the three herbivorous prey to triage.

Here, we present data from a long-term study (seven years) on carnivorous vertebrates and their vertebrate prey at a semi-desert locality in north-central Chile. The salient features of this example are that the species interactions are undoubtedly strong (predator/prey) while the predator species (hawks, falcons, owls, foxes) are simultaneously "vulnerable," "umbrella," "charismatic," and perhaps even "keystone." The data, obtained in a markedly seasonal environment over the span of several years and including at least one major climatic disturbance (the 1987 El Niño/Southern Oscillation), permit us to address the following issues: (1) Is the biodiversity at a given site a stable attribute or a time-dependent, idiosyncratic variable? (2) If the latter, what are the ranges of variation involved? (3) Does the relative degree of variation differ between functional groups (predators versus prey species)? (4) Are certain species consistently "redundant" or interchangeable with others, at least with regard to trophic position?

We also address conservation implications. Decisions among alternatives for sites to conserve are often made on the basis of species lists plus some natural history observations on the species in question (e.g., home ranges of individuals of "flagship" species). The data base is often generated by "snapshot" inventories or "RAP" – Rapid Assessment Profile – approaches (Roberts 1991). Such data bases may be inadequate when some species of interest are migratory. When this is obviously the case, such as with tropical-temperate migrant birds, decision makers attempt to allow for seasonal changes in habitat use. Even in such cases, however, at most annual migration is considered. Basing decisions about biodiversity preservation on snapshot or RAP species surveys, even on year-round species lists, could possibly be dangerous when the environments concerned experience longer-term abiotic and biological fluctuations. This is the case for many dryland areas, such as the western coast of South America, heavily influenced by multi-annual El Niño Southern Oscillation (ENSO) events (Fuentes and Campusano 1985). Our study site lies in this region, and below we illustrate the significance of such long-term events to basic ecology and to possible management decisions.

22.2 Methods of Study

The study area is in the Reserva Nacional de Fauna Las Chinchillas, at Aucó, near Illapel, north-central Chile (details in Jaksic et al. 1992; Jiménez et al. 1992). In this topographically rugged region, the nature of the semi-desert scrub vegetation is highly heterogeneous, varying in relation to slope, exposure, elevation, soil type, and land use history. Based on 20 years of climatic data from nearby Illapel, annual precipitation averages 175 mm, with marked year-to-year fluctuations (Fig. 22.1). Here we report on the first 7 years of the study (early 1987–early 1994), breaking each year up into two biologically meaningful seasons (Jaksic et al. 1993a): the non-breeding season, spanning March through August (i.e., autumn and winter), and the breeding season for both predators and prey, spanning September through February of the following year (i.e., spring and summer).

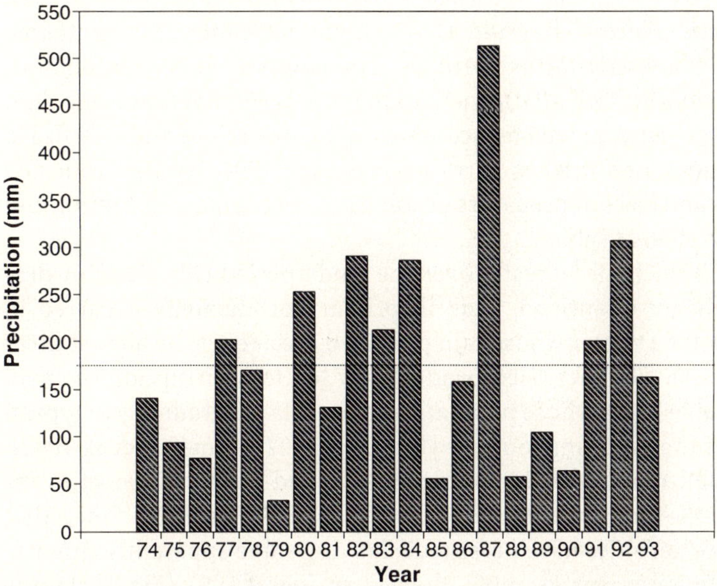

Fig. 22.1. Annual precipitation at Illapel, 15 km S of Aucó (beeline), through successive years. Horizontal line indicates annual average over 20 years (175.6 mm)

The predator/prey system that we are investigating in this ongoing study involves both avian and mammalian carnivores that consume small mammals, among other foods (Jaksic et al. 1992). Since 1987, small mammals have been live-trapped on a monthly basis with medium-sized Sherman traps arranged in grids (see Jiménez et al. 1992 for details of trapping design). Pellets (of avian predators) and feces (of mammalian predators) are collected on a monthly basis and their contents analyzed to the finest level of resolution possible, usually to the species level (vertebrate prey, plant seeds) or the genus level (invertebrate prey). Minimum number of prey individuals per pellet or feces is calculated based on the contents. Visual and aural observations of predators provide accessory data, but we consider a given predator species as present in the study site during a given season only when at least five pellets are obtained.

22.3 Predator and Prey Diversity, Abundance, and Climatic Variation

To date we have recorded eight species of small mammals at the site, the rodents *Abrocoma bennetti*, *Akodon longipilis*, *Akodon olivaceus*, *Chinchilla lanigera*, *Octodon degus*, *Oryzomys longicaudatus*, *Phyllotis darwini*, and the marsupial *Marmosa elegans*. These species have been preyed upon at various times by six falconiforms (*Buteo polyosoma*, *Elanus leucurus*, *Falco femoralis*, *Falco sparverius*, *Geranoaetus melanoleucus*, *Parabuteo unicinctus*), four strigiforms (*Athene cunicularia*, *Bubo virginianus*, *Glaucidium nanum*, *Tyto alba*), and two mammals (*Pseudalopex culpaeus*, *Pseudalopex griseus*), as well as occasionally by one snake and one lizard species not discussed further here. By February 1994, we had collected 14 170 pellets and feces of predators (Table 22.1) that contained a minimum of 55 386 prey items (Table 22.2).

Rainfall has fluctuated greatly over the study period (Fig. 22.1). A dramatic, though not quantified, peak in primary productivity occurred in spring 1987 after a winter whose rain exceeded the average by almost three times (Fig. 22.1). The increased productivity led to an irruption of small mammals (Table 22.3), whose populations then relaxed gradually to typical densities over the following four dry years (Table 22.3; Jiménez et al. 1992). In 1991 rainfall rebounded to average levels, and mammal densities increased the following breeding season (Fig. 22.1, Table 22.3). Since that time moderate densities have persisted, showing some apparently climate-induced fluctuations; for example, the wet spring of 1992 (Fig. 22.1) was followed by an increase in mammal densities (Table 22.3), whereas the dry

Table 22.1. Numbers of pellets and feces collected at Aucó, by season, plus general information on predator species concerned. Note that feces of the two *Pseudalopex* species were not distinguished; therefore, we have split evenly the records among these equally common species

Predator species[a]	Mass (g)[b]	Activity[c]	W 87[d]	B 87	W 88	B 88	W 89	B 89	W 90	B 90	W 91	B 91	W 92	B 92	W 93	B 93	Total
Falco femoralis (F)	334	D	#0	#0	#0	#0	#0	#0	#0	#0	#0	43	#0	#0	#0	#0	43
Elanus leucurus (F)	302	D	#0	#0	#0	#0	#0	#0	#0	#0	#0	#0	#0	#0	23	#0	23
Parabuteo unicinctus (F)	876	D	15	8	9	#0	#0	#0	#0	#2	#0	19	#0	#2	#0	24	79
Buteo polyosoma (F)	975	D	#0	40	19	69	#0	#0	#0	#0	#0	#0	#0	#4	#0	33	165
Geranoaetus melanoleucus (F)	2378	D	#4	#0	28	27	11	#0	#0	#0	#0	#0	#0	#0	#0	#0	70
Falco sparverius (F)	116	D	15	38	5	5	#0	#2	#0	31	#0	#0	#0	#0	#0	89	185
Tyto alba (S)	307	N	51	104	181	71	21	#0	56	#0	13	88	207	302	378	1237	2709
Glaucidium nanum (S)	81	DCN	100	76	129	118	81	68	73	23	#1	#0	38	15	67	56	845
Athene cunicularia (S)	247	CN	124	214	25	101	59	103	217	154	63	#0	60	686	159	504	2469
Bubo virginianus (S)	1227	N	184	194	260	433	62	64	71	10	10	309	217	360	604	229	3007
Pseudalopex culpaeus (C)	4317	DCN	42	88	402	189	242	158	143	124	163	271	176	167	66	57	2288
Pseudalopex griseus (C)	2495	DCN	42	88	402	189	242	157	143	125	163	271	176	167	65	57	2287
Number of pellets/feces			577	850	1460	1202	718	552	703	469	413	958	874	1703	1362	2286	14170

[a] F, Falconiform; S, Strigiform; C, mammalian carnivore.
[b] Mass is mean obtained from field studies or from literature sources.
[c] Activity: D, diurnal, N, nocturnal, C, crepuscular. B, breeding season, W, wintering season.
[d] #, designated "not present," based on criterion given in Section 22.2.

Table 22.2. Minimum numbers of prey items counted in predators' pellets/feces collected at Aucó. See Table 22.1 for explanations

Predator species	W 87	B 87	W 88	B 88	W 89	B 89	W 90	B 90	W 91	B 91	W 92	B 92	W 93	B 93	Total
Falco femoralis	0	0	0	0	0	0	0	0	0	40	0	0	0	0	40
Elanus leucurus	0	0	0	0	0	0	0	0	0	0	0	0	68	0	68
Parabuteo unicinctus	30	25	15	0	0	0	0	5	0	41	0	6	0	35	157
Buteo polyosoma	0	141	42	165	0	0	0	0	0	0	0	6	0	53	407
Geranoaetus melanoleucus	8	0	38	32	14	0	0	0	0	0	0	0	0	0	92
Falco sparverius	119	130	6	34	0	3	0	128	0	0	0	0	0	328	748
Tyto alba	98	193	330	107	32	0	88	0	31	108	291	404	525	1577	3784
Glaucidium nanum	336	204	234	269	188	247	170	62	11	0	131	57	89	104	2102
Athene cunicularia	1225	1505	195	543	634	1936	1902	1497	521	0	698	3902	1100	1784	17442
Bubo virginianus	635	422	583	760	123	78	104	37	13	372	296	519	795	288	5025
Pseudalopex culpaeus	148	505	1625	1137	948	1142	837	886	1443	2823	904	437	80	71	12986
Pseudalopex griseus	147	505	1626	1136	948	1142	838	885	1442	2823	905	437	80	71	12985
Number of prey in diet	2746	3630	4694	4183	2887	4548	3939	3500	3461	6167	3225	5768	2737	4311	55836

Table 22.3. Estimated density (no.ha^{-1}) of small mammals on trapping grids at Aucó, by season (as in Table 22.1)

Mammal prey species[a]	Mass (g)	Diet	W 87	B 87	W 88	B 88	W 89	B 89	W 90	B 90	W 91	B 91	W 92	B 92	W 93	B 93	Mean
Abrocoma bennetti (R)	182	Folivore	n.d.	1.1	0.9	0.5	0.4	0.2	0.7	0.0	0.0	0.0	0.0	0.4	0.4	0.0	0.3
Akodon longipilis (R)	61	Omnivore	n.d.	1.4	0.5	0.9	0.4	0.0	0.0	0.0	0.0	0.4	0.0	0.0	0.4	1.4	0.4
Oryzomys longicaudatus (R)	27	Granivore	n.d.	1.4	5.9	2.7	1.1	0.0	0.2	0.0	0.7	4.1	1.0	0.4	2.2	1.0	1.6
Chinchilla lanigera (R)	306	Folivore	n.d.	5.9	5.3	4.6	2.5	1.8	1.1	0.0	0.0	0.0	0.0	0.0	0.0	0.0	1.6
Octodon degus (R)	123	Folivore	n.d.	11.5	9.3	4.8	2.5	1.1	0.2	0.0	0.0	1.4	0.0	2.8	5.0	18.6	4.4
Marmosa elegans (M)	26	Insectivore	n.d.	6.0	8.8	5.0	2.3	1.6	2.0	2.5	1.8	6.8	11.8	5.4	10.4	2.8	5.2
Akodon olivaceus (R)	28	Omnivore	n.d.	30.0	37.7	14.5	3.4	2.5	1.4	0.5	0.0	1.6	7.2	22.2	28.8	9.0	12.2
Phyllotis darwini (R)	47	Granivore	n.d.	164.9	140.3	65.2	17.6	17.3	10.9	19.3	13.4	59.1	36.8	55.2	50.6	44.4	53.5
Density (number/ha)			n.d.	222.2	208.7	98.2	30.2	24.5	16.5	22.3	15.9	73.4	56.8	86.4	97.8	77.2	79.2
Number of mammal species			n.d.	8	8	8	8	6	7	3	3	6	4	6	7	6	3–8

[a] R, rodent, M, marsupial. Mean mass per species obtained from field records. n.d., no data obtained that season.

winter of 1993 (Fig. 22.1) was followed by a slight decline in densities (Table 22.3).

Likewise, mammalian prey diversity on the trapping grids remained constant at eight species until the wintering season of 1989, then declined steeply over the following two years to only three species, to begin an unsteady recovery by the breeding season of 1991 (Table 22.3). Thus, small-mammal diversity has ranged almost threefold from year to year - although it is likely that the "missing" species have actually persisted in the region, just not on the grids per se. Actually, four abundant species (*P. darwini, A. olivaceus, M. elegans*, and *Octodon degus*, in decreasing order) have been present throughout most or all of the study period. These are primarily a granivore, an omnivore, an insectivore, and a folivore, respectively (Jiménez et al. 1992). Of the other four species, one granivore (*O. longicaudatus*) has been present throughout most of the study period (Table 22.3), while one omnivore (*A. longipilis*) and two folivores (*A. bennetti* and *C. lanigera*) have disappeared at times from the trapping grids, the last species perhaps permanently.

Predator diversity peaked at ten species during the small-mammal population peak at the beginning of the study and declined to five, as defined above during the breeding season of 1989, with five to six species present until the breeding season of 1992 (Tables 22.1 and 22.4), when the system apparently started recovering. During the winter of 1993, seven species were recorded (including one probably accidental species), and nine by the breeding season of that year. Thus, there has been a two-fold difference between peak and crash predator diversities (Table 22.4). Actually, six species have resided continuously or nearly so at the site: the foxes *P. culpaeus* and *P. griseus* (which are omnivorous, consuming fruits and invertebrates in addition to vertebrate prey) and the owls *B. virginianus* (carnivorous), *A. cunicularia* (omnivorous), *G. nanum* (omnivorous), and *T. alba* (carnivorous). Two falconiforms have occurred only rarely (the carnivorous *F. femoralis* and the omnivorous *E. leucurus*). Three of the other four falconiforms at the site (*B. polyosoma, G. melanoleucus*, and *P. unicinctus*, all large and carnivorous) have occurred, as occurrence is defined here, only during the early stages of the study and most recently, while the omnivorous *F. sparverius* has persisted for a more prolonged period (Table 22.4). These four species may all be labeled as transients, as opposed to residents. Predator biodiversity has been the highest with small-mammal abundance at or above 100/ha (cf. Tables 22.4 and 22.3), when transient and even "accidental" species have entered to bolster the sedentary core of predators (Table 22.5).

Every falconiform species can be crudely matched diet-wise with a corresponding owl species (Table 22.5; see details in Jaksic et al. 1992,1993a).

Table 22.4. Residence status of predators at Aucó, based on data in Table 22.1 and as defined in text (at least 5 pellets or feces encountered)

Predator species	W 87	B 87	W 88	B 88	W 89	B 89	W 90	B 90	W 91	B 91	W 92	B 92	W 93	B 93	Total
Falco femoralis										X					1/14
Elanus leucurus													X		1/14
Parabuteo unicinctus			X	X	X										3/14
Buteo polyosoma		X	X	X										X	4/14
Geranoaetus melanoleucus	X	X	X							X				X	5/14
Falco sparverius	X	X	X	X				X						X	6/14
Tyto alba	X	X	X	X	X		X		X	X	X	X	X	X	12/14
Glaucidium nanum	X	X	X	X	X	X	X	X			X	X	X	X	12/14
Athene cunicularia	X	X	X	X	X	X	X	X	X		X	X	X	X	13/14
Bubo virginianus	X	X	X	X	X	X	X	X	X	X	X	X	X	X	14/14
Pseudalopex culpaeus	X	X	X	X	X	X	X	X	X	X	X	X	X	X	14/14
Pseudalopex griseus	X	X	X	X	X	X	X	X	X	X	X	X	X	X	14/14
Number of predator species	8	9	10	9	7	5	6	6	5	6	6	6	7	9	5–10

Table 22.5. Overall diets of predators at Aucó from wintering 1987 through breeding 1993

Prey in Diet	Transients				Accidentals	
	P. unicinctus	*B. polyosoma*	*G. melanoleucus*	*F. sparverius*	*F. femoralis*	*E. leucurus*
Small mammals	62.5–100.0%	41.1–69.8%	33.3–62.9%	0.0–16.7%	0.0%	20.6%
Other Vertebrates	0.0–37.5%	16.7–30.9%	22.0–48.0%	11.8–100.0%	57.5%	17.7%
Invertebrates	0.0–0.0%	3.8–36.9%	0.0–34.1%	0.0–82.3%	42.5%	61.7%
Number of Prey	92	407	157	748	40	68

			Residents		
Prey in Diet	*T. alba*	*G. nanum*	*B. virginianus*	*A. cunicularia*	*Pseudalopex* spp.
Small mammals	92.0–100.0%	8.8–73.0%	69.1–100.0%	0.3–25.8%	6.2–92.5%
Other Vertebrates	0.0–8.0%	1.9–30.4%	0.0–5.7%	1.1–7.7%	1.7–14.6%
Invertebrates	0.0–1.0%	41.9–90.9%	0.0–30.9%	70.6–98.4%	0.0–92.1%
Number of Prey	3784	2102	5025	17442	25971
% Feces with fruits					7.9–75.0%

In general, though, the three largest falconiform species prey more strongly than do their owl "counterparts" on *O. degus*, the only strictly diurnal/crepuscular small mammal at the site (Table 22.3), which displayed pronounced fluctuations in density. Neither the falconiforms nor the six resident predator species (four owls and two foxes) keyed in on the most abundant mammalian prey, *P. darwini* (see Jaksic et al. 1992). Indeed, they consumed this prey either less than expected based on its abundance in the field or barely at the expected rate. Instead, the resident predators often showed preferences for species that were never very abundant and that even disappeared temporarily from the trapping grids, such as *A. bennetti* and *O. longicaudatus* (Table 22.3; Jaksic et al. 1992).

22.4 Interpreting Biodiversity in a Fluctuating Environment

22.4.1 Biodiversity and Redundant Species

In reference to the general questions posed in the Introduction, some answers are straightforward: for example, biodiversity at a given site is obviously not a fixed attribute. To date, at the spatial scale of our sampling methodology, biodiversity has ranged two-fold (predators) to three-fold (mammalian prey), although due to the fact that both "missing" predators and prey may have persisted nearby we would hesitate to ascribe any special significance to that difference in relative variation. More interesting is that populations of redundant prey, in terms of trophic position, have failed to respond identically over time. For example, although one granivore (*P. darwini*) grew rapidly in number early in the study, another (*O. longicaudatus*) did not (Table 22.3). The more heavily granivorous of the two omnivores, *A. olivaceus*, increased in number but its more insectivorous congener *A. longipilis* did not, even though the truly insectivorous *M. elegans* also increased in density (Jiménez et al. 1992). Elsewhere, surprisingly, *O. longicaudatus* has been shown to show population growth more strongly than its granivorous counterpart *P. darwini* (Péfaur et al. 1979; Fuentes and Campusano 1985; Meserve and Le Boulengé 1987; Iriarte et al. 1989; Meserve et al. 1995).

In contrast, predators have displayed strong ecological redundancy both in diets and in timing. Aside from the fact that falconiforms as a group have responded quite differently from their approximate counterparts among the strigiforms, at the level of resolution reported in Table 22.5, redundant predators have displayed quite similar population trends over the years

(Jaksic et al. 1993a). *Bubo virginianus* and *T. alba* are highly redundant, preying on essentially the same prey species at roughly the same frequencies. The same occurs with the two *Pseudalopex* foxes, and with the owls *A. cunicularia* and *G. nanum* (with whom the falcon *F. sparverius* was frequently redundant). The three large falconiforms (*B. polyosoma, G. melanoleucus, P. unicinctus*) had quite similar diets as well. Another study conducted over two years in Fray Jorge National Park (Jaksic et al. 1993b), ca. 100 km to the north, demonstrated high redundancy between *B. virginianus* and *T. alba*, and somewhat less between *Pseudalopex culpaeus* and *A. cunicularia* (see also Jaksic et al. 1981). At a finer level of prey resolution, however, redundancy is not so clear-cut: species redundant with one another for several years often switch diet-wise to other guilds, or display unique diets, in other years (Jaksic et al. 1996).

It is worth mentioning that the system studied at Aucó includes many of the features frequently cited to justify conservation efforts (Noss 1990). The three large falconiforms are charismatic species (feathers, big fierce eyes) that are used by Chilean government agencies to stress the importance of conservation (CONAF 1988). The same species, because of their extensive home ranges (cf. Robinson and Wilcove 1989; Thiollay 1989), could act as umbrella species: any land tract extensive enough as to allow their survivorship will end up conserving numerous smaller or less mobile organisms. Two of the large falconiforms (*B. polyosoma, G. melanoleucus*) and one of the prey species (*C. lanigera*) are officially considered to be vulnerable or endangered in Chile (Jaksic and Jiménez 1986; CONAF 1988). Although it would be difficult to determine if any of the predator species plays a keystone role in the system, the possibility cannot be ruled out. Jaksic (1986), Jaksic and Simonetti (1987), Simonetti (1989), and Lagos et al. (1995) have stressed the probable selective pressure that these and other vertebrate predators exert on microhabitat selection by small mammals. At least one prey species seems to act as a keystone for some predators: the presence of the three largest falconiforms seems to be linked to densities of *O. degus*. In this context it is interesting that humans themselves may have played the fundamental keystone role: Simonetti (1988) suggests that the structure of the entire predatory assemblage is an artifact of human activities that have increased the abundance of *O. degus* (but see Meserve 1988).

22.4.2 Significance for Conservation Biology

The Aucó data serve to provide two warnings to decision-makers in biological conservation. First, in dryland sites such as Aucó, samples taken in a "typical" year may imply that the site supports relatively low biodiversity and may not really reflect the importance of the site in maintaining high regional biodiversity through sporadic support of transient flagship species. Differences between seasons within the same year may be so slight as to suggest that seasonal migration is not important, whereas long-range movement over long time scales may be exceedingly important.

Second, information on home ranges of vertebrate species may or may not indicate their habitat needs over long time frames. For those species that are truly residents in Aucó, estimates of home range available in the literature or estimates from a single season's data (e.g., Jiménez 1993) might provide reasonable approximations of long-term habitat needs, but obviously such estimates would be highly biased relative to long-term needs of the four to six transient predator species that, at Aucó, appear to respond to resource flushes over a much larger scale.

Indeed, the category of transient species might provide more useful "umbrella" species than does the category of residents, because to conserve highly mobile transients we must manage extensive and physiognomically diverse areas (Hansen et al. 1993). Unfortunately, the very nature of transient behavior makes it exceedingly difficult to estimate, for practical purposes, the true extent of the umbrella involved. In the case of Aucó, we do not know the diversity or extent of habitats that would have to be protected for any one of the transient species. Nevertheless, it seems logical to protect regions with high topographic diversity in which mammalian fluctuations in different elements of the vegetation mosaic (on different slopes, with different exposures, at different elevations, etc.) might be out of phase with one another, as it is apparent that spatio-temporal variability in the mammalian prey base is at the root of much of the dynamics of the predatory guild.

Whatever are the details, the Aucó data suggest that care must be taken in making management or preservation choices among sites based on inventories of their species richness taken at one point in time. While this may be particularly true for north-central Chile or similar xeric landscapes, it is likely to be true as well for any landscape known to experience erratic or supra-annual fluctuations in the physical environment.

Acknowledgments. This study has been supported at different times by grants DIUC 87-094 from the Dirección de Investigación, Pontificia Universidad Católica de Chile, INT 88-02054 and 92-14085 from the US National Science Foundation, BES 90-760 from the British Ecological Society, from the Frank M. Chapman Memorial Fund (USA), FONDECYT 90-0725, 92-0038, 193-0639 and 196-0319 from Fondo Nacional de Desarrollo Científico y Tecnológico (Chile), 91-4502 from National Geographic Society (USA), and from Fundación Andes/CONICYT (Chile), and by a Presidential Chair in Science. Numerous people have contributed greatly to the data set; we especially wish to thank Jaime Jiménez, Christian

Muñoz, Boris Saavedra, Enrique Silva, and Elier Tabilo. We also thank Sergio Castro, Carl Marti, Pablo Marquet, Enrique Silva, Sergio Silva, and Hugo Torres, for help with data analysis. Special recognition is owed to the Program of Sponsored Research in the System of Protected Wildlife Areas under auspices of Corporación Nacional Forestal (CONAF) of Chile. Juan Cerda of CONAF-IV Region, and Jorge Silva of CONAF-Illapel, kindly allowed us to work in the Las Chinchillas National Reserve and provided logistic support.

References

CONAF (1988) Red list of Chilean terrestrial vertebrates. Chilean Forest Service, Ministry of Agriculture, Santiago, 67 pp

Franklin JF (1993) Preserving biodiversity: species, ecosystems, or landscapes? Ecol Appl 3:202–205

Fuentes ER, Campusano C (1985) Pest outbreaks and rainfall in the semi-arid region of Chile. J Arid Environ 8:67–72

Hansen AJ, Garman SL, Marks, Urban DL (1993) An approach for managing vertebrate diversity across multiple-use landscapes. Ecol Appl 3:481–496

Iriarte JA, Contreras LC, Jaksic FM (1989) A long-term study of a small-mammal assemblage in the central Chilean matorral. J Mammal 70:79–87

Jaksic FM (1986) Predation upon small mammals in shrublands and grasslands of southern South America: ecological correlates and presumable consequences. Rev Chile Hist Nat 59:209–221

Jaksic FM, Jiménez JE (1986) The conservation status of raptors in Chile. Birds Prey Bull 3:95–104

Jaksic FM, Simonetti JA (1987) Predator/prey relationships among terrestrial vertebrates: an exhaustive review of studies conducted in southern South America. Rev Chile Hist Nat 60:221–244

Jaksic FM, Greene HW, Yáñez JL (1981) The guild structure of a community of predatory vertebrates in central Chile. Oecologia 49:21–28

Jaksic, FM, Jiménez JE, Castro SA, Feinsinger P (1992) Numerical and functional response of predators to a long-term decline in mammalian prey at a semi-arid Neotropical site. Oecologia 89:90–101

Jaksic, FM, Feinsinger P, Jiménez JE (1993a) A long-term study on the dynamics of guild structure among predatory vertebrates at a semi-arid neotropical site. Oikos 67:87–96

Jaksic, FM, Meserve PL, Gutiérrez JR, Tabilo EL (1993b) The components of predation on small mammals in semiarid Chile: preliminary results. Rev Chile Hist Nat 66:305–321

Jaksic, FM, Feinsinger P, Jiménez JE (1996) Ecological redundancy and long-term dynamics of vertebrate predators in semi-arid Chile. Conserv Biol 10:252–262

Jiménez JE (1993) Comparative ecology of *Dusicyon* foxes at the Chinchilla National Reserve in north-central Chile. Master's Thesis, University of Florida, Gainesville

Jiménez JE, Feinsinger P, Jaksic FM (1992) Spatiotemporal patterns of an irruption and decline of small mammals in north-central Chile. J Mammal 73:356–364

Lagos VO, Contreras LC, Meserve PL, Gutiérrez JR, Jaksic FM (1995) Effects of predation risk on space use by small mammals: a field experiment with a Neotropical rodent. Oikos 74:259–264

Mares MA (1992) Neotropical mammals and the myth of Amazonian biodiversity. Science 255:976–979

Meserve PL (1988) Are predator diets a consequence of human disturbance in central Chile? A reply to Simonetti. Rev Chile Hist Nat 61:159–161
Meserve PL, Le Boulengé E (1987) Population dynamics and ecology of small mammals in the northern Chilean semi-arid region. Fieldiana Zool 39:413–431
Meserve PL, Yunger JA, Gutiérrez JR, Contreras LC, Milstead WB, Lang BK, Cramer KL, Herrera S, Lagos VO, Silva SI, Tabilo EL, Torrealba MA, Jaksic FM (1995) Heterogeneous responses of small mammals to an El Niño Southern Oscillation event in north-central semi-arid Chile and the importance of ecological scale. J Mammal 76:580–595
Noss RF (1990) Indicators for monitoring biodiversity: a hierarchical approach. Conserv Biol 4:355–364
Péfaur JE, Yáñez JL, Jaksic FM (1979) Biological and environmental aspects of a mouse outbreak in the semi-arid region of Chile. Mammalia 43:313–322
Roberts L (1991) Ranking the rainforests. Science 251:1559–1560
Robinson SK, Wilcove DS (1989) Conserving tropical raptors and game birds. Conserv Biol 3:192–193
Simonetti JA (1988) The carnivorous predatory guild of central Chile: a human-induced community trait? Rev Chile Hist Nat 61:23–25
Simonetti JA (1989) Microhabitat use by small mammals in central Chile. Oikos 56:309–318
Soulé M (1991) Conservation: tactics for a constant crisis. Science 253:744–750
Thiollay JM (1989) Area requirements for the conservation of rainforest raptors and game birds in French Guiana. Conserv Biol 3:128–137
Tracy CR, Brussard PF, Wilcove D, Franklin JF (1994) Preserving biodiversity: species in landscapes. Ecol Appl 4:205–209
Walker BH (1992) Biodiversity and ecological redundancy. Conserv Biol 6:18–23
Wilcove D (1993) Getting ahead of the extinction curve. Ecol Appl 3:218–220

Part VI

The Future

23 Politics and Environment in Mediterranean-Climate Regions

F. di Castri

23.1 Introduction

The intercontinental analysis of mediterranean-climate ecosystems started in the 1960s with dual comparisons of the soil fauna of Chile and Italy (di Castri 1963), and above all the vegetation: that of California and Israel (Naveh 1967), France, California and Australia (Specht 1969a,b) and California and Chile (Mooney and Dunn 1970a,b).

The ecological comparisons were extended to embrace the five regions of the world with a mediterranean-type climate, particularly thanks to the successive Mediterranean Conferences (called MEDECOS). These were held in Valdivia (Chile) in 1971 on the origin, structure and convergent and divergent evolution of mediterranean ecosystems and biota (di Castri and Mooney 1973), in Stanford (California) in 1977 on fire and fuel management (Mooney and Conrad 1977), in Stellenbosch (South Africa) in 1980 on the role of nutrients (Kruger et al. 1983), in Perth (Australia) in 1984 on resilience (Dell et al. 1986), in Montpellier (France) in 1987 on time scales and water stress (di Castri et al. 1988), in Crete (Greece) in 1991 on plant-animal interactions (Arianoutsou and Groves 1994), and in Reñaca-Viña del Mar (Chile) in 1994 on land use and landscape disturbance (this volume). The importance of the human dimension for the five mediterranean-climate regions was stressed since the first MEDECOS (Aschmann 1973a), and remained as an underlying theme for all the others. In this volume, it becomes a central consideration when discussing land use conflicts and the future of these regions.

A first typology of mediterranean-climate regions, taking into account ecological but also a few political factors, was proposed by di Castri (1981), and refined later (di Castri 1991). Political considerations have also been inserted in the history of biological invasions, with special attention to mediterranean-climate regions (di Castri 1989a). Volume 5 on "Mediterrànies" of the encyclopedia *Biosfera* (Fundació Enciclopedia Catalana 1993) has a wealth of historical, social and economic facts on the

Ecological Studies, Vol. 136
Rundel et al. (eds.) Landscape Degradation and Biodiversity in Mediterranean-Type Ecosystems

five mediterranean-climate regions. The scope of this Chapter is to attempt a first intercontinental political typology, also taking into account the economic and social factors, as related to the environment. Presumably, it will remain as rough and preliminary in nature as the first intercontinental ecological comparisons were in the sixties.

The word "politics" in its broad meaning was preferred in the title of this chapter to the more accurate one of "geopolitics" (political geography), because of the multiple misleading interpretations of the latter term (see Encyclopaedia Britannica). However, geopolitics will be used in the text in so far as it means comparison among regions and countries. In my research work, I define geopolitics as "the discipline which analyzes why political decisions have been taken, are taken and are likely to be taken, and their consequences on regional planning and landscape dynamics, as related to the decisions taken by other countries, other regions, and other administrative units or private sectors." This definition facilitates a bridge between natural and political sciences. Its geographical comparableness is particularly appropriate in this era of worldwide connectedness of economies, technologies, communications, and climatic and biological systems (the so-called globalizations).

23.2 Global Driving Forces and Shift of Disturbance Regimes

When the intercontinental comparisons mentioned above were first undertaken, most disturbance regimes in mediterranean-climate regions were still determined primarily by local driving forces. In several regions, the rule was still that of economic protectionism, of subsidized agriculture, and of a rigid State-based planning. The situation is no longer the same, with a global opening of trade and markets, an end of the political blocks, and the long-distance decision-making in real time over national boundaries. Together with traditional landscapes and territories defined by ecological and/or administrative boundaries, new larger "virtual landscapes" appear with moving frontiers shaped by flows of capitals and communications, or conversely by lack of investments. Natural resources are no more the key factor for development. For the better or worse, the mediterranean-climate regions have inevitably joined the global market economy, with a few vanishing spots of subsistence economy in the poorest areas and some remnants of central-State economy in the south of the Mediterranean Basin.

The human-induced disturbance regimes have shifted accordingly, with very diversified new trends determined by the historical trajectory of the systems and the impact on them of global forces. I will take, as a microcosm, a sector of the central zone of Chile, along the Rio Tinguiririca from the Andes to the coast, where I worked intensively in 1950–1953, and have periodically monitored up until May 1996. Grazing by cattle, sheep and goats, extraction of firewood and charcoal production, that is to say, the past prevailing patterns, have almost completely disappeared because of lack of economic incentive. Matorral and sclerophyllous forest are much more dense, but they are now exposed to great fire hazards because of fuel accumulation, and they have become affected by tourism. The fertile and irrigated lands are intensively utilized for export production of fruits and vegetables, with great economic benefits, but also serious pollution of soil and water by pesticides and fertilizers. The coastal zones, and particularly the wetlands, are increasingly degraded. Rural exodus and land abandonment with no management for regeneration are common features.

Something similar could be said as regards the situation in southern France some 30–40 years ago (Debussche and Le Coz 1985), in mediterranean Spain in the 1970s, and much earlier in California, while some rural parts of Chile and south of the Mediterranean Basin have not yet reached this status. There is, therefore, a temporal shift of the disturbance regime in accordance with the prevailing economic, political and social factors. Retrospectively, early intercontinental comparisons of mediterranean-climate regions did not take into due account these "diachronic" (through time) steps, and the results may have been biased to a certain extent. It would be interesting, for instance, to compare sequentially the current landscape dynamics in Chile with what has happened in Mediterranean France since the 1960s. The steps are likely to be of a similar nature, but the time elapsed from one step to the next will presumably be much shorter in Chile, because of the accelerated transformations provoked by the recent trends of globalization.

More information on these topics can be found in Reich (1991) and in di Castri (1989b, 1995a, 1996, 1997b), with special emphasis on the interaction of different globalizations and on the "winners-losers" syndrome as a consequence of these processes.

23.3 A Geopolitical and Environmental Comparison of the Five Regions

It is useful to attempt to compare geopolitical and environmental viewpoints from the five regions of the world with a mediterranean-type climate: California and Baja California, central Chile, western and south

Australia, the Cape province of South Africa, and the Mediterranean Basin *sensu lato*, including lands with a mediterranean climate – such as Portugal or the Atlantic coasts of Morocco – that do not border the Mediterranean Sea. A qualification is needed as regards method and results. Geopolitical indicators (Table 23.1) also include economic and social factors, since they are at present the main elements for political decision. Among environmental indicators (Table 23.2), consideration is given also to topics such as air pollution, energy consumption or waste treatment, that do not relate strictly with the state of mediterranean-type ecosystems.

Going from the economic to social to environmental indicators, data are becoming progressively gross approximations. Even the way of measuring economic and social indicators (economic growth, inflation, unemployment, rate of poverty, demographic pressure, etc.) may vary from country to country. Furthermore, mediterranean-climate regions do not correspond to administrative regional or State units where this kind of data is collected. "California" corresponds to the major part of the State of California in the United States and northwestern Baja California in Mexico. The Mediterranean Basin covers part of the territories of some 26 Nation-States, some of them with still ill-defined boundaries (e.g. Slovenia, Croatia, Bosnia, Serbia, and Macedonia that have emerged from the breakup of Yugoslavia). Something similar could be said of the other three regions, which nevertheless stay within the boundaries of only one Nation-State (but with one Federal government and four State governments in the case of Australia). Many extrapolations have been needed, therefore, to reach a sensible average, and often it will be necessary to dissociate data according to other groupings (for instance, north, western and south Mediterranean countries). Some important indicators have been discarded (for instance the external national debt), precisely because of the impossibility to differentiate the state of a mediterranean-climate region from that of the overall country. Finally, because of space constraints, it will be impossible to discuss each one of the 28 indicators chosen for this comparison in more depth.

Accordingly, I will restrain from giving systematically figures as applied to each one of the indicators and regions. This would induce an impression of precision that is far from being achieved at this stage. Rather than exact figures, an order of magnitude is required in this case. In addition, several indicators, for instance political stability and security or environmental awareness, are intrinsically of a qualitative nature.

I have to use rank analysis here, as the most appropriate and commonly applied method in similar situations. For each indicator, a rank from 1 to 5 is given to the five regions, 1 to the region that is doing best on a particular subject and 5 to the region that is doing worst, with intermediate ranks for the others. For instance, rank 1 as regards economic growth or international competitiveness shows that the region has the best performance on these subjects, but rank 1 concerning unemployment or air pollution implies that the region has the best opportunities of employment or least polluted air conditions.

As a source of data, I have consulted the usual yearly statistical compilations (Généreux 1993, World Resources Institute 1994, Brown 1996, Le Monde 1996, Atlaseco 1997), but the most significant amount of information comes from preliminary statistical analyses I have access to (mostly from within the framework of the United Nations), and from very numerous personal communications of economists, political scientists and industrialists of the five regions and elsewhere.

Table 23.1 presents the geopolitical indicators (from the political to the economic up to the social ones) for the five regions. Those regions totaling a smaller number of points are considered to be the most healthy and stable

Table 23.1. Geopolitical, economic and social indicators

	California	Chile	Australia	South Africa	Mediterranean Basin
Geopolitical security	1	3	2	4	5
Position as compared to neighbor regions	2	3	4	1	5
Political stability (ethnic conflicts, terrorism, regional wars)	3	2	1	4	5
Economic growth	2	1	3	5	4
Gross product per capita	1	4	2	5	3
Inflation	2	3	1	5	4
International competitiveness	1	2	3	5	4
Unemployment	1	2	3	5	4
Differential of unemployment within the region	2	3	1	5	4
Social stability (poverty and social exclusion)	2	4	1	5	3
Cultural diversity	4	3	5	2	1
Demographic pressure	3	2	1	4	5
Immigration flows	5	1	2	3	4
Urbanization and slums	2	3	1	5	4
Total	31	36	30	58	55

from this point of view. In spite of the inherent approximation of a rank analysis, it is unquestionable that Australia, California and Chile stand in a much better position than the Mediterranean Basin and South Africa. They are also consistently the first three regions, in different orders, as regards the most important indicators of economic growth, inflation, international competitiveness, rate of employment, demographic pressure, and social and political stability. California (USA) would become the first in absolute terms, when taken in isolation from Baja California. Conversely, the Mediterranean Basin would maintain the same rank position, even if only the more developed northern Mediterranean zones were considered alone.

The case of Chile merits special attention since this country has experienced the most rapid and unprecedented change, with its mediterranean-climate region being the most dynamic portion of the country. Chile's gross product per capita is still much lower than that of California, Australia, and most north Mediterranean regions, but it is slightly higher than that of South Africa and is much better than those of Croatia, Serbia, Macedonia, Albania and all of the southern Mediterranean countries with the exception of Israel and Libya. In the ranking of the *Global Competitiveness Report* of 1996, issued by the Geneva-based World Economic Forum, Chile appears well before Australia (while its neighbor New Zealand is now the new economic star), France, Portugal, Spain, Italy and South Africa. The specific criteria of the World Economic Forum reward "dynamic countries that throw open their frontiers to foreign trade and boast lean governments, flexible labor markets, advanced technology and stable political systems" (Serril 1996). While recognizing the importance of these criteria, a greater comparative attention is given in this chapter to social, cultural and environmental dimensions.

In general terms, California, Chile and Australia have the comparative advantage of being situated in the most dynamic and stable economic area of the Pacific, while South Africa and the Mediterranean Basin are neighboring the most underdeveloped and politically less stable areas of the world, Africa south of the Sahara and Eastern Europe.

Another geopolitical comparison is to explore whether the mediterranean-climate regions are doing better or worse than their neighboring regions or countries. A rank analysis has been made, following the same criteria. Particularly, Chile (with special emphasis on its mediterranean-climate region) and South Africa have a much better performance as compared with their South American and African neighbors. Conversely, the mediterranean-climate regions of Spain, France and particularly Italy are in a much worse situation as compared with other regions of the same countries or with northern European countries. Nevertheless, it would be totally arbitrary to impute success or failure to the existence of a mediterranean-type climate. Historical factors and current political and social situations are better suited for providing explanations.

A last comparison refers to the adaptive response of the five mediterranean-climate regions to the recent interconnected globalizations. The main adaptive principles (di Castri 1995b) are those of specificity (tailored responses with emphasis on innovation rather than on imitation), diversification (economic, cultural and biological), flexibility (with special attention to the removal of bureaucratic and administrative rigidities, thus leading to institutional and productive efficiency), proactiveness (capacity of anticipating change, preventive approach, incorporation of environmental factors in the early phases of the production processes, large investments in research and development activities), and reactiveness (short time of reaction to change). After another rank analysis, California stands at the first place (7 points), followed by Chile (11 points); Australia has an intermediate position (14 points), and South Africa and the Mediterranean Basin are the last with more than 20 points. This was an expected result since the two rank analyses are intimately linked with some cause-effect relationships.

Finally, I have not included one of the most important factors of internal political stability, that of the level of democracy and respect of human rights, in the rank analysis of this table. It would be too arbitrary to discriminate among regions that have reached a satisfactory level, after the end of a dictatorship in Chile and of apartheid in South Africa. Very serious problems still remain in parts of the southern and eastern Mediterranean.

Table 23.2 presents environmental indicators for the five regions. The level of precision is lower than in the first table and several indicators are only qualitative in nature. Australia and California keep the first ranks. Demographic pressure is lower in Australia than anywhere else, and in both regions the level of environmental awareness and responsibility is comparatively greater. In addition, these countries are at a stage of economic and industrial development where the impact on many natural resources for extractive processes decreases, because it is no longer viable economically. South Africa has an intermediate position, while Chile and the Mediterranean Basin form the last ranks. It is certain that environmental awareness and scientific and technical know-how on conservation of natural resources have a long tradition in the Cape Province of South Africa. In addition, the Mediterranean Basin, because of political, social, historical and demographic reasons, cannot implement a coherent network of protected areas and national parks of an extension and efficiency comparable to those of Australia, California, South Africa and also, to a minor extent, Chile.

Table 23.2. Environmental indicators

	California	Chile	Australia	South Africa	Mediterranean Basin
Soil degradation and desertification	1	4	2	3	5
Deforestation	1	5	2	4	3
Degradation of freshwater resources	1	4	2	3	5
Floods	2	4	1	3	5
Fire hazards	4	2	1	3	5
Landscape disruption	2	5	1	3	4
Coastal degradation	4	3	1	2	5
Mountain degradation	2	4	1	3	5
Biodiversity loss and biological invasions	1	5	2	4	3
Genetic diversity of landraces and domesticated plants	4	2	5	3	1
Energy consumption	5	1	4	2	3
Air pollution	4	5	1	2	3
Waste treatment	1	5	2	3	4
Environmental education and awareness	1	4	2	3	5
Total	33	53	27	41	56

The situation in Chile again merits special attention, since it is the only mediterranean-climate region where good geopolitical and economic indicators play against the state of the environment. In fact, the first phase of development, as in the case of Chile, with a violent take-off of quantitative competitiveness based on low prices and on extractive processes implies, almost unavoidably, environmental (and social) costs. Hopefully, in the next phases of development, characterized by a greater qualitative diversification and then of innovation, Chile will be able to devote more attention to environmental concerns. It seems to be already happening. In any event, when comparing Chile with other countries of the old Third World that are economically stagnant or regressing (mainly in Africa and South Asia), the environmental degradation is much more accelerated there. Poverty and lack of development, so often intimately associated with demographic pressure, are a major threat to the environment.

As a whole, the main causes of environmental degradation in the mediterranean-climate regions are the impact on coastal ecosystems (a worldwide phenomenon, since more than 60% of human population is concentrated in coastal areas), loss of biodiversity and biological invasions, deterioration of freshwater resources, fire hazards, and catastrophic floods due in part to unwise land-use planning. Air pollution is very serious in cities such as Athens, Santiago or Los Angeles, where specific physical and meteorological conditions aggravate this phenomenon. In the poorest zones, soil degradation and desertification (Thomas and Middleton 1994), deforestation, and inefficient waste treatment also represent very serious threats. Conversely, in several areas that are now void of economic attractiveness, mediterranean shrublands and forests are recovering and regenerating. For instance, the extension of matorral and forest has never been as large in Spain in historical times than it is at present. As a consequence of this phenomenon, however, fire frequencies are now very high.

Are mediterranean-type ecosystems more fragile than the other ecosystems of the world? Usually, it is said that they are more fragile than temperate forests and grasslands, but less fragile than tropical forests. It is probably so, but only long-term monitoring could assess this point. With a historical perspective and in monitoring their reaction to the current development patterns, it is my feeling that mediterranean-type ecosystems are relatively resilient to heavy human impact and regenerate after abandonment. Their biological diversity is undermined, but the thresholds that should not be crossed to sustain their functioning, stability and productivity are still unknown.

This conclusion may seem to be less gloomy than those of the chapters dealing with land use conflicts in the various mediterranean-climate regions. There is no contradiction, but different scales and levels of perception. Undoubtedly, the environmental situation has degraded during recent years, with great changes in landscape dynamics and erosion of biodiversity in all mediterranean-climate regions. New and more frequent land use conflicts have inevitably emerged in this era of international competitiveness. A different issue is that of exploring whether the situation is better or worse, when comparisons are made within the mediterranean-climate regions, between them and with other regions of the world. In addition, some case studies from these Chapters refer, in a very accurate and worthy way, to mediterranean microcosms. The heterogeneity of mediterranean-climate regions is such, not only from an ecological, biogeographical and environmental viewpoint, but also from a geopolitical, economic, social and perceptional one, that there are other microcosms within a single region that could give different highlights.

It would be perilous to attempt a comparison between Table 23.1, with the political, economic and social indicators, and Table 23.2 with the environmental ones. The only sensible conclusion is that there are no linear relations, either in a positive or a negative way. Only a complex web of interactions can be visualized. As expected, economic development is not in itself antithetical to the environment, contrary to what is frequently said in the media, but such development can harm some environmental components in the absence of appropriate planning and mitigation measures. Small human populations with a subsistence economy can live in harmony with their environment and their resources, but there are only few examples of this in mediterranean-climate regions (and elsewhere in the world), and their situation is not tenable in the long-term surrounded as they are by a world of global market economy. On the other hand, there is ample evidence that the central State-mastered economies (such as the ones which were under the Soviet rule) have provoked the worst results with respect to environmental degradation.

The establishment of a positive relationship between development and environment would imply policy decisions, in order to favor the preventive approach and the proactive efficiency that, after an initial investment, is less costly than a non timely reaction leading to degradation and later to restoration of resources (see Fig. 23.1). In this way, the care for the environment does not impair the economic competitiveness. By introducing environment considerations during the first phases of production processes, energy consumption, water use and industrial waste are reduced, and the efficiency of use of resources is greater. These are the so-called *win-win strategies*. Both development and environment are winners. Unfortunately, there is an initial cost, a need for a new technological generation and for investments in Research and Development. Several countries from the mediterranean-climate regions do not yet have the possibility of behaving in a proactive manner. Even Italy and Spain have low research and development spending (Chui et al. 1996). Italy, for instance, spends only 6% of the total investment in environmental protection on preventive and proactive actions. This proportion is higher in California and is now improving in Chile. The promotion of this approach and of this technology would represent a more fruitful and innovative form of international cooperation on the environment than the current patterns that have already shown their inefficiency and inapplicability.

No direct relation can be found between human population density and environmental health. For instance, in several southern Mediterranean countries population density is lower than in the north, but soil degradation and deforestation are much more acute phenomena. Conversely, it could be said that mediterranean-climate zones with a high human popu-

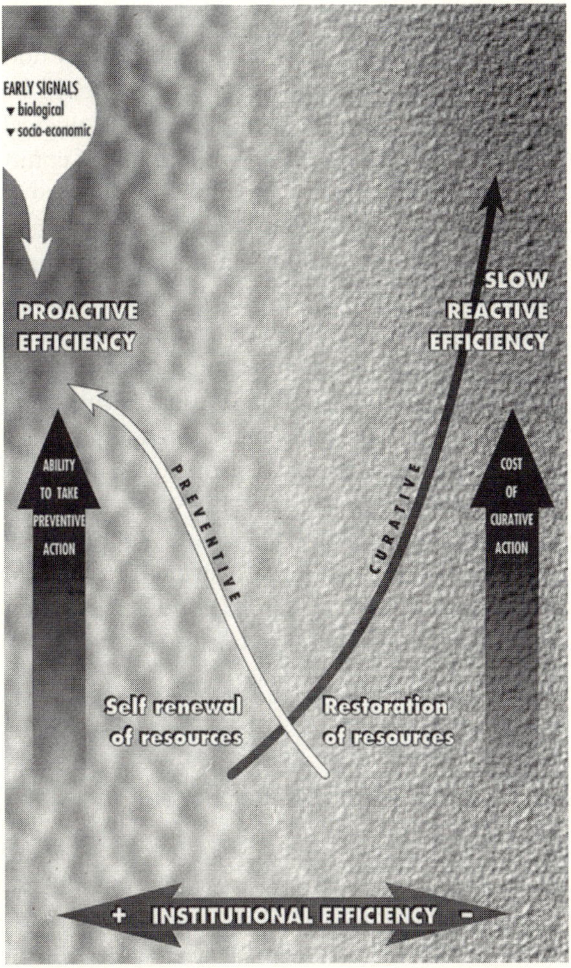

Fig. 23.1. The shift from a curative to a preventive approach in resource use, as related to the increase of institutional efficiency. The preventive approach implies also a proactive technology with consideration and internalization of environmental aspects, since the early phases of the production processes and cycles (from di Castri 1997a and IIASA, personal communication, modified)

lation growth rate have always induced a precarious state of the environment. Nevertheless, this is likely to be a Panglossian statement. High population growth and underdevelopment are so intimately linked that it is impossible to differentiate the cause and the effect. Together with birth control measures, only a take-off towards development can effectively decrease birth rates.

In general terms, it seems inappropriate to attempt to correlate ecological or climatic factors with potential for development, at least in the framework of recent globalized economic trends. For instance, California or Chile are at the top of development rankings, while zones with very similar ecological situations in North Africa and the Middle East or in Mediterranean Europe are in a very precarious state. Appropriately trained human resources, political stability and institutional efficiency are more important discriminants. Similarly, the old myth that tropical countries would be intrinsically inept at development because of their human and ecological characteristics, has been debunked by the explosive development of countries such as Malaysia, Thailand or Singapore.

The last point is to visualize how mediterranean-climate regions stand comparatively in a global perspective, when considering at the same time political, economic, and environmental factors. California and Australia stand very high, Chile is rapidly progressing in the same direction, and South Africa, in spite of great historical constraints, now has favorable perspectives. Conversely, if the geopolitical situation does not evolve, the Mediterranean Basin constitutes, politically and environmentally speaking, one of the most serious "tectonic faults" of the world with its great fractures between the north and the south and towards the West (Balkans). However, if a (not impossible) happy geopolitical "surprise" happens, the Mediterranean Basin with 600 million people and 30% of the gross world product (Rota 1996) could again acquire the crossroads and key position that it held in ancient times and maintained up to a few centuries ago under Venetian economic rule (Braudel 1979, Zorzi 1996).

23.4 The Special Case of the Mediterranean Basin

There are more diverse geopolitical and environmental situations in the Mediterranean Basin than in all the other four mediterranean-climate regions taken together. In addition, the Mediterranean Basin embraces some 60% of all mediterranean-climate lands of the world (di Castri 1981) if the mediterranean climate is taken in a restrictive meaning (Aschmann 1973b), and much more if a wider definition of this climate is accepted (UNESCO-FAO 1963). In addition, there are as many as 26 Nation-States in the Mediterranean region against only five in all the others. Some qualifications are, therefore, needed.

It is usually stated that in the northern Mediterranean region there are few and relatively wealthy people and in the southern Mediterranean there are too many poor people. It is also said that land degradation is more accelerated in the south, because of the assault on fragile natural resources by economically marginalized populations, while pollution is more serious in the north because of industrialization. This implies a strong immigration pressure towards the north, in a similar way than Mexican and other Latin Americans are being attracted by California.

Table 23.3. Population in countries of the Mediterranean Basin

	Total number of inhabitants	%
North of the Mediterrancan in 1950	140 million	66.7
South of the Mediterranean in 1950	70 million	33.3
North of the Mediterrancan in 1990	190 million	50.0
South of the Mediterranean in 1990	190 million	50.0
North of the Mediterranean in 2025	195 million	35.8
South of the Mediterranean in 2025	350 million	64.2

Table 23.4. Population ratio of 15–24 years/55–64 years in countries of the Mediterranean Basin in the year 2000

North of the Mediterranean	1.3
South of the Mediterranean	3.3

This is a gross overgeneralization. Human population density is generally lower in the South than in the north. This is partly, but not exclusively, because in the south there are many more arid lands unfavorable to human settlement. For example, one of the highest population densities of the south is in Israel, where the gross national product per capita is greater than those of all the other countries of the south (and also of Greece, Portugal and Spain). By most indicators, the major exception being its precarious geopolitical security, Israel more closely resembles the northern mediterranean-climate regions. In addition, the average economic growth of the countries of the south (some 4.5%) doubles that of the Mediterranean countries of the European Union (some 1.8–2%). Furthermore, when referring to the north, it is indispensable to differentiate countries of the European Union from the western Balkan countries (from Croatia to Albania), that are more comparable to countries of the south in spite of their much more favorable climatic conditions. Finally, air pollution and, above all, lack of waste treatment, slum conditions or car traffic congestion are more widespread in the south than in the north. These few examples underline the difficulty of making generalizations as regards the Mediterranean Basin countries (see also CIHEAM 1993).

The best discriminant, taken alone, is again the rate of human population growth (together with the age pyramid). Tables 23.3 and 23.4 show the population dynamics and the age relations in the northern and the southern Mediterranean Basin. Most data have been taken from Grenon and Batisse (1988). Data, unlike those in previous examples, refer only to countries bordering the Mediterranean Sea (therefore excluding Portugal), and to their entire territory. For instance, all statistical data for France are considered (including Paris, the Atlantic coasts, the northern and the central regions) and not only those of the relatively small mediterranean-climate zones of the south. There was no other way to proceed in strictly quantitative terms. Table 23.3 indicates a dramatic time shift in the number of inhabitants from the north to the south of the Mediterranean Basin, thus implying an even greater impact on resources and environment in the South. A simple scenario might follow that immigration pressure from the south towards the north will increase accordingly, and with no possibility of stopping it in spite of all restrictive legal measures for entry. Only development in partnership of the north and the south could remedy this situation.

Table 23.3 shows that population is much younger in the south than in the north, and this trend will further accentuate in the future. Again, a dynamic young population, if adequate employment opportunities are not provided, has only the option of migrating elsewhere. The low rate of population growth in the European countries of the northern Mediterranean, thus reaching senescent patterns in only three decades, is certainly not an asset in this case. In addition, new generations will be unable to cope with the burden of health and pension expenses of the more numerous inactive older people. The current Social Security and welfare system of these countries is totally unsustainable, as compared with the more realistic ones of the other mediterranean-climate regions. Great immigration pressure, combined with lower social security and high unemployment, are likely to be the origin of social disruptions and blowups.

A similar demographic trend can be found within a single country, Italy, when comparing the non-mediterranean northern regions with the mediterranean-climate south. In the framework of a scenario of national demographic recession (from 57 millions inhabitants to only 46 million in less than 50 years), 60% of the population is expected to shift from the north to the south. An old population in the north would face a young population in the south. It should be underlined that the level of environmental awareness and of environmental protection, is at present much more implanted in the north than in the south.

Given the extreme complexity of the situation in the Mediterranean Basin countries, several other rank comparisons have been undertaken. The results of only a few of them are given here in a brief manner. Within the mediterranean-climate Europe, two clusters exist. These are the mediterranean-climate zones within the European Union (Portugal, Spain, France, Italy and Greece) and those in the Balkans. The former have better indicators than the latter from both a geopolitical-economic and an environmental viewpoint. Greece, located in the Balkans, but a member of the

European Union, has an intermediate position closer to that of the European Union than of the Balkans. Nevertheless, in the context of only the European Union, the mediterranean-climate zones as a whole are the poorest, most economically marginalized (and often subsidized), and have more serious environmental degradation of their landscapes. In addition, it is in some mediterranean-climate regions of Europe that the highest peaks of unemployment are reached. Spain as a whole has a current rate of unemployment of 24%, while the rate is almost 48% in Andalusia. As an intercontinental comparison, unemployment is only 5.5–6% in the mediterranean-climate zone of Chile.

In Italy, the fifth wealthiest country of the world (and member of the G7), national unemployment is 12%, but it goes from 6% in the northeast to 22.5% in the mediterranean-climate south, where the unemployment of young people entering in the labor market jumps up to 56% (Ingrao 1996). The population classified as poor in Italy is 6.7 million (the percentage of poor has increased from 8.3% in 1980 to 10.6% in 1995); however, only one family out of 20 is poor in the north, while one family out of five is poor in the south. These enormous gaps are the origin of serious political troubles; hopefully, they will give rise to a Federal organization of the Italian regions, that would be in accordance with historical traditions, while the strong North League party is claiming a secession of the north from the rest of Italy.

The situation is equally heterogeneous when the countries of the south Mediterranean Basin are considered. It becomes evident, as already stated by di Castri (1995b), that a globalized economy, with its winners and its losers, increases gaps and divergence among countries and societies. We are very far from a "convergent evolution" or "unique thinking" in terms of countries and economies, as advocated by many politicians. A wide-spread pattern in the south is that of geopolitical insecurity, still weak or absent democracies, dictatorial or autocratic regimes, lack of recognition of the human rights of entire peoples (the Kurds at present, the Palestinians until a short time ago, the Jewish even now by a few countries, etc.). In the whole region, some advance in development goes in hand with better environmental awareness and protection (see for instance the case study of Le Floc'h et al. 1995), but industrialization and urbanization are too often environmentally careless. Sedentarization in arid zones is at the origin of major landscape change.

In the rank analysis, Israel would be first if geopolitical security would improve. Morocco and Tunisia are doing relatively well, while the situation of Algeria, Turkey and, to a lesser extent, Egypt is deteriorating further.

When comparing the four countries of the Mediterranean Basin with the largest extension of mediterranean-climate zones, Spain and Italy in the north, Morocco and Turkey in the south, the rank in order is Italy, Spain, Morocco and Turkey. When only the mediterranean-climate zones of these countries are concerned, the rank remains the same, but there are less sharp differences.

On a more positive side, cultural development and diversity are improving all over the Mediterranean Basin, in spite of the globalization of communications. The revival of old national, regional, local and ethnic identities is a common pattern of the region, with positive effects or negative consequences as in the Balkans. It is astonishing to observe how resilient these identities are over the centuries. Strong cultural identities are a positive factor of political stability and development, in so far they go hand in hand with respect and interest in the diversity and the identity of others. At the same time, such cultural motivation may degenerate towards intolerance, discrimination, exclusions, racism, fundamentalism, terrorism and regional wars. Countries and peoples of the Mediterranean Basin should realize that they have a common destiny, or they have none.

23.5 Prospects for Development and Sustainability

In this era of interconnected globalizations and of new complexities, all prospects should be taken in a very cautious way. A great degree of unpredictability is an inherent feature. There is no more room for linear projections and scenarios, but only for "surprise-rich scenarios" (Brooks 1986). The second qualification is that sustainability does not refer only to the environment, but to the interrelated four dimensions of development, the economic, social, cultural and environmental ones (di Castri 1995b).

Prospects for California are good from an economic viewpoint. Also the economic potential of the mediterranean-climate region of northwest Baja California is excellent in the Mexican context, in spite of a "wild" urbanization. An innovative proactive technology can be implemented in California to minimize environmental impacts, but the fate of its coastal ecosystems is gloomy (see chapter 12) and biological diversity is under threat in spite of the good organization of national parks and protected areas. The major limiting factors towards sustainability are of a social and cultural nature. Because of the relatively healthy status of the economy and of current job opportunities, the United States accepts more legal immigrants than all other industrialized nations combined, and 40% of these immigrants are in California, primarily from Mexico and Central America. In addition, the attractiveness of California is high even within the United States because of its climate and technological and economic advances. The problem is how to assimilate or integrate these massive new arrivals of immigrants in a cohesive way. Thresholds, in number and mostly in time, are not the same as for the previous old immigration waves. Because of ethnic and social instability, riots (such as the ones in Los Angeles) are likely to occur with great tensions within Latino communities, Black communities, and entrepreneurial immigrants from Asia. It is difficult to stop immigrants through simple restrictive legal measures. The long-term sustainability of California will therefore depend to a large extent on the economic development of its southern neighbors (mostly Mexico and Central America). It remains to be seen if NAFTA (the North American Free Trade Agreement) will be an appropriate tool to that effect.

Within all mediterranean-climate regions, Chile has accomplished the most impressive economic boom during recent years. The first phase was characterized by high international competitiveness, with little added value in the use of resources. The next phase of economic diversification and quality of the products is well under way, but it is doubtful if Chile will be able to reach the more stable phase of economic and technological innovation of other emergent countries like South Korea or Taiwan. Investment in research and development is still relatively small in Chile compared with the above countries. The single most important indicator of social stability, that is to say the rate of employment, is good, but poverty and social gaps are still too important (Olivares 1996). The step to democracy, after the dictatorship, went very well, and Chile has a political stability very appreciated by foreign investors. Similarly, political friction with its neighbors, particularly with Argentina, has decreased. No major immigration flows from abroad exist in Chile at present (unlike California), but rural exodus is important. In spite of the immigrant waves of the past, Chile has reached a great national cohesion.

Environmental degradation is important in central Chile, mostly in the coastal zones and in the large cities, with the severe and increasing air pollution in Santiago as a prime example. Slums (*poblaciones callampas*) persist, but in a less dramatic way than in the past. Insufficient waste treatment represents a major threat. Conservation of biodiversity is an important issue that should be treated not only in national parks and protected areas, but in the territory as a whole (di Castri 1994). Environmental awareness is rapidly improving, as can be realized observing the very numerous popular articles on environmental themes in local newspapers and magazines. Ecological know-how on conservation aspects is adequate, but environmental engineering is not satisfactorily applied, mostly because of economic considerations. Environmental economics and law are particularly well developed (Figueroa 1994). As a whole, the mediterranean-climate zone of central Chile stands better than the other productive zone of southern Chile, where temperate rain forests (Valdivian forests) are badly affected by deforestation and widespread plantations of rapid-growth exotic trees (*Pinus radiata* and *Eucalyptus*). In the mediterranean-climate region, vast areas of matorral and forest in the hills are being abandoned because of their lack of economic interest, given the current international rules of a globalized economy. Natural regeneration is possible. The main challenges for sustainability in central Chile are now to step across to a new phase of a more diversified and innovative economy, with greater added values, to become more actively involved in regional economic agreements such as Mercosur (*Mercado Común del Sur*) or NAFTA (Figueroa and Schaper 1995), to reduce social gaps, and to pay greater attention to environmental aspects, as far as possible through a preventive approach. Even from the viewpoint of international competitiveness, Chile should not face the risk of being accused of economic, social or environmental dumping. No one of these challenges represent an impossible task, and Chile is in such a geopolitical situation that it can rely on its own policy.

The situation in Australia is the clearest one, because this region is intrinsically situated in the most stable geopolitical juncture, followed by Chile and California, while South Africa and the Mediterranean Basin are in the worst. Most Asian countries within the sphere of influence of Australia are at present in a period of full economic growth. Immigration is partly controlled, but now up to 60% comes from Asia. Social and political stability is better than in any other mediterranean-climate region, but social disparity has increased as in all industrialized countries. After the terrible drought of 1994, the economy has recovered, but the budget shows a long-standing serious deficit and external debt has reached a historical record. Human density is the lowest among mediterranean-climate regions, but sheep and cattle density are the highest with a considerable pressure on natural resources. In spite of relatively good indicators on most aspects, Australia gives an impression of potential stagnation and institutional inefficiency, even more when a comparison is made with its most dynamic neighbor, New Zealand. Within a large regional context, Australia does not have the leadership position that California holds in North America and Chile in South America. A better integration within the Pacific, Southeast Asian and Indian economic regions will be a key factor for Australia as it reaches a future crossroads.

Together with Chile, South Africa is the mediterranean-climate region that has experienced the most extraordinary advance in its political situation during recent years, through the historical event of the end of apartheid. Progress in human rights and democracy has been made in an almost magical way. The favorable prospects for political stability and the end of international isolation have also attracted considerable foreign investments. In addition, the wealth of mineral resources remains one of the best of the world, although South Africa lacks oil resources. Many heavy geopolitical problems are inevitably still there. Most neighboring countries in southern Africa (Namibia, Angola, Mozambique, Malawi) are in a very bad economic and social position and lack stable political and economic infrastructures. As a whole, Africa south of the Sahara represents the most depressed region of the world. While Botswana has a relatively prosperous situation, the economy of Zimbabwe is deteriorating. Within South Africa itself, the unemployment rate is about 30%, but this percentage would increase up to 50% if underemployment is included (Fritscher 1996). Disparity is still very great. Difference between the gross national product per capita is in the proportion of 1 to 9 between a black and a white. The unemployment rate is about 8% among the white population and 37% in the black communities, leading to problems of criminality. Ethnic tensions between the Zulu ethnic population and other African groups are strong, but the Cape Province is relatively less affected by these problems. From a positive point of view, the creation of the economic area of the *Indian Ocean Rim*, from African countries up to India and Australia, could further boost the South African economy and also reestablish the traditional economic and cultural links between South Africa and Australia.

The South African (and the Cape Province) environment is mostly affected by the social problems mentioned above. Rural exodus, urbanization and poor slum conditions are at the basis of the major threats. Wilderness areas in the Cape region are in relatively good shape, thanks to excellent know-how in conservation and forestry. Environmental awareness is comparatively high, but erosion of biodiversity and biological invasions constitute a matter of great preoccupation.

In spite of the heavy economic and social obstacles mentioned above, and of the fact that disparities could only be attenuated in a lapse of a decade or more, South Africa is destined to play a leading role in the development, environmental protection and economic stability of Africa.

It is in the Mediterranean Basin, with due consideration to its heterogeneity, where the worst indicators can be found with respect to geopolitical security, economic health, social cohesion and environmental protection. Discrimination, exclusion and terrorism are common features. Geopolitical fractures exist towards the south and the west. Unemployment is absurdly high, including in the richest countries, with great disparities even within the context of a single country. Vast migration movements occur, and are likely to increase in the future. Economic growth is slow in the north where, conversely, gross national product per capita and inflation show satisfactory indicators. Soil degradation and ill-managed fresh water resources are a more acute problem than in any of the other mediterranean-climate regions. Establishment and management of protected areas are intrinsically more difficult than in other regions, because of historical and social obstacles. Environmental awareness is comparatively low. In addition, it appears that too many institutions in the Mediterranean Basin, including some university and research institutions, are anachronistically organized, stagnant, thereby lacking adaptive potential and response vis-à-vis the current selective forces of globalization. No one Mediterranean Basin zone stands high in international competitiveness within a global perspective.

From a more positive side, the state of the Mediterranean Sea itself is not as bad as it is often claimed in the media (where it is called a "dead sea") and biological regeneration is taking place. However, coastal zones and wetlands are badly impaired. Cultural diversity is much greater, because of the larger scales of space and time for its development, than in any other mediterranean-climate region. The concentration of people and the gross economic product of the Mediterranean Basin as a whole would allow the establishment of an important economic space in itself, in the absence of the current geopolitical obstacles, cleavages and conflicts. Prospects for changing this situation are gloomy unless Mediterranean peoples and countries join forces through a new deal of conviviality and symbiosis, with a common design and common objectives. There are few signs of this new solidarity, but the resilience of civilization in the Mediterranean Basin has been consistently high over time.

The last point of this chapter will refer briefly to political and geographical obstacles when facing global climate change. Geographically speaking, regions with a long latitudinal extension, like California or Chile, can better ensure shifts of ecosystems and migration of biota, as a long term adaptation, than poleward "truncated" continents as in the case of South Africa and Australia. The political situation in the Mediterranean Basin would again represent a great obstacle for wide-spread adaptive solutions. Too many frontiers and likely conflicts could undermine joint ventures. In addition, most protected areas in the Mediterranean Basin are so small and so affected by human pressure, that they could become pitfalls with increased rates of extinction in a situation with climate change. Populations of plants and animals are in general too small, too disconnected and have too low genetic variability for evolving and adapting in situ, and too isolated and surrounded by vast human modified areas to undertake long migrations. In addition, compared with the other mediterranean-climate regions, the Mediterranean Basin sensu lato has very large deltas (those of the Guadalquivir, Ebro, Rhône, Po, Nile and Danube rivers); sea water infiltration and soil salinization might take place, because of the rising sea levels, with large social and biological impacts.

23.6 Inter-Mediterranean Affinities from Biogeography to Geopolitics

The first generation of inter-Mediterranean comparisons described earlier in the introduction to this chapter put emphasis on biogeographical and ecological affinities among the five regions. In this chapter, political (including social and economic) and environmental similarities are emphasized. To bridge the two series of research, Figure 23.2 shows how affinities shape different clusters among mediterranean-climate regions,

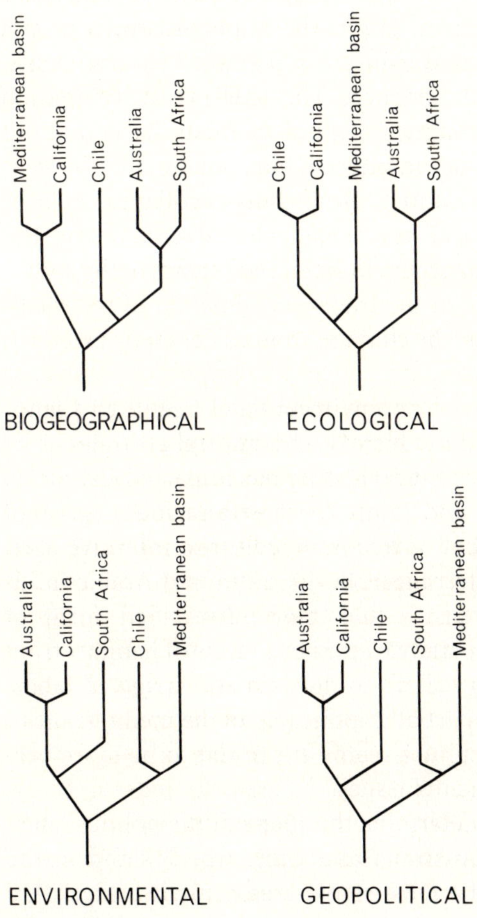

Fig. 23.2. Biogeographical, ecological, environmental and geopolitical (including economic and social factors) clusters, shaped by different levels of affinity among the five mediterranean-climate regions. Affinities have been measured according to Sørensen' quotient (1948)

according to the point of discrimination of the analysis. Biogeographical and ecological data for the clusters have been taken from di Castri (1981, 1991), di Castri and Vitali-di Castri (1981) and Fundaciò Enciclopèdia Catalana (1993). Those data on geopolitical and environmental state have been already been cited earlier in this chapter. There is a strong historical gradient from biogeographical, ecological, environmental and political factors, from ancient to recent times.

Old geological relations primarily, but not exclusively, determine the biogeographical cluster. Southern regions with Gondwanaland and Paleoantarctic contacts (South Africa, Australia and Chile) stand at one side; northern regions from old Laurasia (Mediterranean Basin and California) stand at the other.

The ecological cluster is primarily shaped by selective factors prevailing at the development of mediterranean-type climates during the Pleistocene. A west-coast cold ocean current and a north-south extension of the mediterranean-climate zone with strong summer drought characterize Chile and California. The climates of South Africa and Australia have a considerable west-east extension and truncation poleward by land distribution, and thereby retain tropical influences. The widespread distribution of nutrient-poor soils is also a common feature to Australia and South Africa. The Mediterranean Basin has an intermediate position, but in general is more similar to California and Chile in these ecological factors. Portugal and Morocco have a typical west coast, cold current climate, but some tropical-like summer rainfall occurs in France and parts of the southern Mediterranean Basin. Because of the transverse position of the Mediterranean, Black, and Caspian Seas, the climatic zone is extremely extended to the east.

The environmental cluster, based on environmental health and landscape degradation, has recent land use history and cultural attitudes as its main discriminants. Anglo-Saxon cultural history has impacted California (after the Spanish rule), Australia, and South Africa with various degrees of impact from extended development of tree monocultures, intensive agriculture and grazing. In the Mediterranean Basin, Latin and Arabic influences have created a labor-intensive and more diversified form of agriculture, with a wealth of local plant varieties in a range of habitats from coastal and mountain terraces to valleys to dehesas and irrigated lands. This pattern of land use is at least partially applicable to the mediterranean zone of Chile. The Anglo-Saxon culture seems at present to be more perceptive to and aware of environmental issues.

Finally, security and economy determine the shape of the political cluster. At one side are California, Australia and Chile which stand in the Pacific region that is at present the most stable area of the world, with a largely open global economy. At the other side are the Mediterranean Basin and South Africa that belong to the Old World with all of its ethnic and racial tensions. Zones of potential political instability occur in Eastern Europe and the Balkans, the Caucasus, North Africa, the Near and Middle east, and sub-Saharan Africa. Under these conditions, development and globalization of economies cannot easily take place.

The two clusters at the bottom of Figure 23.2 on environment and geopolitics are remarkably similar. They are determined by short scales of time, while those of the upper half of this figure have a longer time scale. Recent affinities are not fixed, and may readily be altered by accelerated transformations that may occur through intensification of global trade exchanges.

There are no linear relationships among these four groups of factors. They are all embedded, one inside the other, as in a system of Chinese boxes, with a complex web of relationships. Each cluster has its own historic trajectory, with differing scales of time, but there is no predetermination or imprinting of one cluster over another, and thus affinities may shift. For example, Chile is biogeographically closer to Australia, but ecologically a mirror image of California. The vigorous economic growth of Chile in recent years is also similar to California, but Chile lies closer to the countries of the southern Mediterranean Basin when environmental degradation is considered. An understanding of the interactions of these groups of factors is vital for a correct understanding, but analogies and simple extrapolations have often led to biased interpretations.

23.7 Conclusions

From a political viewpoint, three Mediterranean-climate regions, California, Chile and Australia are situated in larger zones with a relatively high stability, while South Africa and the Mediterranean Basin are in a worse position. Economically, California and Chile hold a leadership position as regards their neighbors, while Australia and South Africa have good prospects in this respect. Conversely, the Mediterranean Basin as a whole has lost the crossroads political and economic position that it held from ancient times up to the seventeenth century, with a power shift towards northern zones. Only by attenuating the current social, ethnic, economic and political conflicts, by overcoming the emerging fundamentalism's and discriminations, and by making good use of its particular cultural diversity and richness, can this situation be remedied. In general terms, the main limiting factor for sustainability in all Mediterranean-climate regions is of a social nature, with the existence or the emergence of too many and blatant disparities.

The environment of the five Mediterranean-climate regions has further deteriorated in absolute terms, but it is comparatively better that the one prevailing in most tropical zones. Soil degradation is the main threat in the poorest Mediterranean-climate zones. The effects of a massive urbanization have had serious repercussions in the five regions, and biological diversity is eroded everywhere.

Biological and cultural diversities represent the main feedback mechanisms of control as regards the prevailing forces of the interconnected globalizations. Diversities are also a major asset for a development leading to economic diversification and for the maintenance of options when confronted with the inevitable surprises and bifurcations of an unpredictable future. The fate of the mediterranean-climate regions will largely depend on their capacity to conserve and make use of their rich biological and cultural diversity as provided by their evolutionary and historical trajectories.

References

Arianoutsou M, Groves R (eds) (1994) Plant–animal interactions in Mediterranean-type ecosystems. Kluwer, Dordrecht

Aschmann H (1973a) Man's impact on several regions with mediterranean climates. In: di Castri F, Mooney HA (eds) Mediterranean-type ecosystems: origin and structure. Springer-Verlag, Berlin Heidelberg New York, pp 363–371

Aschmann H (1973b) Distribution and peculiarity of mediterranean ecosystems. In di Castri F, Mooney HA (eds) Mediterranean-type ecosystems: origin and structure. Springer-Verlag, Berlin Heidelberg New York, pp 11–19

ATLASECO (1997) Atlas économique mondial. Editions O.C., Paris

Braudel F (1979) Civilisation matérielle, économie et capitalisme, XVe–XVIIIe siècle. Tome I: Les structures du quotidien: le possible et l'impossible. (Published in English by Collins with the title: The structure of everyday life). Tome II: Les jeux de l'échange. (Published in English by Collins with the title: The wheels of commerce). Tome III: Le temps du monde. (Published in English by Collins with the title: The perspective of the world). Armand Colin, Paris

Brooks H (1986) The typology of surprises in technology, institutions and development. In: Clark WC, Munn RE (eds) Sustainable development of the biosphere. IIASA, Laxenburg and Cambridge University Press, Cambridge, pp 325–347

Brown LR (ed) (1996) State of the world 1996. WW Norton, New York

CIHEAM (1993) Etat de l'agriculture en Méditerranée. The situation of agriculture in Mediterranean countries. Institut Agronomique Méditerranéen, Zaragoza

Chui M, Levine P, Pearlman J, Sentance A (1996) Innovation, imitation and growth in a changing world economy. London Business School, London

Debussche M, Le Coz J (eds) (1985) Etude écologique, sociologique et économique d'une zone méditerranéenne médiane: les Garrigues du Montpelliérais. De la marginalisation à la périurbanisation. PIREN-CNRS, Montpellier

Dell B, Hopkins AJM, Lamont BB (eds) (1986) Resilience in mediterranean-type ecosystems. Junk, The Hague

di Castri F (1963) Estado biológico de los suelos naturales y cultivados de Chile Central. Bol Prod Anim Chile 1:101–112

di Castri F (1981) Mediterranean-type shrublands of the world. In: di Castri F, Goodall DW, Specht R (eds) Mediterranean-type shrublands. Ecosystems of the world 11. Elsevier, Amsterdam, pp 1–52

di Castri F (1989a) History of biological invasions with special emphasis on the Old World. In: Drake JA, Mooney HA, di Castri F, Groves, RH, Kruger FJ, Rejmánek M, Williamson M (eds) Biological invasions: a global perspective. John Wiley, Chichester, pp 1–30

di Castri F (1989b) Global crises and the environment. In: Marini-Bettolo GB (ed) A modern approach to the protection of the environment. Pontificia Academia Scientiarum, Vatican City State, pp 1–33

di Castri F (1991) An ecological overview of the five regions with a mediterranean climate. In: Groves RH, di Castri F (eds) Biogeography of mediterranean invasions. Cambridge University Press, Cambridge, pp 3–16

di Castri F (1994) Medio Ambiente y Territorio. Agric Téc (Chile). 54:366–370

di Castri F (1995a) 1492–1992. Searching for signals and consequences of globalisation. In: Turner BL, Gómez A, González, F, di Castri F (eds) Global land use change. A perspective from the Columbian encounter. CSIC, Madrid, pp 109–130

di Castri F (1995b) The chair of sustainable development. Nat Resour 31 (3):2–7

di Castri F (1996) La risposta delle scienze forestali ai cambiamenti globali. L'Ital For e Montana 51(2):77–111

di Castri F (1997a) The challenge of development and environment in a globalized world. Can institutions adapt? In: Hadley M (ed) Integrating conservation, development and research: scientific responses to the environment-development challenge. UNESCO, Paris and Parthenon, Carnforth

di Castri F (1997b) Mediterranean diversity in a global economy. In: Mediterranean Biodiversity. Ministry of the Environment and ENEA, Rome: 21–30

di Castri F, Mooney HA, (eds) (1973) Mediterranean type ecosystems: origin and structure. Springer-Verlag, Berlin Heidelberg New York

di Castri F, Vitali-di Castri V (1981) Soil fauna of the Mediterranean-climate regions. In: di Castri F, Goodall DW, Specht R (eds) Mediterranean-type shrublands. Ecosystems of the world 11. Elsevier, Amsterdam, pp 445–478

di Castri F, Floret C, Rambal S, Roy J. eds (1988) Time scales and water stress. Proc 5th Int Conf on Mediterranean ecosystems. IUBS, Paris

Figueroa E (ed) (1994) Políticas económicas para el desarrollo sustentable de Chile. Universidad de Chile, Santiago

Figueroa E, Schaper M, (eds) (1995) Chile ante el NAFTA y otros acuerdos comerciales. Una perspectiva ambiental. CEPAL and Universidad de Chile, Santiago

Fritscher F (1996) Afrique du Sud. In: Le Monde, Bilan du Monde. L'année économique et sociale 1995, Paris, pp 87–88

Fundació Enciclopèdia Catalana (1993) Biosfera. 5. Mediterrànies. Enciclopèdia Catalana, Barcelona

Généreux J (1993) Chiffres clés de l'économie mondiale. Editions du Seuil, Paris

Grenon M, Batisse M, eds (1988) Le Plan Bleu. Avenirs du Bassin Méditerranéen. ECONOMICA, Paris

Ingrao P (1996) Précaire recomposition politique en Italie. Le Monde Diplomatique 43 (505):13

Kruger FJ, Mitchell DT, Jarvis GUM (eds) (1983) Mediterranean-type ecosystems: the role of nutrients. Springer-Verlag, Berlin Heidelberg New York

Le Floc'h E, Neffati M, Chaïeb M, Pontanier R (1995) Un essai de réhabilitation en zone aride. Le cas de Menzel Habib (Tunisie). In: Pontanier R, M'Hiri A, Akrimi N, Aronson J, Le Floc'h E (eds) L'homme peut-il refaire ce qu'il a défait?, John Libbey Eurotext, Paris, pp 139–160

Le Monde (1996) Bilan du Monde. L'année économique et sociale 1995, Paris

Mooney HA, Conrad CE (eds) (1977) Proceedings of the symposium on the environmental consequences of fire and fuel management in mediterranean ecosystems. USDA Forest Service, Washington DC

Mooney HA, Dunn EL (1970a) Convergent evolution of mediterranean-climate evergreen sclerophyll shrubs. Evolution 24:292–303

Mooney HA, Dunn EL (1970b) Photosynthetic systems of mediterranean-climate shrubs and trees of California and Chile. Am Nat 104:447–453

Naveh Z (1967) Mediterranean ecosystems and vegetation types in California and Israel. Ecology 48:445–459

Olivares E (1996) Chili. In: Le Monde, Bilan du Monde. L'année économique et sociale 1995. Paris, pp 73–74

Reich R (1991) The work of nations. Alfred A. Knopf, New York (Published in French by Dunod, Paris, with the title: L'économie mondialisée, 1993)

Rota O (1996) Mediterraneo. Il più ricco mercato del mondo. Historia. 40 (6):24–31

Serrill MS (1996) Unlock the shackles. A competitiveness survey finds that lean government and free trade favor economic growth. Time. 147 (24):49

Sørenson T (1948) A method of establishing groups of equal amplitude in plant sociology based on similarity of species content and its application to analysis of the vegetation on Danish commons. Vidensk Selsk Biol Skr 5:1–34

Specht RL (1969a) A comparison of the sclerophyllous vegetation characteristic of mediterranean type climates in France, California, and southern Australia. I. Structure, morphology, and succession. Aust J Bot 17:277–292

Specht RL (1969b) A comparison of the sclerophyllous vegetation characteristic of mediterranean type climates in France, California, and southern Australia. II. Dry matter, energy, and nutrient accumulation. Aust J Bot 17:293–308

Thomas DSG, Middleton NJ (1994) Desertification: exploding the myth. John Wiley & Sons, Chichester

UNESCO-FAO (1963) Carte bioclimatique de la zone méditerranéenne. Notice explicative. Recherches sur la Zone Aride, 21. UNESCO, Paris

World Resources Institute (1994) World Resources 1994–95. A guide to the global environment. People and the environment. Oxford University Press, New York

Zorzi A (1996) La République du lion. Histoire de Venise. Payot & Rivages, Paris

Subject Index

Species Index

Ecological Studies

Volumes published since 1992

Volume 95
Palms in Forest Ecosystems of Amazonia (1992)
F. Kahn and J.-J. DeGranville

Volume 96
Ecology and Decline of Red Spruce in the Eastern United States (1992)
C. Eagar and M.B. Adams (Eds.)

Volume 97
The Response of Western Forests to Air Pollution (1992)
R.K. Olson, D. Binkley, and M. Böhm (Eds.)

Volume 98
Plankton Regulation Dynamics (1993)
N. Walz (Ed.)

Volume 99
Biodiversity and Ecosystem Function (1993)
E.-D. Schulze and H.A. Mooney (Eds.)

Volume 100
Ecophysiology of Photosynthesis (1994)
E.-D. Schulze and M.M. Caldwell (Eds.)

Volume 101
Effects of Land Use Change on Atmospheric CO_2 Concentrations: South and South East Asia as a Case Study (1993)
V.H. Dale (Ed.)

Volume 102
Coral Reef Ecology (1993)
Y.I. Sorokin

Volume 103
Rocky Shores: Exploitation in Chile and South Africa (1993)
W.R. Siegfried (Ed.)

Volume 104
Long-Term Experiments With Acid Rain in Norwegian Forest Ecosystems (1993)
G. Abrahamsen et al. (Eds.)

Volume 105
Microbial Ecology of Lake Plußsee (1993)
J. Overbeck and R.J. Chrost (Eds.)

Volume 106
Minimum Animal Populations (1994)
H. Remmert (Ed.)

Volume 107
The Role of Fire in Mediterranean-Type Ecosystems (1994)
J.M. Moreno and W.C. Oechel (Eds.)

Volume 108
Ecology and Biogeography of Mediterranean Ecosystems in Chile, California and Australia (1994)
M.T.K. Arroyo, P.H. Zedler, and M.D. Fox (Eds.)

Volume 109
Mediterranean-Type Ecosystems. The Function of Biodiversity (1995)
G.W. Davis and D.M. Richardson (Eds.)

Volume 110
Tropical Montane Cloud Forests (1995)
L.S. Hamilton, J.O. Juvik, and F.N. Scatena (Eds.)

Volume 111
Peatland Forestry. Ecology and Principles (1995)
E. Paavilainen and J. Päivänen

Volume 112
Tropical Forests: Management and Ecology (1995)
A.E. Lugo and C. Lowe (Eds.)

Volume 113
Arctic and Alpine Biodiversity. Patterns, Causes and Ecosystem Consequences (1995)
F.S. Chapin III and C. Körner (Eds.)

Volume 114
Crassulacean Acid Metabolism. Biochemistry, Ecophysiology and Evolution (1996)
K. Winter and J.A.C. Smith (Eds.)

Volume 115
Islands. Biological Diversity and Ecosystem Function (1995)
P.M. Vitousek, L.L. Loope, and H. Adsersen (Eds.)

Volume 116
High Latitude Rainforests and Associated Ecosystems of the West Coast of the Americas: Climate, Hydrology, Ecology and Conservation (1996)
R.G. Lawford, P. Alaback, and E.R. Fuentes (Eds.)

Volume 117
Anticipated Effects of a Changing Global Environment on Mediterranean-Type Ecosystems (1995)
J. Moreno and W.C. Oechel (Eds.)

Volume 118
Impact of Air Pollutants on Southern Pine Forests (1996)
S. Fox and R.A. Mickler (Eds.)

Volume 119
Freshwater Ecosystems of Alaska (1997)
A.M. Milner and M.W. Oswood (Eds.)

Volume 120
Landscape Function and Disturbance in Arctic Tundra (1996)
J.F. Reynolds and J.D. Tenhunen (Eds.)

Volume 121
Biodiversity and Savanna Ecosystem Processes. A Global Perspective (1996)
O.T. Solbrig, E. Medina, and J.F. Silva (Eds.)

Volume 122
Biodiversity and Ecosystem Processes in Tropical Forests (1996)
G.H. Orians, R. Dirzo, and J.H. Cushman (Eds.)

Volume 123
Marine Benthic Vegetation. Recent Changes and the Effects of Eutrophication (1996)
W. Schramm and P.H. Nienhuis (Eds.)

Volume 124
Global Change and Arctic Terrestrial Ecosystems (1997)
W.C. Oechel (Ed.)

Volume 125
Ecology and Conservation of Great Plains Vertebrates (1997)
F.L. Knopf and F.B. Samson (Eds.)

Volume 126
The Central Amazon Floodplain: Ecology of a Pulsing System (1997)
W.J. Junk (Ed.)

Volume 127
Forest Decline and Ozone: A Comparison of Controlled Chamber and Field Experiments (1997)
H. Sandermann, A.R. Wellburn, and R.L. Heath (Eds.)

Volume 128
Productivity and Sustainability of Southern Forest Ecosystems (1997)
R.A. Mickler and S. Fox (Eds.)

Volume 129
Pelagic Nutrient Cycles: Herbivores as Sources and Sinks (1997)
T. Andersen

Volume 130
Vertical Food Web Interactions: Evolutionary Patterns and Driving Forces (1997)
K. Dettner, G. Bauer, and W. Völkl (Eds.)

Volume 131
The Structuring Role of Submerged Macrophytes in Lakes (1998)
E. Jeppesen et al. (Eds.)

Volume 132
Vegetation of the Tropical Pacific Islands (1998)
D. Mueller-Dombois and F.R. Fosberg

Volume 133
Aquatic Humic Substances: Ecology and Biogeochemistry (1998)
D.O. Hessen and L.J. Tranvik (Eds.)

Volume 134
Oxidant Air Pollution Impacts in the Montane Forests of Southern California (1998)
P.R. Miller and J.R. McBride (Eds.)

Volume 135
Predation in Vertebrate Communities: The Białowieża Primeval Forest as a Case Study (1998)
B. Jędrzejewski and W. Jędrzejewski

Volume 136
Landscape Disturbance and Biodiversity in Mediterranean-Type Ecosystems (1998)
P.W. Rundel, G. Montenegro, and F.M. Jaksic (Eds.)

Printing: Saladruck, Berlin
Binding: Buchbinderei Lüderitz & Bauer, Berlin